Tesla

His Tremendous
and Troubled Life

Marko Perko
Stephen M. Stahl

Prometheus Books

Guilford, Connecticut

⒫ Prometheus Books

An imprint of Globe Pequot, the trade division of The Rowman & Littlefield Publishing Group, Inc.
4501 Forbes Blvd., Ste. 200
Lanham, MD 20706
www.rowman.com

Distributed by NATIONAL BOOK NETWORK

British Library Cataloguing in Publication Information Available

Library of Congress Cataloging-in-Publication Data

Names: Perko, Marko, author. | Stahl, Stephen M., 1951– author.
Title: Tesla : his tremendous and troubled life / Marko Perko, Stephen M. Stahl.
Description: Lanham : Rowman & Littlefield, [2022] | Includes bibliographical references and index. | Summary: "The authors Marko Perko and Stephen M. Stahl, M.D., Ph.D., D.Sc., propose a 'new-style Biography' entitled TESLA: His Tremendous and Troubled Life. They will examine Nikola Tesla in a manner that has yet to be accomplished in publishing history, asking and answering the seminal question: Who was the real man with an extremely complex psyche/personality, who lived with obsessive compulsive disorder (OCD) and a hyperthymic temperament spilling over at times into high flying bipolar mania and then crashing into devastating depression—and not simply the iconoclastic scientist who invented the modern world?"—Provided by publisher.
Identifiers: LCCN 2021032788 (print) | LCCN 2021032789 (ebook) | ISBN 9781633887725 (cloth) | ISBN 9781633887732 (epub)
Subjects: LCSH: Tesla, Nikola, 1856–1943—Health. | Electrical engineers—United States—Biography. | Inventors—United States—Biography. | Obsessive-compulsive disorder—Patients—United States—Biography. | Manic-depressive illness—Patients—United States—Biography.
Classification: LCC TK140.T4 P37 2022 (print) | LCC TK140.T4 (ebook) | DDC 621.3092 [B]—dc23
LC record available at https://lccn.loc.gov/2021032788
LC ebook record available at https://lccn.loc.gov/2021032789

To my darling wife, Heather Mackay;
my wonderful son, Marko III;
and my beautiful daughter, Skye Mackay
—MP

To Shakila Marie
—SMS

Nature and Nature's laws lay hid in night,
God said, "Let Tesla be," and all was light.

—Dr. B. A. Behrend (scientist/engineer–1917)

Contents

Prologue

On a winter day marked by subfreezing temperatures, the local electric company shut off my electric power, and that of my neighbors, for some nine hours. The very moment the electric power was shut off my world was instantly flipped upside down. I could not prepare bacon and eggs for breakfast; I could not turn on a light to read the newspaper; I could not turn on the forced-air heater in my home (I live a mile high in the mountains with two feet of snow on the ground); I could not switch on the television to check the weather conditions or listen to the news; and the radio was not working. I could not stop food from spoiling in the refrigerator; I could not dry my hair while getting ready to leave the house; I could not open the garage door to get to my car to go to an important doctor's appointment for an MRI and then to the chiropractor for transcutaneous electrical nerve stimulation to help heal my muscle strain. And I could not send an email, because there was no Wi-Fi, hence, no access to the internet; nor could I make a cellphone call because the battery was lifeless. And that's just for starters.

As the daylight hours slowly came to an end, and my electric power was finally restored, all I could say to myself was: "Thank you, Nikola Tesla for your life-changing gift of alternating current, which powers and lights the world today; and thank-you for all your other miraculous inventions and discoveries that make life easier for all of us. It is indeed your time!"

—Anonymous

*N*ow, ponder this: If you never recited a Keats poem or heard a Beethoven symphony or stood before a Rembrandt painting; if you never understood the theories of relativity or read a Tolstoy classic, how would your life be different? Probably not at all, because most people have never had these experiences, and yet they live generally good lives. But try living without AC (alternating current) electricity for nine hours, or even nine minutes, and your life becomes instantly very difficult and frustrating, if not utterly unbearable.

∼

Some years ago a very select group of the world's greatest thinkers, scientists, and Nobel laureates were asked a most stimulating, yet provocative question: What was the greatest invention of the last two thousand years, and why? One said the computer, another answered the internet, and on and on it went. Then one clever thinker hit upon the most important invention of all, the "harnessing of electricity."[1] For without such an invention the world would be very different: no turning night into day; no computer, no internet, no cellphone, would be possible. And a Tesla automobile would still be nothing more than a car designer's pencil sketch.

∼

The enigmatic Nikola Tesla—stalked by his ever-present inner demons—invents the modern world. His astonishing story is that of a new-age god, a genius, a Zeus, a wonderful Wizard, yet a deeply troubled one. He tames the mysterious force called "electricity"; he dazzles the world with his endless inventions and discoveries; he blazes new paths in science that profoundly impact our daily lives; he turns fantasies into realities; his thought experiments disrupt scientific norms; he gives us many of the indispensable tools we use today. And famous actresses and chanteuses clamor for his attention, as powerful men desire to be his friend . . . all before an astonished world. Yet all the while he keeps his own counsel, as he simultaneously struggles with the challenging consequences of *bipolar disorder:* flights of manic energy alternating with depths of great despair—his personal alternating current. He shuns the clichés of a quotidian life, while forever seeking to "lift the burdens from the shoulders of mankind." It would become his lifelong leitmotif, but at what cost to him?

What follows is an examination of the life of Nikola Tesla utilizing a "new-style biography" format. Marko Perko and Stephen M. Stahl, MD, PhD, DSc., will examine Nikola Tesla in a manner that has yet to be accomplished in publishing history—asking and answering the seminal question: Who was the real man behind the extremely complex psyche/personality, who lived with obsessive-compulsive disorder (OCD) and a hyperthymic temperament that spilled over at times into high-flying bipolar mania and then crashed into devastating depression who invented the modern world?

What was it like to be Tesla, having both off-the-wall ideas about science that were eventually proven to be true as well as bizarre habits and bursts of irrational mania? Could he tell the difference? Could others? And how was he able to invent, create, and discover under the weight of bipolar disorder? No one has addressed the role of bipolar disorder in his long life and how it affected him at all levels . . . until now.

In *Tesla: His Tremendous and Troubled Life* the authors have mined the depths of the available sources and utilized their expertise to answer additional questions that have mystified chroniclers and gone unanswered for more than a century: Was Tesla's tortured genius "normal?" What is "normal" genius anyway? What is the relationship between bipolar disorder and immense creativity? How did it apply to Tesla? What was it like for him to have original scientific insights that no other individual had ever conceived of or experienced before? Were his strange beliefs and sensory experiences the product of a bipolar state of mind? Was his lifelong struggle with what we call today more specifically bipolar I disorder (commonly referred to as bipolar disorder) an asset or a liability when considering the profound output of his remarkable mind?

Moreover, this biography will address the age-old question of whether incredible creativity can only occur in the context of mental illness, particularly bipolar disorder as applied to the case study of Nikola Tesla. As such, this "new-style biography"—blending the best of timeless biographical techniques with medical case studies and cutting-edge introspective methods found in psychobiography—will plumb the depths of the "tremendous and troubled life" of Nikola Tesla. The authors will unrelentingly apply their decades of experience both as professional writers and researchers as they investigate, assess, and reveal the "complete" life of Nikola Tesla.

From scientific measurements and electronics to twenty-first-century automobiles and streets that bear his name, Tesla is fast becoming a "worldwide brand." More than a century after he "lit" the world, little is actually known about the man who willingly gave us the products of his unbridled genius: rotating magnetic field, alternating current, induction motor, radio, wireless transmission of power/data, magnetic resonance, neon/fluorescent lighting, the laser, remote control, robotics, X-rays, radar, particle beam accelerator (death-beam), the Tesla coil, and yes, even Star Wars—the Strategic Defense Initiative. This greatest of scientists is still a mystery to the growing millions who now immediately recognize his name but know little to nothing about him and his life-altering achievements. In addition, his lifelong struggle with mental health issues has gone unaddressed in any serious way by previous biographers.

This biography will try to satisfy the insatiable hunger for knowledge and understanding of this man, as more and more people encounter the astounding, mysterious, hidden world of Nikola Tesla, and how he harnessed his incredible gift while shouldering the crushing burden of bipolar disorder.

Regarding author Marko Perko, it is important to know that he brings a truly exclusive perspective and great empathy to this biography, for like Tesla himself, he is a fellow Serb. He was born and raised in the same ethnic/cultural/religious milieu as Tesla. He has the "lived experience" to get inside Tesla's mind as a fellow Serb.[2] He understands what many of Tesla's earliest influences were and how they informed and shaped his multidimensional character. As such, this biography will show the "true" Tesla: the man who always valued "principle over profit"; the man who sacrificed personal success for the success of others; and the man who focused his genius to make the world a better place to live.

Stephen M. Stahl, MD, PhD, DSc, brings to this biography a lifetime's worth of extensive psychiatric experience and a worldwide reputation in the fields of psychiatry and psychopharmacology of the highest order. He will dissect the Tesla psyche and reveal its stunning truths.

The book's subtitle says it all: *His Tremendous and Troubled Life* points to Tesla's long, productive career, which was all the while deeply troubled by personal demons. He was the man who, "in his time," changed mankind for "all time." So now it is the twenty-first century, and *"it is his time"* for the world to know the man—his tremendous and troubled life—and not just his name.

~

The panorama of human evolution is illuminated by sudden bursts of dazzling brilliance in intellectual accomplishments that throw their beams far ahead to give us a glimpse of the distant future. . . . Nikola Tesla, by virtue of the amazing discoveries and inventions which he showered on the world, becomes one of the most resplendent flashes that has ever brightened the scroll of human advancement.

—John J. O'Neill, friend and early biographer of Nikola Tesla

Note to the Reader

INCONSISTENCIES

$\mathcal{D}$ates, locations/places, names, quotations, and so on, can be troubling when writing a biography or any book that contains a myriad of them. The authors made every effort to get all this correct. However, the authors would often find from one reference source to the next that the dates or names regarding a specific event did not agree. The same can be said for places and quotations. To resolve such discrepancies, the authors looked for consistencies—that is to say, several reference sources that did agree, understanding that such agreement usually resulted in getting the facts correct.

BOOK FORMAT

Each chapter's subhead gives a location and date that represent the starting point for the subject matter of the chapter.

BOOKS/NEWSPAPERS/JOURNALS/
OTHER REFERENCE SOURCES

Tesla's autobiography entitled *My Inventions* has appeared in multiple editions over the decades; hence, the "only" complete reference source of his childhood has proved to be inconsistent at times. There are versions wherein Tesla recalls to others about his childhood that do not line up with another published version of his autobiography. The same can be said for other sources.

Often Nikola Tesla would recall his autobiographical story in interviews and publications with slight changes; hence, various versions of his autobiography exist. Then there is the issue of each publisher taking liberty with the text.

DATES

There were times when one reference source said that Tesla did something at a certain age, while another reference source gave a different age for the same event.

At times, especially in his early years, there is an almost "dependable" inconsistency among references as to when Tesla did something.

Because of the inherent difficulties in the story's timeline, dates cannot always be given as definitive.

PUNCTUATION

All quotations and references retain their "original" punctuation unless changes were deemed to be needed for clarity.

QUOTATIONS

All quotations (spellings, grammar, syntax, etc.) appear as in the original in all reference sources unless otherwise noted for clarity.

SCIENTIFIC DEVICES

Whether it be called an oscillator or resonator, as with numerous other devices, they are basically the same electrical class; and most reference sources have a tendency to use the terms *AC motor* or *induction motor* or *AC induction motor* interchangeably, and they vary from one reference source to the next.

DISTANCES

The distances between locations/places vary from one reference source to another.

SOURCES/NOTES

Page numbers in reference sources vary, as different editions of the same publication are cited by different sources. For example, there are U.S. and British versions of several technical publications.

SPELLING/SYNTAX/GRAMMAR

Direct quotations were left in their "original" form. The same for spelling and syntax (Serbian or Servian).

TIMELINES

Chronology of events is not always certain, as is the case with much of history, but the authors have made every effort to get each and every date in the timeline correct. The sequence of events sometimes varied from reference source to reference source and even in Nikola Tesla's autobiography.

EVENTS

Examples of inconsistencies of certain events: Those involved in the purchasing of the Tesla patents vary from one reference source to the next, as does the amount paid for the Tesla patents by Westinghouse. Tesla's meeting with Sarah Bernhardt may or may not have happened, and if it did, different versions are given. Likewise, Edison did not comment on the Nobel Prize fiasco in one source and another source said he did.

LOCATIONS/PLACES

Exact locations/places and whether Tesla was in a particular geographical location or building sometimes differed from one reference source to another.

MEASUREMENTS

Oftentimes measurements from one reference source to the next do not agree—the height of the tower in Colorado Springs, for example. Also, several measurements at the Wardenclyffe project do not agree from one reference source to the next.

NUMBERS

Sometimes the numbers referring to a specific event do not agree from reference source to reference source. If egregious, the authors point out the discrepancy.

PERIODICALS

The names of various periodicals change from one reference source to another, requiring further research to ascertain what other authors were referring to when they quoted a reference source.

NAMES

Various names are used interchangeably such as wireless transmission and wireless transmissons, Serbian and Montenegrin; AC polyphase and alternating current, etc.

CURRENCY INFLATION CALCULATIONS

All currency inflation calculations were made at the time of the book's writing.

WEBSITES

All website addresses were active at the time they were cited.

Cast of Characters

Edward Dean Adams—investor

Leland I. Anderson—technical writer/electrical engineer

Muriel Arbus—secretary

John Jacob Astor IV—industrialist

B. A. Behrend—friend, scientist, and engineer

Sarah Bernhardt—French stage/film actress whom Tesla was greatly interested in

A. S. Brown—principal of Western Union

James D. Carmen—longtime confidant and administrator for Tesla

Samuel L. Clemens (Mark Twain)—dear friend, author

Arthur H. Compton—American physicist, winner of the Nobel Prize

Sir William Crookes—British physicist

Colman Czito—friend and lab assistant

Antonín Leopold Dvořák—composer

Thomas A. Edison—inventor

Albert Einstein—physicist

James Franck—German physicist, winner of the Nobel Prize

Hugo Gernsback—journalist, editor, and publisher

Heinrich Hertz—electrical engineer

Richmond P. Hobson—U.S. Naval Admiral and Congressman

Robert U. Johnson—editor and very close friend and protector

Katharine Johnson—very close friend and protector

Rudyard Kipling—author

Sava Kosanović—nephew and Yugoslav Ambassador to the United States

Fritz Lowenstein—laboratory assistant

Guglielmo Marconi—electrical engineer

T. C. Martin—editor of *Electrical World* and author

Dame Nellie Melba—Australian opera star Tesla fancied

Marguerite Merrington—she attracted Tesla's eye and was his regular dinner companion

Robert A. Millikan—American physicist, winner of the Nobel Prize

J. P. Morgan—financier and banker

John Muir—naturalist/environmental philosopher

John J. O'Neill—personal biographer of Tesla

Ignace Paderewski—pianist/composer

Charles Peck—patent lawyer

Eleanor Roosevelt—First Lady

Franklin D. Roosevelt—U.S. President

George Scherff—secretary and office manager

Dorothy Skerrit—secretary

Charles Steinmetz—GE scientist

Kenneth Swezey—journalist

Anton Szigeti—assistant and confidante

William Terbo—grandnephew

Angelina Tesla—sister

Dane Tesla—older brother

Djouka Tesla—mother

Marica Tesla—sister

Milka Tesla—sister

Milutin Tesla—father and priest

William K. Vanderbilt—industrialist

George S. Viereck—poet

George Westinghouse—engineer/entrepreneur

Stanford White—architect

Timeline of Significant Events

1856	Born at the stroke of midnight between July 9th and 10th in the village of Smiljan, in the province of Lika, Austrian Empire
1861	The accidental death of his older brother Dane, which haunts him all his life
1862	Family moves to the town of Gospić
1866–1870	Attends the gymnasium where he first learns of Niagara Falls
1870	Moves to Karlovac to attend the higher gymnasium to study physics, math, and languages
1874	Contracts cholera and uses it to convince his father to let him study engineering
1875–1878	Attends the Polytechnical University at Graz, Austria, where he sees a demonstration of the Gramme dynamo—it could work without a commutator
1879	Leaves the university without a degree and gains work in Maribor
1880	Studies in Prague; father dies
1881	Moves to Budapest to work at the telegraph company; first recognized nervous breakdown; envisions "rotating magnetic field"
1882	Moves to Paris to work for the Continental Edison Company; continues working on his AC induction motor
1883	Works in Strasbourg to rebuild Edison's failing DC power system for Germany; he is cheated by Edison the first of many times
1884	Returns to Paris; moves to New York City to work for another Edison company; fixes dynamos on a ship to save Edison from embarrassment; and he is cheated again

1885	Quits Edison after having been cheated again; invents several new devices
1886	Westinghouse forms Westinghouse Electric Company; Tesla receives "first" patent for a Commutator for Dynamo Electric Machines, issued January 26, 1886 (U.S. Patent 0,334,823); and patents electric arc lamp
1886–1887	Works as a day-laborer digging ditches; develops polyphase system particulars; forms Tesla Electric Company with Peck and Brown; and builds his prototype polyphase AC induction motor; explores the mysteries of X-rays
1888	Delivers his famous, groundbreaking lecture before the American Institute of Electrical Engineers; receives five patents for AC motors; works on his Tesla coil technology; and joins with Westinghouse, who purchases his patents, thus making him rich for the moment; the War of the Currents begins
1889	He makes the boldest of moves and relocates for the next eight months in Colorado Springs where he develops his Tesla coil—magnifying transmitter; and William Kemmler is put to a gruesome death in New York by means of an AC-powered electric chair
1891	Becomes a naturalized U.S. citizen; first Tesla coil patent issued (U.S. patent No. 454,622) with subsequent patents issued; and demonstrates early version of wireless energy transmission at the AIEE
1892	General Electric is formed with J. P. Morgan's financial backing; lectures in London and Paris on his latest research; learns of his mother's death and returns home to see family
1893	Alternating current is utilized by Westinghouse to light the World Columbian Exposition in Chicago; meets his "protectors" Robert and Katharine Johnson; delivers a lecture in St. Louis on the fundamentals of radio (wireless) transmission; and invents the first wireless system; Marconi utilizes wireless equipment just like Tesla's
1894	Granted some dozen patents ranging from electrical transmission of power to engine technologies; and works on high-frequency electromagnetic radiation
1895	His Fifth Street laboratory is completely destroyed in a fire; depressed; but picks up the pieces and moves on with his work and transmits alternating current 500 miles
1896	Electrical power generation begins at Niagara Falls; destroys his contract with Westinghouse at a cost of millions in royalties to

	him; and Marconi transmits radio waves but is denied a patent because of Tesla's primacy
1897	Successfully transmits radio signal NYC to West Point
1898	Demonstrates his telautomaton—wireless—remote-controlled boat in Madison Square Garden before an astonished crowd of the Institute of Electrical Engineers Electrical Exposition; and moves into the Waldorf-Astoria where he lives for some two decades
1899	Moves research to Colorado Springs, Colorado. There he erects a massive magnifying transmitter—Tesla coil—that produces magic lightning and electrifies the earth for miles with millions of volts; and claims to have received signals from Mars; he writes his comprehensive "diary" on his research that year
1901	Marconi uses some seven Tesla patents to transmit the letter "S" across the Atlantic Ocean; Wardenclyffe project begins
1903	Wardenclyffe Project was Tesla's intent to give wireless power to the world for "free"; Morgan ends his funding of the project
1904	Marconi is granted the patent for the invention of the radio (reversed June of 1943 in favor of Tesla)
1907	Inducted into the New York Academy of Sciences
1907	Develops VTOL technology and is granted two patents in 1928
1909	Invents the bladeless turbine engine, inspired by his early childhood experiments, and upon the boundary-layer effect; receives first patent for the device
1909	Marconi awarded the Nobel Prize for the invention of the "radio"
1912	Tesla experiments with his turbine at the Edison Waterside Station
1914	Apparatus for transmitting electrical energy patent granted (U.S. patent 1,119,732)
1915	Files patent infringement suit against Marconi
1916–1928	Granted numerous ingenious patents
1917	His Wardenclyffe Project demolished; ironically awarded the Edison Medal
1922	Continues work of new inventions not yet revealed
1925	Katharine Johnson dies, one of his protectors
1931	On the cover of *Time* magazine for his 75th birthday and begins his yearly press conferences; Edison dies
1934	Pursues particle beam research that leads to the military's Star Wars technology

1937   Robert U. Johnson dies, one of his protectors
1943   On January 7, Serbian Christmas, Tesla vanishes into the ether;
                  his funeral service in New York City is attended by more than
                  2,000 mourners; all his effects are seized by U.S. authorities;
                  and the U.S. Supreme Court rules Tesla as the inventor of the
                  "radio"

Timeline of Psychiatric Events

Clinical Notes

1856	Born at the stroke of midnight between July 9 and 10 in the village of Smiljan, in the province of Lika, Austrian Empire
1856–1866	Precocious, grandiose, likely in an ongoing hypomanic and occasionally psychotic state, first invention of frog hook, first boom moment stroking the cat, invented a popgun, visual hallucinations, OCD symptoms of counting steps, needing to have things divisible by 3 or something bad would happen, aversion to earrings, bracelets, pearls, hair, peaches, camphor, germophobe
1861	Accidental death of his older brother Dane; first major depressive episode and grief reaction
1862	Family moves to the town of Gospić
1866–1870	Attends the gymnasium where he first learns of Niagara Falls; aloof, socially withdrawn, schizoid, self-centered, yet probably enduring hypomanic and occasionally psychotic state, creative, with visual hallucinations, visualizing solutions to problems and developing mental editing skills; imaginative, thinking outside the box
1870	Second major depressive episode, prostrated in bed; then "miraculous recovery"
1873	In bed for nine months; possible cholera, more likely his third major depressive episode
1876	Phronesis, hyperthymic leading to first full manic episode; diminished need for sleep, studying twenty hours/day, gambling; womanizing, drinking, smoking

1877–1881   First full manic episode of 1876 can see seen as continuing through 1881 as a vacillation between hypomanic/hyperthymic mood state and full mania/psychosis with auditory hallucinations; mind running fast, using OCD chants in attempt to control and slow down thoughts; creative, inventive, schizoid isolation; manic/psychotic when claims to hear thunderclaps 550 miles away; to hear ticking of watch three rooms away; to hear crackles of fire of neighbors' burning house far away, likely auditory hallucinations; to hear/feel heart racing to 260 beats per minute, which if real would be an unlikely supraventricular tachycardia if not a sensory hallucination; felt sure he could mentally engineer things in his head which may not have been delusional; toward the end of this period, began having shorter and less severe cycles of obsessive relentless work/intellectual gluttony for weeks, then utter exhaustion with depression where he was certain he was on the threshold of death consistent with rapid cycling bipolar mood swings

1882   Mania/hypomania persists with his now famous frenetic, boom moment, drawing with a stick in the dirt his AC induction motor invention rather than calmly going back to the laboratory and putting it on paper

1884   Depressed, no investor interest in his AC system; dysphoric and possibly major depressive episode in Europe; however, by the time of arrival in New York had cycled back into hypomania; despite his conflict with Edison after arrival, scientific output continued while hypomanic in 1884 and 1885 resulting in his first patents, six in the U.S. and three in Britain, filed during 1886

1886   Now depressed, feeling cheated and defrauded, work in 1886 led to only one patent filed in 1887

1886–1901   Most productive period of his life, when he filed essentially all of his important patents; most of the time was hypomanic punctuated by episodes of depression then recovery to hypomania

1887–1891   New York City showman, charismatic, hypomanic; work this year led to a record fourteen U.S. patents plus two British patents in 1888; twelve U.S. patents plus four British patents in 1889; six U.S. patents in 1890; and eleven U.S. patents plus two British patents in 1891, the single most productive period of his life

1888   Initially hypomanic at his electrical engineer lecture, success, and move to Pittsburgh, but became dysphoric, deflated, and eventually depressed and left Pittsburgh searching for the discovery that would drive his euphoria—or was he searching for the euphoria that would drive his discovery

1889   Ups and downs; back to New York; dealt with adverse publicity about electrocution and how his invention killed and was dangerous

1890–1891   Financial hardship; became a U.S. citizen; another big lecture with charisma, showmanship, charm, demonstrations

1892   Returns to Europe; prolonged painful ordeal and illness, likely an extended depressive episode linked to grief reaction over his mother's death; famous, was a wizard, but had patent infringements to fight; felt it tired him out, sought electrotherapy tonic for his exhausted nervous system; visited Paris in a tired/exhausted state, had "peculiar sleeping spells" from what he thought was prolonged exertion of the brain, and which may have represented the increased sleep associated with a depressive episode compared to the lack of sleep during hypomanic and manic episodes; only one U.S. patent in 1892 and only two in 1893

1893   Returns to New York City, still not over his depression; haggard, overworked. Was at Chicago exposition but exhausted; visited Niagara Falls but "big machines bothered his spine." Partial recovery but mostly wrote patents based on earlier work, with eleven published in 1894 from pushing himself likely in a normal to moderately depressed mood state but not hypomanic

1894   Although joined the glitterati crowd, he was exhausted, exasperated, lonely, completely worn out, felt he had an inventor's fragile health and Johnsons agreed his "health" was poor probably indicating major depressive episode; he felt it was the endless mental and physical strength of decades of work now depleted. Likely in a mixed state of hypomania with some elements of depression, and cycling up and down

1895   The fire that burned down his laboratory in New York City, devastating event. Broken in spirit, "spent"; "no more to give"; tried "electrotherapy as cold fire to excite the brain" as an attempt to self-treat his depression; betrayed by Edison and Westinghouse. No patents in 1895. Still had some flights of fantasy, unstable mood and psyche, in a state of denial about his losses, including his laboratory, his business partner, and betrayed by Edison and Westinghouse. Mostly depressed and unproductive

1896–1987   Mostly hypomanic but some cycling, from up periods with Mars fever, showing how buildings oscillate, trusting his internal compass and intuition, yet bizarre Niagara Falls speech out of character and somewhat dejected without the usual aplomb and bombast and charisma; sense of inferiority and defeat crept in at times of his lows, and Katharine detects his suffering; Tesla and friends feel he is being treated unfairly; Tesla has bizarre idea that the external forces of nature which cause him pain are cosmic and those who inflict them on him then come to grief themselves due to these forces from the medium of an automaton

1898   Successful Madison Square Garden radio-controlled boat demonstration as well as reports of Marconi failures buoyed his spirits; under fire, envy of competitors abounds, with scientific gossip; bouts of melancholy "poor health," worn out physically and psychologically; spent, lonely, mostly depressed with some bouts of hypomania mixed with depression

1899   Colorado Springs, began in hypomanic state, and progressed to depression by the time he left disappointed and failing to meet his goals; had vision in early part of the year for wireless communication, wireless energy, transmission through the earth, but by the end physically and mentally spent; no patents this year

1900   He began exhausted, no hope, financial pressures, emerging from failure, depressed and had no charisma and no notable hypomania and thus unable to raise financing

1901   Has idea for building a tower for transmission of energy through the earth; got initial seed funding successfully from J. P. Morgan as Tesla emerged from depression into his usual productive hypomania; wrote a treatise a bit grandiose and bombastic

1902–1943   The rest of his career, when he was rarely hypomanic and at most had some mixed features of mania while depressed. He also exhibited insufficient charisma to raise significant funds in order to complete any large-scale projects ever again; during his productive period of 1886–1901, he filed ninety-five patents or about six per year for an absolutely amazing scientific output and with only one year having no filings when he was depressed; otherwise, during this productive period he spent most of his time hypomanic with some bouts of rapid cycling to depression and back to hypomania and some periods where his mania was mixed with a tinge of depression. But, from 1902 until his death, he only filed fifteen patents, about one every three years, and was never again continuously manic or hypomanic with most of the time spent either depressed with a few symptoms of mania in mixed state, or just depressed, with increasing irritability, resentment, lacking social contacts or social grace, lonely, and alone

1901–1902   Famous Wardenclyffe tower built; Marconi successfully sends a wireless message, beating Tesla to this landmark accomplishment and causing disappointment and resentment by Tesla; mental state frayed; felt betrayed by science and colleagues and financial patrons when he did not get sufficient support or credit despite having had the courage to reject received wisdom and reversing standard scientific practices and challenging existing norms; unfortunately he could not make these notions believable to others when the high moods abandoned him

1903   Mental breakdown; chaotic chicken scratch letters showing emotional fraying and distress; strained the "finest fibers of my brain"; felt he was in a complete collapse; thinking and behaving irrationally according to potential financers, "giving away free electricity"; no longer at this time considered a genius and a showman, becoming an aging crank

1904–1905   Early patents expiring with the loss of the income from them; many were for inventions that were not developed in his lifetime; financial problems; reclusive; unpredictable; began to have flashbacks to seeing dead family members, possibly a form of psychotic depression; bedridden from exhaust and depression

1906–1908   Tesla's tower scrapped; nervous breakdown continued; no major projects for the next thirty-seven years; still had ideas but lacked the charisma or energy to carry them out or get them funded; lacked business savvy as always; ahead of his time even with his new ideas presented in this emotional state; invented a turbine

1909   Marconi gets Nobel prize; Tesla has a last gasp with the Tesla Propulsion Company; couldn't understand why industrialists invest money to make money when Tesla invested money to make new technologies; Westinghouse, John Muir, Mark Twain, Astor all died. His friend Johnson had a scandal; lots of stressors

1910–1915   J. P. Morgan dies in 1913; Morgan's son gave initial but not ongoing finance for Tesla's turbine but then WWI intervened; meanwhile, Tesla reduced to prophesizing and not to producing; financial problems and lawsuits

1916–1920   Awarded Edison medal in 1917 with some financial help; still alive was his dream for wireless communication, conceived of radar. Imagineering never seemed to tire, but mounting stubbornness to new ideas that weren't his own; egotism; obstinacy; writing for some income, more and more time spent with pigeons; mounting financial problems; rival Pupin died; met a nineteen-year-old science writer

1921–1929   Closed his offices, no secretary; locked into old ways of thinking and unable to accept new ideas; Katharine Johnson dies in 1925

1930–1933   *Time* magazine cover; Teslascope conceived for interplanetary communication; Edison dies; the world acknowledges Edison as a genius in perfecting existing inventions; Tesla as a genius with brilliant inventions, many at that time not perfected for commercialization; irked Tesla who thought conceiving new inventions was superior to perfecting previous inventions, but the world thought otherwise; some bizarre visions; rude; curmudgeon, refused to shake hands, used gloves once yet consorted with pigeons; began yearly public relations appearances on his birthday

1934–1943   Conceived death-beam, constant beam of tungsten or mercury, to neutralize enemy airplanes; serene, courtly, conceived of Star Wars defense shield that was eventually implemented by the Reagan administration; frail; hit by taxi; walked with a cane; Robert Johnson dies; at the end Tesla was bedridden, irrationality alternating with lucidity and early signs of senility; OCD persisted to the end with his childhood-onset fascination with the number 3 so always stayed at an address or in a room divisible by 3, with his last residence room 3327 at the New Yorker Hotel

I

DREAMING

Lightning and Thunder

Precocious

SMILJAN, PROVINCE OF LIKA, AUSTRIAN EMPIRE, 1856

The blackest nighttime sky was suddenly shot through with powerful bolts of lightning that brightened the heavens and thundered like a lion making its presence known. While the violent electrical thunderstorm raged on, displaying the undisputed power and mystery of nature, a young mother labored in a small, one-window room with wood-planked flooring to birth a child who would be named Nikola. As she shook the bed, writhing in pain and joy, gale-force winds mercilessly battered manmade structures outside, while the midwife beside her was so frightened by the moment that she nervously said, "He'll be a child of the storm." His mother responded insistently, "No, of light."[1] (History suggests they were both correct, as the life of Nikola Tesla would prove to ride the storm of bipolar disorder while lighting the world both literally and figuratively with his genius.)

It was precisely the midnight hour, between July 9 and July 10, 1856 (Julian Calendar).[2] The quaint village of Smiljan ("the place of sweet basil" is east of the Velebit Mountains and the Adriatic Sea), in a humble house, which stood steadfastly next to the local Serbian Orthodox Christian church presided over by his father Reverend Milutin Tesla, was where the family welcomed its second son and fourth of five children. Little did anyone know that in the years to come, the lightning and thunder that introduced Nikola to his family would prove to be a harbinger of tremendous and troubled times to come for the newborn. And little did anyone know at the time of his unheralded, simple birth that in the decades to follow he would create the modern world by harnessing nature's power, unlocking many of its greatest mysteries, and seeing what no others saw.[3] Yet, despite their happiness for the arrival of their second son, the family insisted on baptizing him at home the day of his birth

3

believing he was weak and frail. Moreover, Austrian law also required that he be enlisted in the military and to enter service from the age of fifteen onward.[4]

~

Born into a Serbian family in a land controlled by Croatians, the young Nikola was part of a racial and religious minority living in the greater Austro-Hungarian Empire of the Hapsburgs, to whose heavy-handed rule the people adapted as best they could.[5] A Serbian's Orthodox Christian faith did not square with the predominately Roman Catholic faith that dominated the empire. The theological dissimilarities proved to be a wedge that often drove the followers of each faith to opposing corners.[6] It has been a centuries-long religious schism that endures to this very day.[7]

~

The young Tesla's Serbian Orthodox Christian faith was all around him as he grew up, because his father and uncles were priests in the Church, as was his mother's father. Like many great men who came before him, throughout his long life, his faith would inform his creative process by instilling the belief that everything in Creation has an underlying foundation.[8]

He was born into a distinct *zadruga*, also known as a rural community, comprised of a Southern Slav's extended family. His family's original surname was Draganić, and it dates back more than two and a half centuries. By the mid-eighteenth century, having ancestors who had journeyed from Western Serbia in the late 1690s, the tightly knit clan settled in Croatia (an Illyrian province at the time) as many other Serbs did, and the Tesla surname became its calling card.[9]

It is said that the family acquired the new surname for two reasons. First, because several of its family members had wide, protruding teeth—Nikola's were not—resembling that of a woodworking tool called an adz or axe. And second, this tradesman's tool is fitted with a broad, steel cutting blade at right angles to the wooden handle used for cutting large tree trunks into squared timbers. In the Serbian language the tool is still called a *tesla*.[10]

The Tesla family's Serbian Orthodox Christian faith was the glue that kept them and all other Serbs together as they progressed over the centuries across the Balkan Peninsula, in search of farmable land, homesteading opportunities, and at times, simply to put distance between themselves and invaders.[11]

One only has to see the land that the Serbs called home to understand the difficult, unrelenting challenges they faced just to survive from day to day. The earth beneath their tired and blistered feet constantly resisted their best efforts to till the soil and feed their families. Later in life Tesla would on occasion recall the many times he heard local Serbs say of the Likan soil, "that when God distributed the rocks over the earth He carried them in a sack, and that when He was above our land the sack broke."[12]

Tesla's comment about the hardness of the land the Serbs occupied at the time has never changed, for once they settled in what would become modern-day Serbia (and Montenegro), the hardened earth continued to fight them at every turn.[13] At first blush the earth appears to be simply a firm soil, but upon closer scrutiny one sees that countless rocks showing only a small portion protruding above ground, like an iceberg that only reveals 10 percent of its total mass above water. The rocks are intractable and often cause one to stumble, if not fall upon contact. Simply put, during Tesla's time making his way along the paths in the province of Lika often proved most trying at the very least—and not much has changed today.

The difficulty of traveling along the paths the Serbs have traveled is emblematic of other hardships they have endured through the ages. It is reflected in their poetry as well, as Tesla said of his people: "Hardly is there a nation which has met with a sadder fate than the Serbians. From the height of its splendor, when the empire embraced almost the entire northern part of the Balkan Peninsula and a large portion of territory now belonging to Austria, the Servian [sic] nation was plunged into abject slavery, after the fatal battle of 1389 at the Kosovo Polje, against the overwhelming Asiatic hordes. Europe can never repay the great debt it owes to the Servians [sic] for checking, by the sacrifice of their own liberty, the barbarian influx."[14]

The Battle of Kosovo Polje, or the Field of the Blackbirds, was a brutal battle fought between Orthodox Christian Serbs and the invading army of Muslim Turks of the Ottoman Empire in 1389. As Tesla's sentiment so aptly suggests, although the Serbs were ultimately defeated—with the leaders on both sides dying and with innumerable deaths sustained by both the Serbs and Turks—they had stamped their national identity as the bravest of warriors. To this very day the battle is celebrated as a national, ethnic, and religious holiday called Vidovdan (Saint Vitus Day in English), which is recognized each year on June 15.[15]

At the zenith of the Ottoman Turks' power in the fifteenth and sixteenth centuries, they were a pernicious and ferocious force of invaders who came up from the southeast and wreaked great havoc in the Balkan Peninsula, causing the Serbs to leave their homeland—which would later become modern-day Serbia and Kosovo—to escape certain death and seek refuge in Croatia (then a protectorate of the Austrian Empire), where the people were also natural enemies of the Ottoman Turks and their Islamic faith. As a consequence,

Croatia was primarily a military region and a source of troops to defend Austria's borders.[16] It is interesting to note that the Serbs were warriors tasked to protect the empire's borders from the Ottoman Turks. So, young Nikola Tesla, the namesake of his grandfather Nikola Tesla the elder, a man of great bravery born in 1789, who fought as a sergeant in Napoleon's French army in the Illyrian provinces 1809–1813, as the unification of the Balkan states was underway, would become filled with the strength of a courageous warrior as well. And as a reward for fighting against the Austro-Hungarian Empire, Serbs were permitted to own their own land, something never afforded the Croats.[17]

~

GOSPIĆ, PROVINCE OF LIKA, AUSTRIAN EMPIRE, 1815

The Napoleonic Wars drew to an end with Napoleon's defeat at Waterloo on June 18, 1815.[18] It was during this time that Tesla's grandfather Nikola, having married the comely Ana Kalinić, the daughter of a decorated military officer, returned to his birthplace in Gospić, transitioning from the French army only to resume his warrior ways in the Austrian Empire's military forces. It was then that the young couple began a family with the birth of their first son Milutin (Mitchell), the inventor's father, in 1819. The family would soon be blessed with another son and three daughters.

Tesla's father, Milutin (1819–1879), was the eldest of the five children and was later remembered in Tesla's autobiography as "a very erudite man, a veritable natural philosopher, poet and writer and his sermons were said to be as eloquent as those of Abraham a Sancta Clara. He had a prodigious memory and frequently recited at length from works in several languages. He often remarked playfully that if some of the classics were lost he could restore them."[19] Despite receiving a military education in Gospić—the only one available—and eventually enrolling in an Austro-Hungarian military academy at his father's behest, he very quickly learned that he did not share his father's love of military life. He summarily rejected what he saw as the trivialities of military regimentation and was ultimately forced to drop out after an incident in which an officer demanded that he keep his brass buttons polished to a fine shine.[20]

Unlike his brother Josip, who ultimately became a military officer and eventually a professor of mathematics at a military academy in Austria, Milutin would take a very different path. He entered the Orthodox seminary in Plaski (Croatia) and graduated at the top of his class in 1845. Now free from the demands and desires of his father, he became a political activist and fancied himself a poet, writing under the pen-name Srbin Pravicich, "Man of Justice." He championed social justice issues, mandatory education for children, and the creation of Serbian schools throughout all of Croatia.[21] Considered a progressive of his time, he soon attracted the attention of the local intelligentsia, and he would live his life of fighting for common causes, including "nationalism" for his people, until the end of his days.

In 1852, Reverend Milutin penned a letter confirming his beliefs, essentially engraving them in stone. In that letter he wrote: "By God! Nothing is as sacred to me as my church and my forefathers' law and custom, and nothing so precious as liberty, well-being and advancement of my people and my brothers, and for those two, the church and the people, wherever I am, I'll be ready to lay down my life."[22]

∼

It must be appreciated that a Serb's faith in Serbian (Eastern) Orthodox Christianity clashed with many aspects of Roman Catholicism at the time—the Roman Catholic faith still dominates modern-day Croatia. Reverend Milutin understood that to keep his people's religion, culture, and ethnic customs alive, he needed to fight for an education that served all Serbs, for it was paramount to the maintenance of their *Serbianness*. He felt compelled to be the Serb's greatest advocate. As a consequence, Serbs were allowed to have their own Orthodox churches and religious practices after the Illyrian Provinces were returned to the Austrian Empire after the First French Empire. However, because of Austria's effort to keep the aggressive Turks at bay and to hold a tight rein on both the Serbs and Croats, a strong military presence still needed to be maintained on the ethnically charged frontier.[23]

It must be recognized that during the seventeenth to nineteenth centuries the diverse demographics, shifting alliances, and conflicting faiths found in Central and Eastern Europe caused great confusion for all parties concerned. Moreover, that confusion led to disputes on all sides, which resulted in continual wars over many decades.

∼

By 1847, Reverend Milutin was ready to marry and begin a family, for unlike priests of the Roman Catholic faith, Orthodox priests were permitted to marry. He took as his wife Djouka (Đuka—Georgina) Mandić (1822–1892), the twenty-five-year-old daughter of Nikola Mandić, the head of a well-known Serbian family and a highly respected priest in his own right. While the Tesla family produced a preponderance of military men in its early years, the Mandić family's course favored the priesthood, with numerous men in her family achieving great prominence in the Serbian Orthodox Church. Also, several of the Mandić family members as well as many in the Tesla family were of the "educated aristocracy" of the Serbian community. This higher level of education spanned generations and Tesla would prove to be the result of good DNA on both sides.[24] Moreover, descendants on the Mandić side of Tesla's family also produced several inventors. As Tesla recalled in his autobiography about his mother's family, "Both her father and grandfather originated numerous implements for household, agricultural and other uses."[25]

After their wedding Reverend Milutin received word from senior clerics that he was to be assigned his first posting in Senj, a picturesque seaside town that hugged the spectacular coastline of the ancient, blue-green Adriatic Sea, where he was prepared to proselytize his faith to all.[26] A primitively made whitewashed stone structure, seemingly plonked on a precipitous bluff, served as the church where the novice priest welcomed some forty families. His task was to increase the congregation and to represent Serbs before "foreign and Catholic persons."[27]

The stark reality of all Serbs at the time was that their way of life was always under assault while the ability to worship in their own way and to preserve their catechism, ethnicity, traditions, and culture were paramount to who they were. The majority of those around them were Croats who followed the precepts of the Roman Catholic Church, a church that recognized a pope as the leader of the flock and made use of the Latin alphabet. The antithesis was Serbian Orthodoxy. It held tightly to the Greek Church's religious ways, adopted the Byzantine patriarch (Ecumenical Patriarch of Constantinople, who is not the equivalent of the Roman Catholic pope), and wrote using the Cyrillic alphabet. History records that the Cyrillic alphabet and language needed to coalesce the many tongues of the Slavic peoples and was created by brothers and saints Cyril and Methodius in the ninth century.[28]

In 1848, the year after the couple's marriage, they welcomed their first son, Dane (Daniel). Two years later came their first daughter Angelina (grandmother to William Terbo, Tesla's great-nephew) and two years after that, in 1852 arrived Milka, daughter number two.[29] With Djouka working day and night to make a good home, Reverend Milutin proved his mettle and was effective in preaching the "word" as he effervescently evangelized to the misbegotten around him in Senj. In fact, one of his sermons, called "On Labor," proved so remarkable that he was awarded the Red Sash by his local bishop. Yet the existential threat of the Serbs' Christian Orthodoxy being taken away from them remained. As an example, when Reverend Milutin asked the military commander in town to permit Serbian soldiers to participate in Sunday services at his church, the Austrian authorities insisted the Serbs continue attending the Catholic church.[30]

Although Tesla appreciated what his father had done for him in his formative years regarding all manner of daily mental exercises—a nineteenth-century mental bootcamp—such as guessing one another's thoughts, discovering defects of some forms of expression, repeating long sentences or performing mental calculations, all of which led to strengthening his memory, and reason, and critical thinking, it was his mother, Djouka, who was his polestar.[31] She was a woman of inherited ingenuity.[32]

Tesla would often broach the subject of his dear mother throughout his lifetime as his inspiration, as his fount of creativeness. He said of her abilities, "She was a truly great woman, of rare skill, courage and fortitude, who had braved the storms of life and passed thru many a trying experience."[33] He went on to say of her inventiveness and tenacity, which he began to mimic as a young boy, "My mother was an inventor of the first order and would, I believe, have achieved great things had she not been so remote from modern life and its multifold opportunities. She invented and constructed all kinds of tools and devices and wove the finest designs from thread which was spun by her. She worked indefatigably, from break of day till late at night."[34]

Was this a model for young Nikola's own disposition and temperament? From the record of his mother provided much later by Nikola himself, he also described her with increased energy and productivity and short sleep patterns which some might call hyperthymia. Others might call this Type A personality—competitive, highly organized, ambitious, impatient, highly aware of time management and/or aggressive. Or possibly this is more simply an adaptive attitude toward the demands of motherhood and management of the home in that era. We will never know for sure.

He went on to add, "She was also creative and inventive, with most of the wearing apparel and furnishings of the home the product of her hands. When she was past sixty, her fingers were still nimble enough to tie three knots in an eyelash."[35] In his autobiography, he also stated that his mother, the oldest daughter of eight children born into a patriarchal society, was at the age of sixteen saddled with ominous responsibilities, including the preparation for burial of an entire family stricken with cholera.[36] This undoubtedly added to her toughness—her iron will—and her ability to deal with future responsibilities of a grave magnitude. Unknowingly, they were just around the corner.

Reverend Milutin found Senj, a place he and his growing family had lived in for eight years, was becoming increasingly a very difficult posting for him. The salary was meager, the congregation small, and the wet coastal weather was detrimental to his delicate health, so he put in for a transfer to a more acceptable area and climate. Finally, in 1852, after his daughter Milka's birth, his request was granted. He was transferred to the church of St. Apostles Peter and Paul in Smiljan, in the province of Lika, and three miles from Gospić.

Thinking that he had found the perfect place to preach the gospel and advance his causes, he quickly realized that the church was isolated, with the nearest neighbors several miles away—its isolation would be a portent of how young Nikola would live most of his life . . . *alone*. Admittedly, he was given a fine house and enough fertile farmland and cows to raise and feed his growing family.[37] However, the place was not just right, but he would bide his time and make the best of it because he knew that his future was bright.

Reverend Milutin began assembling an impressive library during the time he and his family were in Smiljan. It contained books in German by Goethe and Schiller, encyclopedic works in French by D'Alembert, and other classics, most likely in English, from the eighteenth and nineteenth centuries. Among the multitude of books he possessed, he cherished a copy of *Sluzhebnik*, a Serbian book of church liturgy printed by Božidar Vuković in Venice in 1519. Tesla would ultimately receive this, his father's greatest possession, and happily carry it with him across the Atlantic Ocean to his new home America.[38] He also carried with him to his New World a love of the liberal arts that his father had instilled in him. And as he became a scientist of great repute, education served him very well. He understood what Ada Lovelace (1815–1852) understood during the Romantic Age: the symbiosis between the humanities and sciences is undeniable, and so too did Tesla understand this intricate relationship, as he would go on to found a new revolution in the science of electricity.[39]

Having a library in his own house would in time lead the young Nikola to a lifelong habit of voracious reading, from the hard sciences to the great authors down through the ages. Oddly though, Reverend Milutin's library was off limits to him. He said of his father, "He did not permit it and would fly into a rage when he caught me in the act. He hid the candles when he found I was reading in secret. He did not want me to spoil my eyes. But I obtained tallow, made the wicking and cast sticks into tin forms, and every night I would bush the keyhole and the cracks and read, often until dawn."[40]

SMILJAN, PROVINCE OF LIKA, LATE 1850s

Although the very young Nikola enjoyed playing with his siblings—essentially being raised by his two older sisters most of the time—his mind was often off thinking of other matters. And although he was fortunate to spend his early years in an idyllic environment surrounded by all manner of farm animals from geese and cows to chickens and local-area birds, he simply looked at the world differently than other children of his young age.[41]

Whether swimming with his friends at a local creek during the warmer months or catching frogs with the same, young Tesla was always thinking on a distinctively different wavelength. Then it happened, in what he described as his "first" invention.

> People have often asked me how and when I first began to invent. This I can only answer from my present recollection in the light of which the first attempt I recall was rather ambitious for it involved the invention of an *apparatus* and a method. In the former I was anticipated but the latter was original. It happened this way. One of my playmates had come into the possession of a hook and fishing-tackle which created quite an excitement in the village, and the next morning all started out to catch frogs. I was left alone and deserted owing to a quarrel with this boy. I had never seen a real hook and pictured it as something wonderful, endowed with peculiar qualities, and was despairing not to be one of the party. Urged by necessity, I somehow got hold of a piece of soft iron wire, hammered the end to a sharp point between two stones, bent it into shape, and fastened it to a strong string. I then cut a rod, gathered some bait, and went down to the brook where there were frogs in abundance. But I could not catch any and was almost discouraged when it occurred to me to dangle the empty hook in front of a frog sitting on a stump. At first he collapsed but by and by his eyes bulged out and became bloodshot, he swelled to twice his normal size and made a vicious snap at the hook. Immediately I pulled him up. I tried the same thing again and again and the method proved infallible. When my comrades, who in spite of their fine outfit had caught nothing, came to me they were green with envy. For a long time I kept my secret and enjoyed the monopoly but finally yielded to the spirit of Christmas. Every boy could then do the same and the following summer brought disaster to the frogs.[42]

Despite the frog-on-a-hook trick employed by the young Tesla in the spring and summer, he was more concerned about how to make the best use of the village's waterwheel. As such, he and his friends looked to dam the seasonal flooding to save the lowlands. And yet Tesla continued to look at the actions of the waterwheel. As was often the case, many years down the line his analysis of the workings of the waterwheel formed the essential concept of his groundbreaking bladeless turbine engine.[43] And his uncontrollable need, a *cacoëthes* if you will, to invent was sparked by such an experience. Blessed with his mother's bent for invention, young Tesla was on his way.

Although he could not articulate his philosophy regarding mankind at the time, and he most likely did not even think about it in specific terms, through the decades in America he would rise to the greatest heights as an inventor and discoverer in his quest to "lift the burdens from the shoulders of mankind." Young Tesla's constant analysis of the world around him was highly usual. His first attempts at invention came fast and furious. Yes, he did act as a young boy, on the one hand, but on the other hand, he was decades ahead of his peers in virtually everything he did, especially when it came to basic physics, mathematics, and critical thinking. He unknowingly demonstrated aspects of "phronesis," an ability that dates back to the time of Plato and refers to a special type of wisdom and judgment. It is rarely encountered among most individuals of any age.

Young Tesla often drew inspiration from his environs in the province of Lika. Observing a snowball gaining size as it rolled down a hill gave rise to the concept of "magnification of feeble action," causing Tesla to later consider using the earth's resistance to amplify the impact of an electrical thrust. Unknowingly at the time, his observations had even made a connection with Sir Isaac Newton's laws of motion. He analyzed nature's sequencing: the order of appearance of lightning, thunder, and torrents. He noticed that the power of falling water could be harnessed and that warm winds melted snow, sending massive amounts of water into rivers that would then destroy property and life.[44] The observation of falling water would lead him to the advantages of hydroelectric power.

Later in life Tesla wrote a letter to a Miss Fotitch.[45] It recalled his hyperthymic mother's indefatigable work ethic—rising before 4 a.m., every morning and laboring nonstop until 11 p.m., for her large family—and several of his childhood experiences. He recalled his happiest times with his family, but there was one unusual family member that was nearest and dearest to him. Although the young Tesla truly loved to frolic with his siblings and friends, his favorite companion was the family's beloved black cat Mačak (Serbian for *cat*), the finest of all cats in the world.[46]

He said the cat was scrupulously clean, had no fleas or bugs whatever, shed no hair and showed none of the objectionable traits and habits of cats as he knew them later.[47] He went on to praise his cat, imbuing it with humanlike characteristics. He spent a good deal of time outside with Mačak, especially in autumn and early winter. It was then that something truly magical happened that demonstrated his critical thinking ability, something that would happen to him countless times throughout his life. But the difference was that most individuals would simply not recognize the moment as significant or would not see it at all. But not the young Tesla. He believed that every encounter, whether it was with a person, an animal, or nature itself served a primary purpose—to advance the human condition.

Tesla often wrote and spoke about this fundamental matter of the inventor's charge: "The progressive development of man is vitally dependent on invention. It is the most important product of his creative brain. Its ultimate purpose is the complete mastery of mind over the material world, the harnessing of the forces of nature to human needs. This is the difficult task of the inventor who is often misunderstood and unrewarded. But he finds ample compensation in the pleasing exercises of his powers and the knowledge of being one of that exceptionally privileged class without whom the race would have long ago perished in the bitter struggle against pitiless elements."[48] Throughout his life, Tesla used his powerful mind to master the material world, but he would prove unable to recognize those times when his mind had mastery over him.

~

One very chilly winter day in Smiljan, young Tesla experienced his first "boom" moment. It was to set him on his way to becoming the greatest of all inventors. As he wrote to Miss Fotitch, he explained the monumental moment, which is worth repeating here:

> It happened that on the day of my experience we had a cold drier than ever observed before. People walking in the snow left a luminous trail behind them and a snowball thrown against an obstacle gave a flare of light like a loaf of sugar hit with a knife. It was dusk of the evening and I felt compelled to stroke Mačak's back. Mačak's back was a sheet of light and my hand produced a shower of sparks loud enough to be heard all over the place. My father was a very learned man, he had an answer for every question. But this phenomenon was *new* even to him. Well, he finally remarked, this is nothing but electricity, the same thing you see on the trees in a storm. My mother seemed alarmed. Stop playing with the cat, she said, he might start a fire. I was thinking abstractly. Is nature a gigantic cat? If so, who strokes the cat? It can only be God, I concluded. You may know that Pascal was an extraordinarily precocious child who attracted attention before he reached the age of six years. But there I was, only three years old, and already philosophizing![49]

This was the moment when the small wheels in his nascent mind were turning at warp speed. He went on to emphasize in his letter to Miss Fotitch: "I can not exaggerate the effect of this marvelous sight on my childish imagination. Day after day I asked myself what is electricity and found no answer. Eighty years have gone by since and I still ask the same question, unable to answer it. Some pseudo scientist of whom there are only too many may tell you that he can, but do not believe him. If any of them knew what it is I would also know and the chances are better than any of them for my laboratory and practical experiences are more extensive and my life covers three generations of scientific research.[50]

Early on, young Tesla's inventive spirit produced a basic version of a modern helicopter. His drive to do so is best described by him in his own words: "In my next attempt I seem to have acted under the first instinctive impulse which later dominated me—to harness the energies of nature to the service of man."[51] Was it part of his ingenious and real inventiveness to think he could harness the energies of nature to serve man, or was it grandiosity? Is it an absurd exaggeration if it foretells what really happened? Distinguishing between the real limits to his perception—when he was wrong on the one hand, or his ahead-of-the-curve thinking on the other hand—would come to plague Tesla over his lifetime.

Believing the contraption could gain lift if it had power, he decided to use live Maybugs for the power source. He found an endless supply of them everywhere in the surrounding area. He recalled, "I would attach as many as four of them to a crosspiece, rotably [sic] arranged on a thin spindle, and transmit the motion of the same to a large disc and so derive considerable 'power.'" All went well until a young boy, the son of a retired Austrian military officer, came along and ate the Maybugs. Young Tesla was so horrified by what had unexpectedly happened that he vowed to never again touch another insect.[52]

He also spent a good deal of time making primitive hand weapons of war, for he was "under the spell of Serbian national poetry (his mother would flawlessly recite Serbian poems and ancient stories to him and his siblings) and full of admiration for the feats of the heroes." He added, "I used to spend hours in mowing down my enemies in the form of corn-stalks which ruined the crops and netted me several spankings from my mother. Moreover these were not of the formal kind but the genuine kind. I had all this and more behind me before I was six years old and had past [sic] thru one year of elementary school in the village of Smiljan where I was born."[53]

In between the invention and construction of other devices, he took to taking apart and assembling his grandfather's clocks, one after another. As he described it: "I undertook to take apart and assemble the clocks of my grandfather. In the former operation I was always successful but often failed in the latter. So it came that he brought my work to a sudden halt in a manner not too delicate and it took thirty years before I tackled another clockwork again."[54]

Soon young Tesla set to inventing an even more sophisticated weapon—a pop-gun. He described it as going "into the manufacture of a kind of pop-gun which comprised a hollow tube, a piston, and two plugs of hemp. When firing the gun, the piston was prest [sic] against the stomach and the tube was pushed back quickly with both hands. The air between the plugs was comprest [*sic*] and raised to high temperature and one of them was expelled with a loud report."[55]

There was no doubt that young Tesla had an overly active imagination. Admittedly, most young children have active imaginations to some degree, but his was well beyond the norm. He was using his imaginative powers to conceive and build new inventions and to discover laws of nature that would "lift the burdens from the shoulders of mankind." When one thinks of his earliest inventions, even one powered by Maybugs, one can only conclude that he was operating with a different frame of mind than that of his peers. Yes, he was normal by all outward appearances, and boyhood adventures were his as well—even some that defied death itself. From falling into a giant kettle of boiling milk, just drawn from the paternal herds to nearly drowning after swimming under a raft as well as being carried over a fast-moving waterfall at one of the nearby dams, he was a frequent habitué of boyhood misadventures.[56] But inside him something very different was going on, something was festering. Yes, something was patently wrong. Since he believed that he essentially willed things to happen in his imagination and controlled his thoughts, it came as a shock when he began to experience visual hallucinations. Were these the product of his brain, which he did not control, or of his mind, which he believed he completely controlled?

In his autobiography Tesla describes his late awakening and why:

In my boyhood I suffered from a peculiar affliction due to the appearance of images, often accompanied by strong flashes of light, which marred the sight of real objects and interfered with my thought and action. They were pictures of things and scenes which I had really seen, never of those I imagined. When a word was spoken to me the image of the object it designated would present itself vividly to my vision and sometimes I was quite unable to distinguish whether what I saw was tangible or not. This caused me great discomfort and anxiety. None of the students of psychology or physiology whom I consulted could ever explain satisfactorily these phenomena. They seem to have been unique altho [*sic*] I was probably predisposed as I know that my brother experienced a similar trouble. The theory I have formulated is that the images were the result of a reflex action from the brain on the retina under great excitation. They certainly were not hallucinations such as are produced in diseased and anguished minds, for in other respects I was normal and composed. To give an idea of my distress, suppose that I had witnest [*sic*] a funeral or some such nerve-racking spectacle. Then, inevitably, in the stillness of night, a vivid picture of the scene would thrust itself before my eyes and persist despite all my efforts to banish it. Sometimes it would even remain fixt [*sic*] in space tho [*sic*] I pushed my hand thru it.[57]

If illusions are distortions of images that are really present, and hallucinations are perceptions of visions that are not really present, these seem like descriptions of visual hallucinations. There is no record that Tesla took hallucinatory drugs that provoke visual hallucinations or experienced auditory hallucinations, more common in mood and psychotic disorders. We do not know what mood state accompanied these visions, whether he was upbeat, high, or hypomanic, or low and depressed, nor whether they correlated with periods of great inventive productivity or with gaps in productivity. They seem to have begun as a child before the age of ten, and this type of experience can be associated with premorbid mood disorders that manifest more symptomatically later in life. What we do know is that Tesla did not like these particular visions, and he tried to use his incredible willpower to control them or at least give himself the impression that he could control them and make them go away. Tesla writing about how he fought "peculiar images" did not end there. He followed up in his autobiography with this:

> To free myself from these tormenting appearances, I tried to concentrate my mind on something else I had seen, and in this way I would of ten [often] obtain temporary relief; but in order to get it I had to continuously conjure new images. It was not long before I found that I had exhausted all of those at my command; my "reel" had run out, as it were, because I had seen little of the world—only objects in my home and the immediate surroundings. As I performed these mental operations for the second time or third time, in order to chase the appearances from my vision, the remedy gradually lost all its force. Then I instinctively commenced to make excursions beyond the limits of the small world of which I had knowledge, and I saw new scenes.[58]

As the young Tesla battled his early demons, he was resolute that willpower was his best weapon, but that proved futile at critical times. He finished his analysis of this matter when he said,

> These [images] were at first blurred and indistinct, and would flit away when I tried to concentrate my attention upon them, but by and by I succeeded in fixing them; they gained in strength and distinctness and finally assumed the concreteness of real things. I soon discovered that my best comfort was attained if I simply went on in my vision farther and farther, getting new impressions all the time, and so I began to travel—of course, in my mind. Every night (and sometimes during the day), when alone, I would start on my journeys—see new places, cities and countries—live there, meet people and make friendships and acquaintances and, however unbelievable, it is fact that they were just as dear to me as those in actual life and not a bit less intense in their manifestations.[59]

These visions seem to have come from his brain and not from his mind, and tormented him as opposed to visions of inventions that he felt that he willed from his own efforts. Visualization of this latter type is often done by others who successfully see themselves hit a baseball, catch a football, win a tennis match as part of contemporary sports psychology. Tesla thus seems to have had two types of visualization, one that may have been associated with his mood and came involuntarily from his brain, and another that came from his mind that he produced with effort, made a conceptual leap to an invention, foresaw its design, and then either tested it visually, or experimentally. Tesla did this until about the age of seventeen when his desire for invention took over his primary thoughts, where he was now able to visualize with the greatest of facility, needing no models, drawings, or experiments.[60]

It is manifestly obvious that something serious was going on in young Tesla's mind. Years later, upon reflection chronicled in his autobiography, he said, "To what extent imagination played a part in my early life I may illustrate by another odd experience. Like most children I was fond of jumping and developed an intense desire to support myself in the air. Occasionally a strong wind richly charged with oxygen blew from the mountains rendering my body as light as cork and then I would leap and float in space for a long time. It was a delightful sensation and my disappointment was keen when later I undeceived myself."[61]

Whether these are a manifestation of the normal imagination and grandiosity of a child or warning signs of magical thinking and premonitions of later mood instability remain an interesting speculation. These manifestations as a young boy grew in time. They proved to be the genesis of further mental thoughts and physical actions out of the norm. In his reminiscence, Tesla offered:

> During this period I contracted many strange likes, dislikes and habits, some of which I can trace to external impressions while others are unaccountable. I had a violent aversion against the earrings of women but other ornaments, as bracelets, pleased me more or less according to design. The sight of a pearl would almost give me a fit but I was fascinated with the glitter of crystals or objects with sharp edges and plane surfaces. I would not touch the hair of other people except, perhaps, at the point of a revolver. I would get a fever by looking at a peach and if a piece of camphor was anywhere in the house it caused me the keenest discomfort. Even now I am not insensible to these upsetting impulses. When I drop little squares of paper in a dish filled with liquid, I always sense a peculiar and awful taste in my mouth. I counted the steps in my walks and calculated the cubical contents of soup plates, coffee cup and pieces of food—otherwise my meal was unenjoyable. All repeated acts or operations I performed had to be divisible by three and if I mist [sic] I felt impelled to do it all over again, even if it took hours.[62]

This strangeness demonized him his entire life. There is little doubt that these strange thoughts, present before the age of seven, are the description of classical obsessions and compulsions characteristic of obsessive-compulsive disorder (OCD), and signs of this affliction can be found in the records of Tesla's behavior for the rest of his life.

By 1861, young Tesla had already been entered into school and completed a year at the primary level in Smiljan before by the age of five.[63] There he studied German, arithmetic, and religion. Quite a feat for what would be considered a preschooler today.

~

GOSPIĆ, PROVINCE OF LIKA, AUSTRIAN EMPIRE, 1863

Unexpectedly, Reverend Milutin received a new assignment as the parish priest—where he would preach for the next sixteen years—in the little city of Gospić, some three-plus miles away.[64] The "onion-shaped dome" of the Church of the Great Martyr George served as the parishioners' place of worship. So the family picked up and moved from Smiljan, a quaint village, to a place where too many people would prove too much for young Tesla.

Although the physical move from Smiljan certainly was not dramatic for most of the family, nonetheless, young Tesla found it very traumatic because he was uprooted from the close proximity to Nature. He loved the countryside and the high mountains where he felt free. The "sudden transition to the artificialities of the city was a very definite shock" to his psyche.[65] Despite its isolation, he so loved his life in quiet Smiljan that when he filed for his first patents in America, he named.Smiljan as his home and not Gospić.[66]

Finally, with great effort, the family had settled into their new surroundings. There was an equipoise of calmness, a sense of lagom took root in Gospić for the family, despite the seven-year-old Nikola's feelings of doom and gloom. Reverend Milutin and Djouka were flush with the joy of having had a good life for several years in the village of Smiljan and five healthy children as well as a promising future, but tragedy was about to hit hard like Thor's hammer struck on an enormous anvil upon which tragedy became case-hardened. Their family equilibrium had been punctured: Dane, the family's oldest son had died in an accident. It was a Black Swan event that would ravage their family's psyche and stalk Tesla all his days. In fact, from then on Tesla would face the greatest test of his life: It was a test that he could never pass, for the seven-year-old Nikola Tesla had witnessed his teenage brother's horrible death.

~

Dane was the family's oldest child—"the favored one"—and male. In the patriarchal society of the Serbs, a first-born "male" held great status in the family and all efforts were focused on him to achieve greatness. Tesla remembered his older brother as the one "who was gifted to an extraordinary degree—one of those rare phenomena of mentality which biological investigation has failed to explain."[67] Moreover, Dane, some seven years older than his younger brother, had "foreshadowed in his early years the strange manifestations which in his surviving brother were a prelude to greatness."[68] Sadly, Dane's death was never accepted by the family, for its best hope was gone forever.

Unfortunately, there is more than one story as to how Dane died. The most accepted story tells of Dane riding atop the family's high-spirited Arabian stallion—a gift from a local Bosnian Turkish pasha to Reverend Milutin for assisting in a matter involving several local Muslims.[69] The previous winter the horse had thrown Reverend Milutin and left him for dead, returning home without its rider. Fortunately, the horse was smart enough to lead a search party back to rescue Reverend Milutin. Ironically, one summer day, the very same horse threw Dane, who died of his injuries shortly thereafter. Some biographers have even postulated that young Tesla spooked the horse Dane was riding.[70] Now here's where the story of how Dane died could be apocryphal to some degree; that is, he did indeed die of an accident, but what accident caused his premature death?

A second version of the story has Dane falling down a flight of cellar stairs, and in his semiconscious state of mind he accused his younger brother of pushing him. He died of a head injury, most likely a hematoma. Unfortunately, no hard details remain to verify the truth of either story of Dane's death. However, what can be said is that the tragic incident changed young Tesla, who anguished over his brother's death forevermore, because he felt personally responsible.[71] It was a monumental event about which Tesla said very little in his autobiography.

However, he did remember with specificity a rather unnerving event just after his brother's death. His mother came to his room and awoke him in the middle of the night after his brother had succumbed to his injuries. "It was a dismal night with rain falling in torrents," he recalled. "My mother came to my room, took me in her arms and whispered almost inaudibly: 'Come and kiss Dane.' I pressed my mouth against the ice-cold lips of my brother knowing only that something dreadful had happened. My mother put me again to bed and lingering a little said with tears streaming: 'God gave me one [Nikola] and at midnight he took away the other one [Dane].'"[72]

～

Overcome with grief, the move to Gospić allowed the family—much to young Tesla's resistance—to change the surroundings that so reminded them of their darling Dane. However, Reverend Milutin's life became more reclusive as time went on. He wrote fewer articles and took up fewer causes. He began to "talk to himself and would often carry on animated conversation and indulge in heated argument," even altering his voice so that it appeared that different people were talking. The death of Dane, the family's favored son, stalked Reverend Milutin all his days, and before he was called *home*, he came to be called "Old Man Milovan."[73] His son's tragic death had aged him so—parents never overcome one of their children preceding them in death. In the case of Reverend Milutin, he may have slid into an episode of grief that spilled into a major depressive episode with psychotic thinking.

The death of young Tesla's brother was undoubtedly the tipping-point in his young life. It was not enough that the family had uprooted itself and moved to Gospić after the tragedy, but the death and move both served to dramatically change the relationship with his parents, and his sensitive mental state was knocked completely off kilter. Today we would say he was traumatized. As Tesla so painfully put it years later, "Anything I did that was creditable merely caused my parents to feel their loss more keenly, so I grew up with little confidence in myself."[74] The seven-year-old also told of running away because of his brother's death and secreting himself in an inaccessible mountain chapel visited only once a year. Nightfall came upon the frightened young boy, so he was forced to spend it entombed within, until day broke. "It was an awful experience," he confirmed.[75] Moreover, young Tesla's parents became forever disconsolate and continued to increasingly idolize their dead son Dane's talents and consistently projected accomplishments as if he had lived.[76] This only served to exacerbate his frail psyche. Sadly, this would not be his only encounter with the church.

Young Tesla now faced what he remembered as possibly his most difficult moment as a child one Sunday morning. As he stated: "But my hardest trial came on Sunday when I had to dress up and attend service. There I meet with an accident, the mere thought of which made my blood curdle like sour milk for years afterwards. It was my second adventure in a church."[77] Tesla was referring to his night in the abandoned church that was an awful experience, but he claimed that the second time was worse.

He went on to remember the event. "There was a wealthy lady in town, a good but pompous woman, who used to come to the church gorgeously painted up and attired with an enormous train and attendants. One Sunday I had just finished ringing the bell in the belfry and rushed downstairs when this grand dame was sweeping out and I jumped on her train. It tore off with a ripping noise which sounded like a salvo of musketry fired by raw recruits. My father was livid with rage. He gave me a gentle slap on the cheek, the only corporal punishment he ever administered to me but I almost feel it now. The embarrassment and confusion that followed are indescribable. I was practically ostracized until something else happened which redeemed me in the estimation of the community."[78]

~

The insecure young Tesla was searching feverishly to find approval elsewhere, for his father Reverend Milutin was never to give his overt approval. After all, he lost his first and ideal son, and it was the Serbian father's way. But an odd sort of salvation came along quite by accident to provide a momentary succor for his fragile mindset and need for acceptance. He explains it best in his own words just how it happened:

> An enterprising young merchant had organized a fire department. A new fire engine was purchased, uniforms provided and the men drilled for service and parade. The engine was, in reality, a pump to be worked by sixteen men and was beautifully painted red and black. One afternoon the official trial was prepared for and the machine was transported to the river. The entire population turned out to witness the spectacle. When all the speeches and ceremonies were concluded, the command was given to pump, but not a drop of water came from the nozzle. The professors and experts tried in vain to locate the trouble. The fizzle was complete when I arrived at the scene. My knowledge of the mechanism was nil and I knew next to nothing of air pressure, but instinctively I felt for the suction hose in the water and found that it had collapsed. When I waded in the river and opened it up the water rushed forth and not a few Sunday clothes were spoiled. Archimedes running naked thru the streets of Syracuse and shouting Eureka at the top of his voice did not make a greater impression than myself. I was carried on the shoulders and was the hero of the day.[79]

Young Tesla's instinctive move to solve the problem of the plugged-up fire-hose was just that . . . instinct, and it would serve him well throughout his long and productive life. Yet his instinct could not overcome his feelings of inadequacy as a young boy.

Tesla recalled a particular event regarding his feeling of inadequacy even though he was a clever boy. "One day the Aldermen were passing thru a street where I was at play with other boys. The oldest of these venerable gentlemen—a wealthy citizen—paused to give a silver piece to each of us. Coming to me he suddenly stopt [sic] and commanded, 'Look in my eyes.' I met his gaze, my hand outstretched to receive the much valued coin, when, to my dismay, he said, 'No, not much, you can get nothing from me, you are too smart.'"[80]

As the Tesla family did their best to settle in, hoping Gospić would provide them with the tranquility that they so yearned for after their great loss, young Tesla struggled every day to make sense of it all. Every effort to fit in just did not work. He said of his circumstance, "The change of residence was like a calamity to me. It almost broke my heart to part from our pigeons, chickens, and sheep, and our magnificent flock of geese which used to rise to the clouds in the morning and return from the feeding grounds at sundown in battle formation, so perfect that it would have put a squadron of the best aviators of the present day to shame."[81]

Reverend Milutin intended that his only son follow him into the priesthood, fearing the dangers of military life and the rigors of intense engineering study. Although the religious prospect "constantly oppressed" young Tesla, who "longed to be an engineer," his father proved to be quite inflexible. Reverend Milutin seemed hell-bent on not letting young Tesla become Nikola Tesla.[82] He would use his powers as a Serbian father to get his way. Often it was in a sign language unknown to others. It would be a Serbian father's look, a micro-gesture, or a simple word said in a very commanding way to his son. As a sensitive young boy, he felt such things had great authority, but he was not just a young boy.

It was now time for the young Tesla to commence his tremendous and troubled journey to becoming Nikola Tesla.

· 2 ·

Halcyon Days

Learning the Basics

GOSPIĆ, PROVINCE OF LIKA, AUSTRIAN EMPIRE, 1866

*Y*oung Tesla had now reached the tender age of ten. His curiosity about nature was overflowing, as were his emotions, unstable as they were at times. He
had self-doubt; he had guilt. He had troubling visions; he had obsessions and
compulsions. As he wrote, "I had neither courage or strength to form a firm
resolve."[1] He had to change, and he knew it. It was time for him to learn the
basics. It was time for the grammar schooler "Niko" to show his family that he
could fulfill the promise of his older brother. So, he began a mid-nineteenth-
century version of the ancient Greek *trivium*: reading, writing, and arithmetic.
Indeed, he had vowed to himself that he would become famous and change
the world, and it all began innocently with his discovery of a certain book.

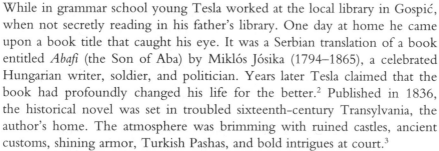

While in grammar school young Tesla worked at the local library in Gospić,
when not secretly reading in his father's library. One day at home he came
upon a book title that caught his eye. It was a Serbian translation of a book
entitled *Abafi* (the Son of Aba) by Miklós Jósika (1794–1865), a celebrated
Hungarian writer, soldier, and politician. Years later Tesla claimed that the
book had profoundly changed his life for the better.[2] Published in 1836,
the historical novel was set in troubled sixteenth-century Transylvania, the
author's home. The atmosphere was brimming with ruined castles, ancient
customs, shining armor, Turkish Pashas, and bold intrigues at court.[3]

    The protagonist, a fictitious lord named Oliver Abafi, emerges as the hero
of the story. He remakes himself through true grit from being disruptive and
frivolous to being noble and disciplined to the extent that he gives himself
over to his prince and country. This compelling story resonated so deeply with

the impressionable young Tesla, that even at such a young age, he decided to focus his vast intellectual powers to control his feelings and actions.[4]

As Tesla said later in life, "My feelings came in waves and surges and vibrated unceasingly between extremes. My wishes were of consuming force and like the heads of the hydra, they multiplied. I was opprest [*sic*] by thoughts of pain in life and death and religious fear. I was swayed by superstitious belief and lived in constant dread of the spirit of evil, of ghosts and ogres and other unholy monsters in the dark. On one occasion I came across a novel entitled 'Abafi.'"[5]

It was a seminal moment for young Tesla, for he saw himself as a new person, as reborn. He wrote in his autobiography how the process of change took place. "This work somehow awakened my dormant powers of will and I began to practice self-control. At first my resolutions faded like snow in April, but in a little while I conquered my weakness and felt pleasure I never knew before—that of doing as I willed. In the course of time this vigorous mental exercise became second nature. At the outset my wishes had to be subdued but gradually desire and will grew to be identical. After years of such discipline I gained so complete a mastery over myself that I toyed with passions which have meant destruction to some of the strongest men."[6]

Still too young to fully process what was happening with his psychological conversion, nonetheless, by developing his self-control and learning to direct his powerful imagination, young Tesla had begun to acquire the mental skills that would serve him well as an inventor. Moreover, he would be able to unearth and explore new ideas in his mind, and also have the discipline and focus he would need to shape and mentally edit these ideas into actual devices.[7] As he developed these mental skills, in time he would be able to challenge, if not reject "received wisdom" at every turn in an effort to blaze new trails in science, to think new thoughts, and to discover new concepts and laws of nature.

Young Tesla was not aware at the time, but he was in the preliminary stages of creating his very own personal "Superman."[8] He was a "superboy" becoming a superman.[9] Was it his prodigious willpower that forged this superman or was he assisted by the mass forces of mania as well? Over the decades he would become purposely impervious to the outside influences of others; thus, in some ways, severing the connective tissue that binds the human race—love. As a grown man he consciously made the hard decision to leave no progeny, hence, no disciples to carry on his magnificent work. He was the architect of his own life, and he was the kryptonite that would eventually end it. He never needed any other scientists to collaborate with because he believed them to be nothing more than mere distractions. It was Sir Francis Bacon who denounced Aristotle's metaphysics (received wisdom) when he published his *New Organon*. And it was Nikola Tesla who denounced the received wisdom of the day to forge a new science. He had dismissed what had come before (distractions) and what Bacon called "idols."[10] Tesla wanted no one to influence his thinking and methodology. So, as his skills increased in complexity, so did his purpose, which can be best expressed by his tireless need to make man's lot in life easier. He would deny himself personal relationships, particularly with women, for his superman could not handle such interactions.[11] Thus, he was destined to live his long life virtually alone.

Because of his exceptionally complex psyche, young Tesla, the boy, understood that he was not normal. He had visions and he had a force within that he still had not harnessed. What amused him and what drew his attention were different than other boys his age, and hence, caused him to isolate himself much of the time. Yet despite these great differences, and the fact that what he did was always greater or better than his peers or unable to be done by them, he was nonetheless still a boy at the molecular level and did what normal boys did on occasion.[12]

By now young Tesla was comfortable with his nickname "Niko" that his family had bestowed on him as an infant, yet he still found his mental health troublesome, his physical health challenging, and his imagining to be too plentiful. When these thoughts consumed him, he sought isolation much of the time. One aunt recalled that when young Tesla was with relatives "he would always like to be alone."[13] He was aloof, some would say socially withdrawn, even schizoid. The modern medical perspective might be that he suffered a bit from Asperger's syndrome, but even that term is out of vogue. For sure, he was unique. He would go off in the morning "into the woods to meditate. He measured one tree to another, making notes, experimenting." Such introverted actions put a fright in most locals. They would approach and say, "We're sorry; your [cousin] seems to be crazy."[14] (Today this is called autism spectrum disorder. Asperger's syndrome is considered a "high-functioning" type of autism spectrum disorder, with mostly lack of social skills or lack of exhibiting emotions but without intellectual disability.)

Tesla's need to be alone, although untenable by most people, is mirrored in other great thinkers throughout the ages. The great Sir Isaac Newton, physicist/natural philosopher extraordinaire, preferred his own company, knowing that he could enjoy the benefits of undisturbed thought—"Truth is the offspring of silence and meditation."[15]

During young Tesla's three years at the Real Gymnasium (the nineteenth-century version of junior high school) in Gospić, he had become enamored of demonstration models of waterwheels and turbines and other scientific apparatuses, both electrical and mechanical.[16] He was so taken by them that he built exact replicas of several models and tested them in a nearby stream. One day he proudly showed his uncle the waterwheels and turbines he had constructed with his own hands, and much like the negative reactions he often received from his father, his Uncle Josip dismissed them out of hand as well, not appreciating the youngster's mental ingenuity and admonishing him for wasting valuable time. Thankfully, he continued to think about turbines, and when he read about the mighty Niagara Falls and then saw a metal engraving of the natural wonder, he dreamed of using a giant waterwheel to capture the power of the falls. Young Tesla told his uncle that he would one day travel to America and do just that. In fact, he recalled: "Thirty years later I saw my ideas carried out at Niagara."[17] He also remembered the significance of his time in the Real Gymnasium's department of physics in Gospić as "undoubtedly a powerful incentive to invention."[18]

While there, he excelled in mathematics, much like his erudite father and Uncle Josip, both of whom were said to have taught at the new institution during the time young Tesla was a student.[19] He performed mathematical feats of quick and accurate complex calculations before his peers and professors.[20] It was his ability to visualize things in his mind's eye—he would go on to demonstrate his abilities of photographic, eidetic, and didactic memories that would serve him well as an inventor and scientific discoverer—that would garner plaudits from all who saw what appeared to them to be mathematical magic. Was his development of this incomparable "mind's eye" the fortuitous product of having more bizarre hallucinatory experiences that he harnessed and learned to use to visualize his inventions, and even build them and test them all in his mind? He also said of his mathematical skills, "Up to a certain complexity it was absolutely the same to me whether I wrote the symbols on the board or conjured them before my mental vision."[21] That said, he simply had no gift for freehand drawing, to which many hours were devoted in school. He attributed his lack of such a talent in this area to his preference for "undisturbed" thought, even though "most of his family excelled in it."[22] Young Tesla also excused himself from the need to be proficient at the subject of drawing because he began life as left-handed—out of necessity he later became ambidextrous. He went on to bemoan the fact that "it was a serious handicap as under the then existing educational system, drawing being obligatory, this deficiency threatened to spoil my whole career and father had considerable trouble in railroading me from one class to another."[23]

By his second year at the Real Gymnasium, he became preoccupied with the thought of producing continuous motion through steady air pressure, hence a flying machine. He had said that the earlier incident with the fire engine pump had "set afire my youthful imagination and imprest [*sic*] me with the boundless abilities of a vacuum."[24] He grappled with the idea of a way to combine a vacuum with the fact that the air on the atmosphere is under pressure at fourteen pounds per square inch.[25] But ultimately he built a model and hoped that he would achieve mechanical flight. It was a dream of his and something that he said made him "delirious with joy."[26]

A further recollection on the subject of flight had him stating emphatically, "Every day I used to transport myself thru [sic] the air to distant regions but could not understand just how I managed to do it. Now I had something concrete—a flying machine with nothing more than a rotating shaft, flapping wings, and—a vacuum of unlimited power! From that time on I made my daily [imaginary] aerial excursions in a vehicle of comfort and luxury as might have befitted King Solomon. It took years before I understood that the atmospheric pressure acted at right angles to the surface of the cylinder and that the slight rotary effort I observed was due to a leak. Tho [sic] this knowledge came gradually it gave me a painful shock."[27] Young Tesla had indicated that he had wished that he could build a real flying machine that would connect his dreams with reality.[28]

At the time, this not only may have appeared fanciful and bizarre to young Tesla, but frankly delusional. Helicopters? Flight? Teleporting around the earth at 1,000 mph? Wireless communication? All completely harebrained, except he made the last one actually happen. Tesla had a hard time telling fancy from genius and impossibility from imminent invention. Is this the benefit yet the curse of manic genius?

Aside from his unquenchable need to invent, by the age of twelve, acts of self-denial and self-mastery marked the leitmotif of Tesla's young life, and he would live with such troubling idiosyncrasies all his days.[29] In addition to struggling with his bizarre thoughts, he fought to maintain his health. From time to time, several relatives offered home cures to help build up his strength. Oddly enough, once he had achieved adulthood, he demonstrated considerable agility and strength.[30] In fact, his physical health improved so much that he lived nearly eighty-seven productive years.

That said, such was not the case by the time he had completed his three-year curriculum at the Real Gymnasium in 1870. He stated in his personal life's story: "I was prostrated with a dangerous illness or rather, a score of them, and my condition became so desperate that I was given up by physicians. During this period I was permitted to read constantly, obtaining books from the Public Library which had been neglected and trusted to me for classification of the works and preparation of the catalogues. One day I was handed a few volumes of new literature unlike anything I had ever read before and so captivating as to make me utterly forget my hopeless state. They were the earlier works of Mark Twain and to them might have been due the miraculous recovery which followed. Twenty-five years later, when I met Mr. Clemens and we formed a friendship between us, I told him of the experience and was amazed to see that great man of laughter burst into tears."[31] Although young Tesla was known to have experienced medical illnesses including malaria and cholera, this one was different. Nobody could diagnose his physical or psychological problems. In retrospect, it seems possible that this was his first major depressive episode.

KARLOVAC (KARLSTADT),
CROATIA, AUSTRO-HUNGARIAN EMPIRE, 1870

Once young Tesla had regained his strength and returned to somewhat normal activities, his father Reverend Milutin sent him off for more intensive studies, this time to the Higher Real Gymnasium (a nineteenth-century version of a middle to high school program of classical education that prepared smart students for university and where all classes were taught in German) on Rakovac Street just outside the city of Karlovac, a swampy area located on a tributary of the Sava River near Zagreb.[32]

The city began as a fortress in 1579 to repel the advance of the Muslim Turks. Some 9.5 miles from Gospić, it was named after Charles II, Archduke of Austria. Needless to say, the move for young Tesla was not easy. Compared to Gospić, moving to Karlovac was akin to leaving a village for the big city, full of all manner of distractions, but blessed with a first-rate gymnasium. However, because of his variable state of physical health, he suffered repeated bouts with malaria in the fetid, swampy land that made up the city of Karlovac. And because of such trying physical times, he had to resort to copious amounts of quinine on a regular basis. Yet this did not discourage or prevent him from developing a passionate interest in "electricity" under the stimulating influence of his physics professor, Martin Sekulić, "an ingenious man and one who often demonstrated the principles by apparatus of his own invention."[33] In fact, "Nikola read all that he could find on the subject . . . [and] experimented with batteries and induction coils."[34]

Moreover, while at the Higher Real Gymnasium, he also studied languages and mathematics with great vigor. He knew he had a facility for complex mathematics, believing that "there is scarcely a subject that cannot be mathematically treated and the effects calculated or the results determined beforehand from the available theoretical and practical data."[35] But now that same ability was manifested in his extraordinary capacity for languages.[36]

〜

The developing young Tesla lived in Karlovac for three years with his Aunt Stanka, his father's sister, and Uncle Branković, a rough-and-ready colonel in the military, whom young Tesla thought of as "an old war horse."[37] Decades later Tesla put it so aptly when he remembered the years he spent with his aunt and uncle:

> I can never forget the three years I past [passed] at their home. No fortress in time of war was under a more rigid discipline. I was fed like a canary. All the meals were of highest quality and deliciously prepared but short in quantity by a thousand percent. The slices of ham cut by my aunt were like tissue paper. Whenever the Colonel would put something substantial on my plate she would snatch it away and say excitedly to him: "Be careful, Niko is very delicate." I had a voracious appetite and suffered like Tantalus. But I lived in an atmosphere of refinement and artistic taste quite unusual for those times and conditions.[38]

Through very hard work, he had been able to reduce four years of schooling into three and began to work out a way to approach his obdurate father with his controversial decision not to enter the ministry but to study engineering instead.[39] He would recollect that "during all those years my parents never wavered in their resolve to make me embrace the clergy, the mere thought of which filled me with dread."[40] He even pleaded with his father to understand, for "it is not humans that I love, but humanity."[41]

Yes, young Tesla did indeed love humanity. It was his most powerful driving force that played out during his entire life, as he focused his unique brilliance on "lifting the burdens from the shoulders of mankind." Indeed, he was interested in mankind, but maybe not in individual men and women, preferring a life of relative isolation especially as the years passed, consistent with a more schizoid demeanor.

"At last, however, my course was completed, the misery ended, and I obtained the certificate of maturity which brought me to the cross-roads."[42]

GOSPIĆ, PROVINCE OF LIKA, AUSTRO-HUNGARIAN EMPIRE, 1873

"Just as I was making ready for the long journey home I received word that my father wished me to go on a shooting expedition. It was a strange request as he had been always strenuously opposed to this kind of sport. But a few days later I learned that the cholera was raging in that district and, taking advantage of an opportunity, I returned to Gospić in disregard of my parents' wishes."[43]

The stench of putrefied corpses, noxious fumes, and floating coal soot hung heavy over the small town, making breathing all but impossible at times for many. "I contracted the awful disease on the very day of my arrival and altho [*sic*] surviving the crisis, I was confined to bed for nine months with scarcely any ability to move."[44] There were times during his forced quarantine that he experienced "dropsy, pulmonary trouble, and all sorts of diseases until finally my coffin was ordered." His daily life was plagued by vomiting, shriveled skin, and diarrhea.[45] But were it not for young Tesla's history of physical health issues, he could have very well avoided the disease altogether.[46]

Now in a desperate state, lingering on the precipice of death, and confined to bed for months on end, he ate little and oftentimes regurgitated what he did eat, while he struggled to breathe, gasping in the poisoned air. "My energy was completely exhausted and for the second time I found myself at death's door. In one of my sinking spells which was thought to be the last, my father rushed into the room. I still see his pallid face as he tried to cheer me in tones belying his assurance. 'Perhaps,' I said, 'I may get well if you will let me study engineering.'"[47] After considerable thought and some obvious soul-searching, young Tesla's protective father fought his strongest beliefs and acquiesced, stating emphatically, "You will go to the best technical institution in the world."[48] One can only imagine that a wide smile must have painted itself on the teenager's ashen face, because he had carried the burdensome weight of uncertainty for so long, believing his father would never let him be an engineer, but now he was convinced his father meant what he said.

With his father's acquiescence about his career and the restoration of his health after being at death's door, young Tesla's future was beginning to look bright. Indeed, perhaps the worst was behind him, but there would be many trials ahead. One of the first he had to face was the unexpected issue of local bullies who now invaded his world. Tesla explained years later, "There was a tough in our town who once thrashed a friend of mine because he parted his hair in the middle, and I henceforth resolved to do the same. I received likewise a thrashing which, for thoroughness, left nothing to be desired." He suffered such a painful indignation and beatings for at least a year. But in the end, he learned to box, and he soon "had the satisfaction of returning the favors." Ironically, before young Tesla could get his desired justice, "the fellow was shot in a brawl."[49] Later in life Tesla befriended many boxers because of his youthful experiences with the practical sport.

The revived and reinvigorated young Tesla now felt free, free to pursue his dreams and to serve humanity. By now he and his parents had decided on the highly acclaimed and prestigious Joanneum Polytechnic School of Graz, Austria, located some 226 miles north of Gospić. But first he faced an existential dilemma that could very well negatively impact his future. He was now of the age when each young man in its (Austria) territories had to serve in the army for three years. Given that his father now feared the loss of his remaining son, he would do what had to be done to have his young Nikola avoid military service. Some believe that inasmuch as Reverend Milutin's family, along with his wife's family, had a history of meritorious military service that young Tesla's avoiding such a duty seems counter to his family's tradition and history—no record of a military deferment exists.[50] Supposition suggests that perhaps because of his history of poor health or through family contacts or even bribery young Tesla was able to avoid conscription even though such deception was a major offense.[51] It is interesting to note that Tesla's autobiography barely mentions the incident other than to say that his father insisted he "spend a year undertaking healthy physical outdoor exercise."[52]

So, from the early fall of 1874 until the summer of the following year, young Tesla spent nine months roaming the rugged mountains and valleys of Croatia "loaded with a hunter's outfit and a bundle of books."[53] The area's impressive biodiversity brought him even closer to nature. As he kept on the move traversing the territory, he became physically stronger both in body and mind. He mentally worked on various dreamed-of inventions, albeit somewhat harebrained for the time. One in particular was a system wherein he could transport letters and packages between continents via a pipe under the ocean. Mail placed in spherical containers would be shot through a pipe using hydraulic pressure. Once he learned that there were difficult physical technicalities to his plan, he was forced to abort his idea and move on.[54]

As the months ticked by, the eager young Tesla hatched another inventive, yet misguided plot. He would attach a ring around the Earth's equator in order to improve passenger travel, speeding travelers along at one thousand miles per hour. It too had to be aborted. But these visionary inventions demonstrate that he wanted to utilize everything the Earth, hence nature, had to offer—power that was forever being dissipated and replenished.[55] Moreover, he was thinking about the future, and his early visionary inventions were prescient, for they were precursors to the transmission of wireless power and the internet.[56]

Evidence of young Tesla's early thoughts on "wireless communication" can be found on a train he was traveling on in September 1870, which was to take him to Karlovac for additional studies. While on the train he talked to another student journeying to another school for the year. He told his companion that "he was thinking about a device which will transmit conversation between America and Europe without wires."[57]

Additionally, his time in the wilderness did reveal to him one of nature's special powers. One day while throwing snowballs with friends next to a mountainside, the concept of "hidden trigger mechanisms" able to unleash immense storehouses of energy was revealed to him. "One [snowball] . . . found just the right conditions; it rolled until it was a large ball and then spread out rolling up the snow at the sides as if it were a giant carpet, and then suddenly it turned into an avalanche . . . stripping the mountainside clear of snow, trees, soil and everything else it could carry with it."[58]

In addition, his curious mind, working freely during his time in the wilderness, proved to him that he could focus it on creating new inventions virtually out of the ether—organic in every sense. Ironically, while the loneliness he endured during his nine-month sojourn in the wilderness aided his power of focusing on inventions, that very same loneliness would become one of his life's subplots that would define his life in many ways. He would later in life say, "I observed to my delight that I could visualize with the greatest facility. I needed no models, drawings or experiments. I could picture them all as real in my mind."[59] This remarkable ability of conjuring up mental images with extraordinary accuracy is known as *hyperthymesia*, or highly superior autobiographical memory (HSAM), and it allowed him to dissect and learn the ideal behind an invention.[60]

Moreover, by now young Tesla's gift of phronesis was also continuing to grow in strength and intensity. In his curiosity and desire to find and understand the "ideal" underpinning of an invention, young Tesla drew from his religious beliefs that he learned from the time he could speak and understand language. His father and uncles were clerics of the Serbian Orthodox Church. And like all Christians, the Orthodox believe in the Trinity, that God is three persons in one: the Father, the Son, and the Spirit. But, unlike in Western Christianity, for Orthodox Christians "the Son of God is the Word" has a much deeper meaning. Hence, for Orthodox Christians, "the material universe" is not simply orderly but everything in it, whether natural or manmade, has an essential principle, a logos that can be discovered by mere humans. So, as an Orthodox Christian, Tesla felt compelled to seek out the ideal in his inventions. It's worth noting that although Tesla was inspired by his faith and its worldview and thus, he learned to look for the ideal in every invention, he was not a practicing Christian in the traditional sense—no records exist of him having attended regular Orthodox Church services (except as a youth in his father's church) or any other form of obvious practice of his faith.[61]

$\sim$

While young Tesla, the explorer, was wandering the mountains and valleys of Croatia, his father Reverend Milutin was busy securing a scholarship for his son from the Military Frontier Administration Authority. The document confirming the scholarship was issued in Zagreb in September of 1876, and it was addressed to the rector of the Joanneum Polytechnic School in Graz, in the province of Styria. The document was very explicit in its request: "Nikolaus Tesla, of Gospić in Lika, the Military District of Otocac . . . in return for scholarship of 420 gulden per year, has pledged that, on conclusion of his studies, he will discharge his military duty and serve a minimum of eight years in the army."[62]

Not focusing on the problematic side of his scholarship acceptance, young Tesla was excited beyond his wildest dreams, forever fantasizing about attending an engineering school. So, that fall, the colorful, embroidered shoulder bag his mother had made for him slung casually across his tall, lean body, he set-off for new horizons, first in Graz. The embroidered shoulder bag and his steadfastness and determination and honor as a Serb would one day accompany him across a vast ocean to America. There he would become world famous, but his bag and his values would never leave him. They were symbols of who he truly was. They were reminders that he should never sacrifice them for any reason. He would ultimately live a life of "principle over profit." However, that noble philosophy would allow opportunists to take advantage of him in the most egregious of ways . . . and they did!

"At the termination of my vacation I was sent to the Polytechnic School in Gratz, Styria, which my father had chosen as one of the oldest and best reputed institutions. That was the moment I eagerly awaited and I began my studies under good auspices and firmly resolved to succeed. My previous training was above the average, due to my father's teaching and opportunities afforded. I had acquired the knowledge of a number of languages and waded thru [sic] the books of several libraries, picking up information more or less useful. Then again, for the first time, I could choose my subjects as I liked, and free-hand drawing was to bother me no more."[63]

This was the moment that the young Nikola Tesla felt truly free. But was he? His bizarre thoughts still encircled him. His wild images came in and out of focus. The inclement weather was no help, given his fragile physical condition. And his insecurities were often predominant. However, at times he was able to mask them with distractions, some of which would get him into trouble all too often while in Graz.

GRAZ, AUSTRIA, AUSTRO-HUNGARIAN EMPIRE, 1875

The excited young Tesla was now an official student at the prestigious school where the likes of physicist and philosopher Ernst Mach (Mach number) had entered years before and where he had been a professor. Renowned experimental physiologist, physicist, and philosopher Gustav Theodor Fechner also taught at the school for several years. It was the school the student Tesla wanted and needed to attend.[64] Moreover, to save on expenses, he roomed

with Kosta Kulishich, a fellow Serb whom he befriended at the Student Society of Serbia. Kulishich subsequently became a highly regarded professor of
philosophy in Belgrade (capital of present-day Serbia).[65]

From the start, the eager student took classes in arithmetical subjects and
geometry from Professor Rogner, a man known for his dramatic presentations.[66] He added integral calculus and differential equations taught by Professor Allé, whom Tesla said "was the most brilliant lecturer to whom I ever
listened. He took a special interest in my progress and would frequently remain for an hour or two in the lecture room, giving me problems to solve, in
which I delighted. To him I explained a flying machine I had conceived, not
an illusory invention, but one based on sound, scientific principles, which has
become realizable thru [sic] my turbine and will soon be given to the world."[67]
Here we now understand that Tesla's early thought experiments with his
bladeless turbine technology began as a very young child while playing with
other boys on a stream near his home in Smiljan, seemingly spontaneously.
His method of increasingly adding to his thought experiments never deserted
him, and many inventions that became reality began as nothing more than a
dreamer's ambition. Even his thoughts of flying harkened back to stories told
to him as a young lad at the knee of his grandfather Nikola about Napoleon's
enjoyment of hot-air balloons, which were used for monitoring enemy troop
movements and for dropping bombs.[68]

Other courses included analytical chemistry, mineralogy, machinery
construction, botany, wave theory, optics, French, and English as well as the
Slavic dialects. Hungry to learn it all, besides the hard sciences, he read widely
the works of Shakespeare, Goethe, Descartes, and Spenser. It has been said that
he could speak at least nine languages fluently as he grew older.[69]

So how did the new student master so many subjects so quickly and
with so much success? It certainly helped that his mind was turbo-charged by
having been gifted by God with photographic, eidetic, and didactic memory.
But there was something else at work. He had a force within himself that
was not simply genius alone. This time in his life is consistent with having an
extended period of hypomania or bouts of mania, with diminished need for
sleep. He was also gambling and womanizing, completely out of character for
him during almost the remainder of his life. And he told us himself decades
later how he did it. He said, "I had a veritable mania for finishing whatever I
began." His mania was so forceful that he took to reading the hundred large
volumes, in small print, of Voltaire. Although the task had cured him of ever
doing another self-appointed assignment, it did nothing to suppress the pattern
of relentless self-denial and self-determination.[70] However, as he said upon
reflection, "It had to be done, but when I laid aside the last book I was very
glad, and said, 'Never more!'"[71]

Of the many great professors young Tesla studied under, one stood out above them all, one who would point him toward the mysteries of "electricity." Professor Jacob Poeschl was chair of the theoretical and experimental physics department. "Prof. Poeschl was a methodical and thoroly [*sic*] grounded German. He had enormous feet and hands like the paws of a bear, but all of his experiments were skillfully performed with clock-like precision and without a miss."[72] It was said of the professor that he wore the same coat for twenty years, adding to his peculiarity. But despite the professor's oddness, he was a classic nineteenth-century lecturer in electricity, who most likely provided a historical overview of the subject beginning with the ancient Greeks and concluding with the most recent developments of dynamos and electric lighting.[73]

It is worth mentioning that all of young Tesla's classes were taught in German, a language that has absolutely no connection to his native language—then called Serbo-Croatian. And to be contextual, the transition from Serbo-Croatian to a non-Cyrillic based language such as German or French or especially the language he would eventually speak for most of his life, English, was an extraordinary feat all its own and worthy of recognition of his unique facility for languages—a bona fide polyglot.

∼

After his freshman year, young Tesla returned home, a brilliant intellectual warrior, and pinned to his coat were medals for all A+s—having passed his freshman classes brilliantly. His expectation was to be hailed a great academic; after all, his first-year professors had already done so. But it was not to be, for his father was insistent that he did not return to school. Why? Because his teachers had secretly written Reverend Milutin, warning that the boy was at risk of injuring his health by his neurotically long and intense hours of study, oftentimes up to twenty hours per day. Adding to his disappointment was the fact that the Military Frontier Authority had been terminated and the invaluable scholarship was no more.[74] This one-two punch, as he later wrote, "almost killed my ambition."[75] The outstanding student, who wanted to be a professor of mathematics and physics, began to doubt, if only for a fleeting moment, whether there was real value in studying so hard.[76]

In the end, the brilliant student Nikola Tesla would have none of the negativity and returned to school.

∼

He was often ridiculed by fellow students for his monastic study habits, his outstanding academic success, and his having been befriended by many professors. His reaction was to go carousing with the best of them. As Professor

Kosta Kulishich, Tesla's former roommate at the Polytechnic School described it in 1936: "At first Tesla was very studious. His professors praised him for this, and the jealousy of other students was aroused. One day, when he was loitering in the halls of the college building, a German student slapped him on the back and said, 'You are wasting your time here. It would be better for you to go home and sit and study, so that your professors could praise you more.' He answered the student: 'I know I can study better than you, but I'll show you that I can also be better at having a good time.' This usually includes a certain amount of drinking, but Tesla drank nothing but coffee. He began to stay late at the Botanical Garden, the students' favorite coffee house, playing cards, billiards and chess. His skill was such that a crowd always gather round to watch when he was playing. Even during this period, however, he seldom missed a lecture, which is a rare record in Continental universities, where students may miss as many as they like."[77]

Adding to Professor Kulishich's insight into his friend Tesla, one evening in 1879, he was in Maribor looking for a teaching position when he ran into Tesla playing cards with two men in the Happy Peasant pub. He asked him, "What's with you Nikola, by God, we've been looking for you everywhere, especially your parents." Tesla responded somewhat harshly, "I like it here; I work for an engineer, receive sixty forints a month, and can earn a little more for every completed project."[78]

It was not the answer his former roommate wanted, and he pushed further and stated, "No, you must return to Graz, to finish your studies." "We'll see," Nikola said and turned away.[79]

It was during young Tesla's sophomore year (1876–1877) at the Polytechnic School that he experienced the first of many watershed moments in his tremendous and troubled life. In the course of a lecture delivered by Professor Poeschl, he was introduced to an example of a direct-current Gramme dynamo recently delivered from Paris to the physics class.[80] As Tesla described it decades later, "having the horseshoe form of a laminated field magnet, and a wire-wound armature with a commutator."[81] "It was connected up and various effects of the currents were shown. While Prof. Poeschl was making demonstrations, running the machine as a motor, the brushes gave trouble, sparking badly, and I observed that it might be possible to operate a motor without these appliances."[82] The customary commutator showed itself to be at the very core of the direct-current versus alternating-current debate. The professor publicly chided him and rejected young Tesla's comments out of hand. Professor Poeschl was simply convinced that the commutator-less motor broke the laws of nature.[83]

In fact, Professor Poeschl's insistence on the indispensability of the "commutator" and Tesla's reaction clearly demonstrated Tesla's confidence that he was right, and no amount of "received wisdom" endorsed by the professor or anyone else was ever going to intimidate him in the long run. He knew he could go the distance. It was many decades later that Larry Page, the co-founder of Google, echoed Tesla's belief in himself, when he said, "There is a phrase I learned in college called, 'Having a healthy disregard for the impossible.' That is a really good phrase. You should try to do things that most people would not do."[84]

Tesla continued to push the envelope with his professor, suggesting that the dynamo was far more complicated than it needed to be and by eliminating its commutator and brushes, much of its functionality would be improved—most true geniuses such as Kepler, Newton . . . and even Steve Jobs have an instinct for "simplicity."[85] With a look that suggested the comment was thoroughly annoying, the professor chastised Tesla before his classmates: "Mr. Tesla will accomplish great things, but he will certainly never do this [alternating current]. It would be equivalent to converting a steady pulling force like gravity into rotary effort. It is a perpetual motion scheme, an impossible idea," the tradition-bound Professor Poeschl replied with a sternness expected of such a man.[86] Although Tesla thought Professor Poeschl a brilliant teacher, and the one who had systematically introduced him to the science of electricity, his statement was received by Tesla as a direct challenge.[87] "For a time I wavered," he wrote, impressed "by the professor's authority, but soon became convinced I was right and undertook the task with all the fire and boundless confidence of youth."[88] Tesla would spend the next several years proving the ossified-in-thought professor wrong.

Tesla's method for tackling the problem was unique. "I started by first picturing in my mind a direct-current machine, running it and following the changing flow of the currents in the armature. Then I would imagine an alternator and investigate the processes taking place in a similar manner. Next I would visualize systems comprising motors and generators and operate them in various ways. The images I saw [in my mind] were to me perfectly real and tangible. All my remaining term in Gratz was passed in intense but fruitless efforts of this kind, and I almost came to the conclusion that the problem was insolvable."[89]

∽

Although Tesla remained undeterred, he was facing existential pressures from all ends. What to do? He returned to Graz in his third year (fall 1877) but stopped attending classes, and school records show he was not registered for the spring of 1878. Obviously, the cancellation of his military scholarship was partially caused by such behavior. Moreover, he was consumed with his not being able to solve the problem of AC—alternating current. The rejection and frustration proved too much for him. Soon he fell in with a group of social miscreants and began gambling, sometimes for twenty-four hours without ever sleeping. Because of his advanced mathematical skills, winning was rarely in doubt. Although his kindness and generosity dictated that he return his winnings to big losers, no one ever reciprocated. However, during one card game he lost his entire allowance, and hence, his tuition money. His father Reverend Milutin was incensed when he learned of the matter, with Tesla swearing he could "stop whenever I please." Later his mother decided to stake him, believing her son would lose it all and end his gambling habit and said to him, "Go and enjoy yourself. The sooner you lose all we possess the better it will be. I know that you will get over it."[90] Ironically, by winning all his initial losses back (and giving winnings back to heavy losers) and returning the remainder to his family, he forswore to never gamble again, thinking he had conquered his dreadful habit. He said emphatically of the incident, "I conquered my passion then and there and only regretted that it had not been a hundred times as strong. I not only vanquished but tore it from my heart so as not to leave even a trace of desire. Ever since that time I have been as indifferent to any form of gambling as to picking teeth."[91] But there would be other times when the temptations simply were too great. Even when he arrived in the United States, he quickly became a very skilled billiard player, and he was known at times to understate his abilities at sport when playing with wealthy socialites, hence, leaving their wallets somewhat lighter at the end of the day. But in truth, he did indeed "gamble" like a libertine with his future.[92]

That senseless act of gambling was perceived of as an affront to his family and a personal betrayal to his father, a man of the greatest integrity. It so troubled Reverend Milutin that he returned home from seeing his son in Maribor, a city where Tesla had been quietly working in a tool and die shop after he dropped out of school (he never graduated) and gambling and playing chess (at this point Tesla recognized his gambling addiction to be a "mania") at night in seedy places throughout the city. In fact, a few weeks later Tesla was arrested for vagrancy and sent home to Gospić, never to return.

GOSPIĆ, PROVINCE OF LIKA, AUSTRO-HUNGARIAN EMPIRE, 1879

Tesla's antics were more than his father could take, and without warning, Reverend Milutin Tesla became very ill. He died of a broken heart at the age of sixty on April 17, 1979 (Old Style). It was said that his funeral liturgy was fit for a saint.[93]

Thankfully, Tesla managed, through sheer willpower, to exorcise the demon of smoking from his life—at one time he was consuming upward of twenty black cigars a day. Very concerned about his already challenged state of health, he also stopped drinking coffee, having destroyed even the urge to indulge. He admitted that his denials preserved his life, but more importantly to him, he "derived an immense amount of satisfaction from what most men would consider privation and sacrifice."[94]

With Tesla forced back to Gospić, having reconciled with his family, he sought the refuge of his father's church. There he tried to wash himself clean of his "womanizing" and "lost weekend" lifestyle for a while. At church was where he met Anna. She was statuesque, beautiful, and possessed mesmerizing eyes. The two spoke of building a future together. Tesla admitted, "I fell in love." Anna was the only non-family member whom he ever loved. But it was not to be—although they kept in touch when he moved to New York, even attending her son's boxing matches until he was killed in the ring—for he was determined to "carry out my father's wish and continue with my education."[95]

The two parted ways in January of 1880, yet a paradox existed: Why would a germophobe, an isolationist, a self-centered, madly driven genius ever enter into such a close, personal relationship?[96] Although Tesla's life was already riddled with paradoxes that would become more complex as he grew older, why he loved Anna and then never again loved another woman still befuddles biographers and serious observers alike. Perhaps he was afraid of such a close relationship and used his father as an excuse to end it. And why did he not say something about her son's death in the boxing ring? After all, he had gone a few rounds with a bully in his youth and had developed a love for the fistic sport.[97]

~

That same January, Tesla left for Prague (Bohemia), where he enrolled in the Charles-Ferdinand branch of the prestigious University of Prague, one of the foremost institutions of higher learning in all of Europe.[98] But in order to have enough funds for tuition and living expenses, he first had to solicit his maternal uncles, Petar and Pavić Mandić, for much-needed assistance. Thankfully, they both graciously agreed to support him.[99]

The evolving student Tesla was now eager to continue his tremendous and troubled journey to becoming Nikola Tesla . . . inventor of the modern world.

On the Continent

Testing the Limits

BUDAPEST, HUNGARY, AUSTRO-HUNGARIAN EMPIRE, 1881

A January day in Budapest was anything but a pleasant place to be. It was teeth–chattering cold and sunless much of the time. People did their best to smile and say, "Hello," but it was oftentimes an effort. The city, whose name is an amalgamation of the antediluvian settlements of Buda and Pest, sat elegantly on the historic Danube River, and in the late nineteenth century it was a beehive of activity. Once protectorates of the Ottoman Empire, their citizens decided in 1873, that they would merge their two cities, thus forming one of Europe's most picturesque capitals. And it was also at this time—the Victorian Age—that man's desire to measure, to calculate, and to purposely change and control "nature" reached a febrile pitch when the transition from a qualitative (subjective) to a quantitative (scientific) understanding of the world was set in motion.

Hansom cabs and four–wheeled carriages stuffed with the well–heeled and pulled by exquisitely groomed, prancing horses clippety–clopped along cobblestone streets marked by splendid examples of High Victorian Gothic architecture. Everything seemed to move about as if it were all under the direction of a seasoned choreographer.

~

As the enthusiastic Nikola Tesla walked with a sense of powerful purpose across the ice–covered Széchenyi Chain Bridge—the first permanent bridge to span the historic Danube River—the sun broke through the thick cloud cover laced with toxic, gaseous wastes emanating from towering, lanky factory smokestacks, as the streets below came alive with a complication of manual laborers, mendicants, and merchants all going about their quotidian routines.

Given his devotion to mankind, one can imagine him reciting as a personal obsessive-compulsive mantra in his head:

> Lift the burdens from the shoulders of mankind.
> Lift the burdens from the shoulders of mankind.
> Lift the burdens from the shoulders of mankind.
> Lift the burdens from the shoulders of mankind!

If he paused too long, he knew his obsessive thoughts would fly out of control, but before that could happen, he was able to remind himself why he came to Budapest: It was fate. He was convinced of it.

You see, it all began months before when he read in a local newspaper while in Prague attending the Charles-Ferdinand branch of the University of Prague (attended/audited unofficially) that Tavidar Puskás, a family friend and member of Transylvanian aristocracy, was given permission by Thomas Edison to build, as his representative, an American Telephone Exchange in Budapest. Puskás enlisted his brother Ferenc, a retired Hussars military officer, to supervise the project. Tesla saw this as an omen and had a family member contact the Puskás brothers for a job.[1]

So on this day, Tesla felt unusually comfortable in his own skin, thoughts under control, feeling confident, and in good—but not those dangerously good—spirits as he moved closer to the building that he believed housed his new job and future.

Tesla stood rod-straight and square-shouldered at six foot three, his only black Callahan frock coat wrapped tightly around his lean, muscled physique. It is interesting to note what others thought of Tesla when first encountering him. Novelist Julian Hawthorne, son of Nathaniel Hawthorne, wrote the following when he chanced upon Tesla as a young man:

> I saw a tall, slender young man with long arms and fingers, whose rather languid movements veiled extraordinary muscular power. His face was oval, broad at the temples, and strong at the lips and chin; with long eyes whose lids seldom fully lifted, as if he were walking in a dream, seeing visions which were not revealed to the generality. He had a slow smile, as if awakening to actualities, and finding a humorous quality in them. Withal he manifested a courtesy and amiability which were almost feminine, and beneath all were the simplicity and integrity of a child. . . . He has abundant wavy brown hair, blue eyes and a fair skin. . . . To be with Tesla is to enter a domain of freedom even freer than solitude, because the horizon enlarges so.[2]

Many others had equally effusive thoughts about Tesla, and all agreed that the power of his physical presence and his charismatic personality were well beyond the norm, and his faultless English and fluency in numerous other languages brought into high relief his ability to communicate without hesitation and with firm conviction. But little did they know, however, of his ongoing and relentless struggle to sort the brilliant ideas from the bizarre ones—which were always buzzing around in his head. From his earliest memories, he found it a never-ending challenge to slow them down, to take control.

~

Tesla's dark, high-top Parker boots were half covered by recently laundered dark wool pants, while his Viceroy high-collar dress shirt and waistcoat had seen better days. A red satin puff tie set off his attire. Although his dress was less than the best, every article of clothing was immaculate and put together in a most fastidious manner: It was something that in the years to come would present itself as an obvious manifestation of the exacting and seemingly senseless repetitive compulsive grooming behaviors he exhibited when his mood was somewhat normal, and as such, he used these as a braking mechanism, a check on his thoughts in order to keep them from winding up and soaring to dizzying heights. Moreover, he felt that being properly dressed was an aspect of his professionalism, and throughout his life, whether in a state of feast or famine, he always made it a point to be well dressed. He was indeed a Beau Brummel of his time, and that would in itself become a visual trademark.

Over time, Tesla's disciplined appearance would prove to be less of a fashion statement and more of an external manifestation of the internal control mechanisms he employed to keep his thoughts, as well as his shirttails, from erupting and catching fire.

~

Today he was in a hurry, no doubt, to confirm his new job as he made his way up the ascending wide steps to the grand Central Telephone Exchange's imposing stone structure and faced the main entrance. Suddenly, and despite his personal mantra, his mind took control of itself and began racing with all manner of thoughts, not just wondering about what it would be like to work at the exchange, but mostly unrelated thoughts coming in rapid sequence—some scientific in nature, but the majority were those "other" thoughts that had always inhabited his inner world: demons and visions that had haunted him since early childhood.

At that moment, he understood that he did not have control of these thoughts, but rather such thoughts had control of him: barraging his mind at an ever-accelerating pace, crowding out all other ideas, now becoming very disquieting—once again his mind was careening nearly out of control as he lost control and obsessions accelerated into racing thoughts. Then momentarily suppressing the thoughts secreted in his inner world by literally shaking his head from side to side as if to command them to settle down, he took a deep breath, forcing his mind back to his compulsive guiding mantra, as he welcomed the quiet within himself. As he opened the enormous door to the Central Telephone Exchange's foyer, he whispered to himself in Serbo-Croatian, "*Ja sam spreman*, I am ready."

Tesla was now ready to embrace the challenge of his new job. It would prove to be a form of basic subsistence while allowing him to begin to do what he really wanted to do with his life, best expressed as his unquenchable wont to better mankind. By now such a belief was branded into his psyche. It would also prove to be much more, and in fact, his "unifying philosophy of life."[3] In essence, it gave him another control mechanism he needed to rule over the pace of his thoughts as best he could, but he would continue to be tested time and again throughout his long life. Moreover, his mantra would not only underlie his endless successes and highs, but his failures and lows as well in the many decades to come. Simply put, this control mechanism came with a high price. It caused him to stay aloof from individuals while substituting a connection to mankind in general being justified in his mind as grandiose benevolence. His schizoid isolation from individuals allowed him the necessity of internal preoccupation using his obsessive mantra as a mental mechanism to control his genius or bizarre mania, whatever his amazing mind threw his way. This preoccupation with his thoughts to the neglect of relationships in the outer world would come to characterize Tesla's adaptation to his *evil twins* of genius and mania.

He was glad to escape the stinging cold as the door closed, vault-like, behind him, and he marveled at the intricately tiled floor beneath his feet. He had now achieved a level of clarity—able to focus on just one thing at a time—and he was duly impressed with what he saw. He continued to scan his surroundings with reflexive contemplation, concentrating on what was before him, all the while keeping his racing thoughts in check—it was when he was at his best.

Once securely inside the impressive building, he saw all manner of activities: young, dark-suited men scurrying about from office to office, some assisted by equally young women with sheaves of clerical papers cradled in their arms, as uniformed technicians crisscrossed the large floor space only to stop before a gargantuan wall apparatus and deliver messages to operators who then pushed and pulled electrical wires into and out of receptors dotting its surface. Then, as he turned to his right, he saw a sign reading: Superintendent of Operations. The eager Tesla surmised that his first real job as an electrical engineer was now just before him. In that moment, he reminded himself that his personal aspirations, as well as economic necessity, had dictated much of his decision to move from the quietude of Gospić, Lika (then part of the Austro-Hungarian Empire) to the city of Budapest. And with the passing of his father, he was now the breadwinner for his family as well, yet nothing was going to stop him from realizing his personal dreams.

When he inquired about his employment status with the superintendent, he became instantly crestfallen to learn, much to his dismay, that the new telephone company was still under construction and was not completely functional, so he was offered instead—what he had hoped would be temporary—a position as a draftsman at the Central Telegraph Office for the Hungarian government. In that moment, he felt deeply dejected and thought to himself, "I came to Budapest prompted by a premature report concerning the telephone enterprise and, as irony of fate willed it, I had to accept a position as a draftsman in the Central Telegraph Office of the Hungarian Government."[4] He was relegated to drafting chores and calculations, but he was called upon at times to climb telephone poles to repair failed equipment.[5] He was not a laborer, he was a creator, nevertheless, he would stay the course until things turned in his favor.

Fortunately, in short order, his talents were acknowledged by the inspector in chief, and he was quickly put in charge of the telegraph company's technical division as its chief electrician. There he would do far more than merely serve as the chief electrician directed to troubleshoot the needs of telegraphy for the government—he also filled his time with the compulsive inventing of the sound adjuster, voice amplifier (loudspeaker), and numerous other devices that would streamline voice transmission, none of which he had even bothered to patent.[6] He was creating needed inventions on the fly—to him it was a mere byproduct of harnessing the wellspring of thoughts that were spontaneously and continuously erupting from his fertile brain. And to think, this all happened even before he began his desired position with the Central Telephone Exchange. Such early experimental and creative times were simply joyful play to him. In 1766, physician and natural philosopher Erasmus Darwin called such playful experimentation "a little philosophical laughing."[7]

Although Tesla was not enamored of his present position in the telegraph office—he was always seeking to solve much more complex scientific problems—or its meager recompense, he was nonetheless determined to soldier on, making the most of the job at hand. Quitting was not an option for him; he had much to do.

Thankfully, on May 1, 1881, he was rewarded with employment at the newly opened telephone exchange, and there he became secure in the knowledge that he now had gainful, permanent employment.

Finally feeling confident in his employment, his mind was suddenly unburdened for a moment, and he allowed it to begin running free again, liberating his imagination to soar, focusing first on what would become one of his many signature inventions, and the one invention that would revolution-ize the entirety of all mankind—the "rotating magnetic field" (AC–alternating current)—and then conceiving the "induction motor" in a flight of ideas, coming at warp speed, one after the other.[8] But before he realized it, he was balancing precipitously on a dangerous cliff's edge faced with the genius in-ventor able to implement his ideas on the one hand and feeling out of control and overly excited on the other hand; hence, he was at times ineffective as an engineer unable to capture and control his ideas as quickly as they appeared. This state of mind only served to progressively frustrate and exhaust him.

By mid-July, Tesla's mind was running too fast again, and because of his overzealousness, he found himself yet again in the throes of a cataclysmic manic high. It was too many brilliant ideas coming too quickly, making him feel like his mind had gone off the rails and blown up into a berserk high, then crashing into a bleak depressive low in which his mind was blank and he was empty and forlorn. Without his usual hyperthymic thoughts, Tesla felt worthless and worried he might never recover his treasured flow of creative thoughts again. Now he had no ideas at all—the ultimate torture for Tesla's unbridled, voracious genius.

At the same time, he had to manage the demands of his prominent posi-tion at the Central Telephone Exchange. On top of this, the mental energy expended—while fully functional between his highs and his lows—during his own time to disentangle and clarify the technical requirements of his alternat-ing current polyphase system—eventually to be known as the Tesla Polyphase System—pushed him to the brink.[9] Could he have slowed himself down so that his mania would not have crashed into a terrible depression? Did his highs and lows have a mind of their own or could he control them? Admittedly, he enjoyed his highs as his mind took flight, but when the cost of the highs was such terrible lows, then was it all worth it? Would he recover? Would his pain ever end? If not, would it happen the next time the ideas began to flow at hyper-speed? He simply did not know.

~

After several months in Budapest, the newly minted, twenty-five-year-old bona fide electrical engineer—having previously completed his extensive studies in physics and mathematics at the University of Prague—was plagued with an ever-increasing level of depressive exhaustion following uncontrol-lable flights of original ideas swarming about in his head. Indeed, he seemed at times to have rapid cycling between too many and too few thoughts. The thoughts no longer made him feel euphoric and exuberant—too many, too quickly, for too long made him wish they would simply stop. He may have thought, "Can't I have controlled mania and a constant flow of genius that is sustained without the disruption of too high of highs and too low of lows," as many bipolar patients ask? He wished for a state of sustained hypomania rather than out-of-control full mania crashing into a devastating and empty lack of thoughts in his state of depression.

This is a common refrain of patients who have experienced hypomania, and the wish to be in an enduring hypomanic state can interfere with their willingness to take medications to bring their mood down to normal. They want to be "a little bit manic." However, hypomania is an unstable mood state that risks flying into full psychotic mania or crashing into a devastating depression and cannot be purposely sustained with or without medications. On the other hand, Tesla's history, as can be the case for many bipolar patients, showed that he naturally experienced long periods of likely hypomania especially as a child, adolescent, and a young man. In reality, bipolar illness can have not only rapid switches between mania/hypomania and depression and back to mania/hypomania, but also mixtures of both mania and depression at the same time, something now called "mixed features." As the patient swings from one mood state to another, they may go from pure mania to a state of mania with a few simultaneous symptoms of depression (mania with mixed features), to a state of major depression with a few simultaneous symptoms of mania (depression with mixed manic features) as the manic symptoms fade and the depressive symptoms accelerate, to a state of pure depression, and back again across the spectrum, up and down, sometimes rapidly.

Then, after a long day's work solving one technical problem after another, he finally collapsed from utter mental exhaustion in his solitary room at a local boardinghouse. What frustrated him was that when the thoughts were exploding one after another in his mind, he was unable to implement them in real time. His mind inevitably ran into a wall of abrupt, hard reality that day, and he did not know what to do as a cacophony of sounds, voices, and images fought for his attention. It was as if a fierce competition of impressions in his head flipped a switch from rapid-fire to frustration and then to darkness, to emptiness, and finally to no thoughts at all. But would he prevail? Would he cycle back to brilliance?

Tesla described his soaring thoughts hitting the wall as if the source of his difficulties arose as physical illness in his body and not that they may have been from his mind as major depressive episodes alternated with irrational mania or hypomania:

Three times in my youth I was rendered by illness a hopeless physical wreck and given up by physicians. More than this, thru [*sic*] ignorance and lightheartedness, I got into all sorts of difficulties, dangers and scrapes from which I extricated myself as by enchantment. I was almost drowned a dozen times; was nearly boiled alive and just mist [*sic*] being cremated. I was entombed, lost and frozen. I had hair-breadth escapes from mad dogs, hogs, and other wild animals. I past [passed] thru dreadful disease and met all kinds of odd mishaps and that I am hale and hearty today seems like a miracle. But as I recall these incidents to my mind I feel convinced that my preservation was not altogether accidental.[10]

These experiences beg the question: Was he avoiding the obvious, thinking his problems stemmed from the physiological rather than the psychological?

In fact, Tesla had indeed crashed and burned into overt mental madness that cold winter day—from the heights of mania to the depths of depression and in a strange city where he had few friends. He had certainly been singed by the fire of manic psychosis on the up side of his mood, but this time it was more disturbing and severe; this time it could only be described as a "nervous breakdown"; it was a testing of the limits of his psyche on the down side of his mood. In modern psychiatric parlance, his situation would be described as a "psychotic break" from a manic episode followed by a cascading hell-ride into deep bipolar depression.

He was in the dire straits of bipolar depression, as coworkers and friends brought him food and flowers and expressed their deep concerns. A renowned physician came to see him and pronounced his "malady unique [and] incurable."[11] There were no treatments for bipolar disorder at this time, and lithium treatment of manic-depressive disorder would not be discovered until five years after Tesla's death. For a time he administered daily sedations of calcium bromide to help reduce his nervous stress, but to no avail.[12]

For several days Tesla lingered precipitously on the verge of death—so he believed, for he had suffered a "complete breakdown of the nerves."[13] Eating proved difficult for him, and even simple hydration had become problematic. Each new episode of cycling ups and downs became more intense as the weeks rolled on. "All the tissues of my body quivered with twitchings and tremors."[14] At times his steely-blue eyes focused on nothing. Something was troubling him; something was stabbing at his psyche. He knew he needed help, but his arrogance prevented him from seeking it. Moreover, his cocksure brilliance— a unique form of charisma—would in time be recognized as Tesla's personality signature. He often had visions of grandeur, believing himself a true "creator," not merely a backyard "tinkerer." All he had to do, he reasoned, was to turn his prodigious mind into slowing down his thoughts when they came too fast, thinking erroneously that his prodigious willpower could control them. The uncertainty of it all was simply the price of genius, he surmised.

~

Actually, many geniuses and experts alike have thought the same. Those in the creative professions like inventors, artists, writers, designers, and even chefs have much higher rates of manic-depressive illness (now called bipolar disorder) than the general population. Why? In part because manic symptoms are associated with creativity and productivity, high energy, fast thinking, even racing thoughts that go in novel and never-before-ventured areas, and in this state, you do not even need to sleep much. Manic traits make for success, especially in lines of work that value novelty. When mania is added to true genius, as has rarely been seen in the history of mankind, you get a Nikola Tesla. Many people have these manic symptoms all the time, as part of their personality—to a mild degree, called hyperthymia—but they are not a Nikola Tesla.

~

So, what must it have been like to have been Tesla? This is something few in history have experienced, so how can we gain insight into who he was, rather than just what he did. The record is consistent; when considering Tesla's baseline temperament between his soaring highs and rare crashing lows, we find hyperthymia tinged with obsessive compulsiveness and a shade of schizoid isolation. Tesla's journey as a hyperthymic bipolar inventor who relieved the burdens of humankind as few did before obviously had some predictable downsides. His high energy could be accompanied by anxiety ("nervous energy") over losing control, with the adaptive mechanism—maladaptive at times—of isolating himself from others and forceful direction of his thoughts toward internal preoccupation with controlling the speed and bizarreness of his thoughts with obsessive mantras, which he called willpower. In Tesla's case, and in Tesla's era, no diagnoses were given, of course, and he proceeded to live a fast-paced, active, colorful life, with occasional severe, emotional instability, and no love life, wife, children, or family as time progressed. In all, the trade-off of genius for mania in Tesla's case was certainly more positive than negative, at least for society, and perhaps for Tesla himself, although at a great personal cost perhaps not recognized by others. How else could a curious and perhaps onetime mixture of genius with practical wisdom—"phronesis" fueled by the energy and drive of bipolar disorder explain the phenomenon known as Tesla? It yielded breakthrough thinking and creativity coupled with the extremely rare ability to discern and make good judgments about the right thing to do from a practical point of view.

As we now understand, Tesla always believed his thoughts were controllable and that he should be able to keep them from getting out of control through sheer willpower. He was convinced his genius was so great that he could control the pace of having ideas, so they would not explode into psychotic mania, but time and again he would cycle into sustained flights of ideas, derailing from one thought to another before he could really process a specific idea or thought. In this state, his mind was not ready to process the present thought, as he was already on to the next thought without really having solidified the previous one. The result was that he became exhausted from the process and his mind eventually shut down.

From his earliest years, Tesla had always been acutely aware of his mind's inner workings and that his senses were complex, but when ideas started to come in an ever-accelerating pace, he became highly sensitive to external and internal stimulants—both real and imagined. Sometimes he could not tell the difference. His ability to perceive what others could not occured at times when his mind was most productive, when he had the most energy, and when he felt vested with an omnipotence—it was both an asset and a liability to him. He was often beleaguered as he struggled to differentiate between ideas born of incandescent genius and the bizarre perceptions that were sometimes more likely the product of psychotic mania.

This time, in Budapest, his genius had all but flown into the sun—he was Icarus redux. He had so surpassed his prodigious limits that he now had plummeted into the depths of frustration and despair. To his great astonishment, what happened during his nervous breakdown exceeded anything that he had experienced before

Throughout his long life, perhaps at moments of manic highs with the waning of his insight, he would often claim to hear thunderclaps of lightning more than 550 miles away, or the faint crackling of flames in neighbors' burning houses, thus warning them and saving their lives.[15] But the Budapest depressive episode was so very different that Tesla was clearly shaken by how it started with wildly bizarre and curious manifestations of creative ideas and endless energy occurring just before crashing into crippling depression, where his thoughts turned slow, leaden, empty, and painful from the eventual absence of brilliance. In 1919, he recalled the episode in Budapest decades before with a noticeable level of angst as he said, "What I experienced during the period of that illness surpasses all belief."[16]

So began a lifetime of ever-increasing, severe manic-depressive episodes, characterized by an obsessive period of relentless work and intellectual gluttony that would continue unabated for weeks on end, eventually throwing

Tesla into a state of utter exhaustion and eventual depressive illness—where he was certain at times that he was on the threshold of death.

~

It cannot be overstated that Tesla's was a strange and tormenting mental disorder whose primary cause was an overabundance of mental capacity, overflowing with endless thought experiments flowing freely in and out of his mind. It bordered on torture at times, almost a form of personal psychologically imposed terror—his evil twins of genius and mania never stopped. At his most original, utterly unheard-of ideas occurred to him with uncanny regularity—the rotating magnetic field, magnetic resonance, wireless transmission, robotics, and so on—while this mental state activated an acute hypersensitivity in all of his sense organs. He would tell others that he could hear the ticking of a watch three rooms away—to him it sounded like the beating of Thor's mighty hammer on an anvil; and the slightest touch of his skin felt like an agonizing body blow from a prize fighter's punch. He would similarly complain at times that his heart rate felt like it raced toward 260 beats per minute as his flesh trembled and shook, causing him great pain and gripping angst. Normal speech sounded like thunderous bedlam; and a beam of sunlight shining on him produced the effect of an internal explosion of a nuclear magnitude.[17] And yet, during these episodes, he continued to feel an inner-calling, as he was driven to recover from what he would later understand to be "manic-depressive illness," and today it is recognized in public mental health parlance and literature as "bipolar disorder."

In response to his early mental breakdowns, Tesla writes: "It is my eternal regret that I was not under the observation of experts in physiology and psychology at the time. I clung desperately to life, but never expected to recover."[18]

There were times when Tesla's mental breakdowns pushed him into the deepest corners of his mind, challenging him to overcome the great forces allied against him. Although mentally destitute, bankrupted of ideas for days on end, as frustration took the place of creativity, and pure madness lurked on the edges of his psyche, his genius would not yield, and in the end, he was convinced he would triumph. Despite the paroxysm of the moment during his first recognized psychotic breakdown in Budapest, he was still able to "intuit" that the solution to his problem with alternating current electricity was coming.

Theoretical physicist Albert Einstein said of intuition: "A new idea comes suddenly and in a rather intuitive way." "But," he hastened to add, "intuition is nothing but the outcome of earlier intellectual experience."[19] Hence, Tesla's years of work on alternating current was the byproduct of intuition grounded in never-ending thought and physical experiments, and they would continue until the questions nagging him were answered and the physical result was achieved. At times, his cycle of experimentation was a pattern he often repeated: genius mixed with bizarre thoughts, but he was unable to tell the difference, thinking everything was a manifestation of his genius—it was all in conflict. Conversely, onlookers sometimes had the opposite reaction: thinking his bizarre genius was a manifestation of his mania. Nevertheless, he would overcome thoughts of failure because, in the end, he was always convinced he would succeed.

Obviously, Tesla's mental abilities were far beyond the norm. It is ironic that the medical disciplines of neurology have no specific word for such a condition of overabundance of normal mental ability other than "genius." However, if one is mentally lacking due to either a physical malady that disturbs mental function or simply born at a lower-functioning mental state, there is a very specific word for it: *deficit*. Furthermore, according to famed neurologist and physician Dr. Oliver Sacks, there is no word for a super or overabundance of function. As Dr. Sacks points out, "A function or functional system—works, or it does not. Thus, a disease which is 'ebullient' or 'productive' in character challenges the basic mechanistic concepts of neurology."[20]

What had occurred on that day in Professor Poeschl's class was life-changing, for Tesla's professor had planted a momentary seed of doubt in his mind, yet the student knew he had to stay alive and be ready when the truth arrived.[21] The moment also solidified in his mind that he would become an electrical engineer and solve the problems demonstrated by the Gramme dynamo and to do other great things. There was no doubt; he would soldier on; it was his raison d'être. Moreover, Tesla was committed to the belief that his life was guided by peerless instinct, about which he said, "I could not demonstrate my belief at that time, but it came to me through what I call instinct, for lack of a better name. But instinct is something which transcends knowledge. We undoubtedly have in our brains some finer fibers which enable us to perceive truths which we could not attain through logical deductions, and which it would be futile to attempt to achieve through any willful effort of thinking."[22] Unlike intuition, whose processes are learned, instinct is something that is "hardwired" from birth—it is innate.

Undoubtedly, there were myriad times Tesla was right: His instinct did indeed transcend knowledge. However, at other times his thoughts were unmanageable and too fragmented to be constructive or useful. Tesla understood direct current's manifold problems and deficiencies, however: the high cost of construction and installation; the inefficient transmission and significant power loss issues when transported over great distances; increased power consumption problems; and great difficulties when using industrial machines, compressors, most lighting systems, electronic devices, et cetera. He also knew that his alternating current polyphase system would address all these various shortcomings of direct current—while Thomas Edison bet his future on "direct current" power transmission.[23]

It is most telling that Tesla had reminded others many times that solving the problems inherent in direct current power transmission and usage was more than merely succeeding: "With me it was a sacred cow, a question of life and death. I knew that I would perish if I failed. Now I felt the battle was won. Back in the deep recesses of the brain was the solution, but I could not yet give it outward expression."[24]

$\sim$

Finally, Tesla emerged from his mental breakdown. For whatever reason, and from whatever disease, the self-described "hopeless physical wreck" eventually recovered his physical health. More than that, he recovered with *abnormal* vigor and continued "testing the limits" of his thoughts and actions.[25] His power of visualization—the ability to see as solid objects before him the things that he conceived in his mind, and which he had considered such a great annoyance in childhood—now proved to be of great aid to him in trying to unravel this problem.[26] He believed he could "mental engineer" anything he thought of because he was not afraid to reverse standard scientific practices . . . to challenge existing norms.[27] Like all great scientists, Tesla faced down received wisdom and was fearless in his desire to go his own way, confident in his ability to blaze a new path.

Einstein had much the same approach: the ability to conduct thought experiments in which he could visualize how a theory would play out in practice. It also helped him peel off the irrelevant facts that surrounded a problem.[28] But unlike Einstein, Tesla had the remarkable ability to turn his thought experiments into physical (tangible) inventions by himself.

In time, Tesla would achieve his goal, built upon a concatenation of scientific experiments and observations that led to his creation of the "alternating current polyphase system," which many would argue is one of the greatest of the greatest inventions of all time.[29] To put a fine point on it, celebrated senior Westinghouse engineer and the subsequent vice president of the American Institute of Electrical Engineers, B. A. Behrend, said that Tesla's invention of alternating current was "the warp and woof of industry. His work marks an epoch in the advance of electrical science."[30]

$\sim$

In what can only be described as an organic incident, which led to a cataclysmic event—"The Boom" moment—occurred on a chilly February day in
1882. It was a late afternoon in the Vàrosliget, Budapest's city park, as sunset
fast approached and a vespertine stillness was in the frigid air. The city on the
Danube River was beginning to close its doors for the day, as mothers called
their children in from their play and tired workers returned home to a warm
meal and a fitful night's sleep, only to rise again the next day to do it all over
again.

A thick grass punctuated by deciduous trees and dense shrubbery—many
still coated in the remnants of an earlier hoar frost—carpeted much of the
frozen ground with their jettisoned leaves. Suddenly, while walking slowly on
a demarcated dirt path in the park, Tesla's thoughts coalesced into a vortex of
images and ideas, each fighting him for his undivided attention, as his confidante and master mechanic Anthony Szigeti, an unusually physically fit man
for the time, followed faithfully behind in lockstep. Tesla began to recite aloud
(in fluent German) Goethe's *Faust*:

> The glow retreats, done is the day of toil;
> It yonder hastes, new fields of life exploring;
> Ah, that no wing can lift me from the soil
> Upon its track to follow, follow soaring!
>
> A glorious dream! though now the glories fade.
> Alas the wings that lift the mind no aid
> Of wings to lift the body can bequeath me.

Inspired by Goethe's brilliant imagery of the sun retreating and racing forward
and of imperceptible wings lifting the mind but not man, Tesla envisioned
the idea of doing away with the impractical commutator and using a "rotating
magnetic field" in his induction motor (Tesla motor).[31] Then, as he continued
to recite *Faust*, one of his favorite classical stories and a harbinger of what was
to come, he suddenly stopped. That was when it happened!

"As I uttered these inspiring words the idea came like a flash of lightning
and in an instant the truth was revealed."[32] Now thoroughly frenetic, Tesla
called out to a concerned Szigeti to hand him a stick. He needed to work out
the answer right then and there.

At first blush, Szigeti had no idea what he was watching, and he must
have questioned himself, "Is it madness or genius that I am watching?" He
attempted to calm the highly animated Tesla and insisted that he sit down on
a park bench and relax. Tesla would have none of it. He called out again for
a stick, as he stood trance-like, supreme focus evident in his steely-blue eyes.
Szigeti hesitantly answered the call. Then, with stick in hand, Tesla changed
science and the world forevermore!

What he had done was to engrave—for the ages—in the damp, coarse dirt of the city park's pathway a series of diagrams that he would show just six years later in his landmark address before the august American Institute of Electrical Engineers in New York City.[33] He would subsequently recall: "The images I saw were wonderfully sharp and clear and had the solidity of metal and stone, so much so that I told him [Szigeti]: 'See my motor here; watch me reverse it.' I cannot begin to describe my emotions. Pygmalion seeing his statue come to life could not have been more deeply moved. A thousand secrets of nature which I might have stumbled upon accidentally I would have given for that one which I had wrested from her against all odds and at the peril of my existence."[34] And so he gloried in the magnificence of "The Boom" moment with his companion. It had been done. He had created a new physics. A new truth!

Curiously, like Tesla, Einstein had his "Boom" moment in 1905, which has been referred to as "The Step." He had been puzzling over certain issues regarding the observations of light under various circumstances with regard to relativity. Then one day while walking to work at the Swiss Patent Office in Bern with his best friend Michele Besso, an engineer, he told him that he could not resolve the problems to his liking, and he said to him, "I'm going to give it up." But as the two discussed it, Einstein recalled, "I suddenly understood the key to the problem."[35] He arrived at the simple, yet profound conclusion that there is "no absolute time."

Much like Tesla, Einstein had achieved a supreme level of clarity in that moment. But unlike Einstein, Tesla would go on to produce inventions and discover laws of nature that directly lifted the burdens from the shoulders of mankind.

The "Boom" moment was undoubtedly an emotional turning point in Tesla's young life, and in that instant, he became certain of his creative powers: "I had carried out what I had undertaken [solving the problem of the Gramme dynamo's sparking motor] and pictured myself achieving wealth and fame. But more than all this was the revelation that I was an inventor. This was the one thing I wanted to be. Archimedes was my ideal. I [had] admired the works of artists, but to my mind, they were only shadows and resemblances. The inventor, I thought, gives the world creations which are palpable, which live and work."[36] Tesla went on to add, "*The progressive development of man is dependent on invention. It is the most important product of his creative brain. Its ultimate purpose is the complete mastery of mind over the material world, the harnessing of the forces of nature to human needs.*"[37]

Many years later well-known electrical engineer B. A. Behrend—one of the few individuals who understood at the time what Tesla had done in The "Boom" moment—would add to his thoughts about what Tesla's alternating current polyphase system of power generation and distribution had achieved when he said, "Not since the appearance of Faraday's 'Experimental Researches in Electricity' has a great experimental *truth* been voiced so simply and so clearly as this description of Tesla's great discovery of the generation and utilization of polyphase alternating currents. He left nothing to be done by those who followed him. His paper contained the skeleton even of the mathematical theory."[38]

But what of Tesla's nervous breakdown? What of its manifestations? What of his visions? Dr. Sacks writes knowledgeably of the experiences of Hildegard of Bingen (1098–1180), a German Benedictine abbess, mystic, visionary, and polymath who experienced countless visions from childhood to her death, just as Tesla had. And like Tesla, she was the great intellect of her time.

Tesla, like Hildegard, also had visions of radiant luminosity, possibly related to disturbed ecstasy or migraine manifestations.[39] Hildegard was considered a rare woman of her time, who consumed knowledge at a startling rate and whose mental acuity matched any man of her time. However, she languished in obscurity for centuries until the 1980s, when women's studies programs and the feminist movement took root.[40] Likewise, until recently, only those in the sciences cared to learn much about Nikola Tesla. Now, at the dawn of the twenty-first century many want to learn of the man who invented the modern world.

Is there a direct, more significant psychological nexus between Hildegard's experiences and Tesla's? Hildegard writes that her experiences were in what she perceived as a normal state of awareness: . . . *I perceive them in open view and according to the will of God.*[41] And yes, both Tesla and Hildegard were affected by mood disorders. There is evidence that suggests that genius and mental disorder go hand in hand. It is not that geniuses want to be unstable; it is just a noticeable fact that they have a tendency for it.[42]

By the spring of 1882, the young electrical engineer was looking for more. He realized that Budapest did not hold his future anymore. The telephone exchange was now fully functional, but Ferenc Puskás had sold the business to a local entrepreneur, so it was time for Tesla to move on. He was convinced that his alternating current polyphase system was the answer to the coming age's need to increase efficiency in commercial industry and to make the daily lives for all mankind better. So, as fate would have it, it intervened in his favor. His was not a grandiose delusion. Mr. Tavidar Puskás would come to the rescue, again. He invited both Tesla and his assistant Szigeti to come to Paris where jobs were waiting for them.[43] Tesla eagerly accepted the invitation and moved with Szigeti in April to Paris.

PARIS, FRANCE, 1882

Mr. Puskás proved to be a man of his word, and the promised jobs were indeed waiting for Tesla and his reliable assistant Szigeti when they arrived in the city pulsing with life around every corner. The warmth of a spring sun told Tesla that great things were to come. The company, Société Electrique Edison of France, was installing incandescent lighting systems—utilizing Edison's direct current power system—throughout the city, and Tesla welcomed the opportunity to work for such a prestigious company. Moreover, Englishman Charles Batchelor, Edison's emissary in France, had established multiple companies in France on Edison's behalf to deal with everything from patents and manufacturing to the installation of Edison products on the Continent. Tesla himself was under the employ of the Société Electrique Edison.[44]

Tesla's first impressions of Paris simply overwhelmed him. He stated, "I can never forget the deep impression that magic city produced on my mind. For several days I roamed through the streets in utter bewilderment of the new spectacle. The attractions were many and irresistible, but, alas, the income was spent as soon as it was received."[45]

Although Tesla had arrived in the City of Light with few personal belongings, he was nonetheless elated by the knowledge that his alternating current polyphase system would in time prove to open up a whole new world of inventions, and he wanted to be part of the coming "electrical age." He felt that Paris, in the midst of the preening atmosphere of La Belle Époque—a time of excited optimism, high culture, and great scientific advancements—held his future in its grasp. He knew the city to be a giant petri dish of opportunity, for it was continental Europe's ground zero for the Second Industrial Revolution.[46]

However, it was while in Paris that Tesla first encountered deceitful and cunning people, against whom he had no defense. He was so preoccupied with his ideas that he often paid scant attention to people and failed to cultivate a sophisticated notion of complex human behavior, and this would prove to be a weakness in his personality that he would never overcome. His naivete was stronger than his desire to be devious himself, and his Serbian upbringing and Orthodox Christian faith had taught him that "principle was more important than profit." As a consequence, he was to constantly and unwittingly deal with rich and authoritarian men who sought to take advantage and exploit him in ways that he had never contemplated, and such a scenario only served to test his emotional insecurities at virtually every turn in his tremendous and troubled life.[47]

Within days of Tesla's arrival in Paris, he assumed his new position. There his work in a busy laboratory brought him into contact with many of Europe's most advanced scientists, all focused on designing and manufacturing electrical equipment for Edison's fledgling company. It was also in the laboratory that he came into contact with American scientists, with whom he was able to share his vision of lighting the world with his alternating current polyphase system.[48] Unfortunately, no one in Europe at the time had the vision to anticipate and appreciate the supremacy of alternating current and all of its manifold manifestations to come.

Although Paris was Tesla's home base, he spent the next several months crisscrossing Europe troubleshooting and solving all manner of electrical problems for Edison's direct current powerplants and lighting systems. Because of his facility for languages, he also served as Edison's technical ombudsman in France, Germany, and wherever else he was needed. He was continuously inventing devices to improve upon Edison's direct current (DC) system of electrical power and in doing so, he was evermore convinced that his alternating current (AC) polyphase system would in time displace Edison's highly inefficient electrical system as the first choice for electrical power around the globe.

On one occasion the German government refused to accept a powerplant it had commissioned from the Société Industrielle & Commercial Edison of France. Edison was facing a serious financial loss and the possibility of bankruptcy. Because of Tesla's fluency in German and his undisputed technical expertise, he was called in to save the day. Tesla stated, "I was entrusted with the difficult task of straightening out matters [in France] and early in 1883 I went to Strasbourg on that mission."[49]

STRASBOURG, FRANCE, 1883–1884

While working for Thomas Edison in Strasbourg, France, Tesla would spend his days occupied not only with fixing and improving the electrical powerplant in the Strasbourg railway station owned by the German government, but the very building it was housed in. He had to rewire and reengineer sections of both. But nighttime was his time; it was spent constructing prototypes of his alternating current induction motor that he had inscribed on the dirt pathway some two years before in a Budapest park.

By late summer, he had completed a working model of his induction motor that he had imagined, built, and tested in his mind for more than a year. Its initial tests proved flawless, and he was now ready to show it to the world.[50] Needless to say, his excitement was growing, for he knew he was edging closer to answers that had alluded him to this point.

In time, Tesla was befriended by a Mr. Bauzin, a staunch supporter of his talents, and the former mayor of Strasbourg. He immediately arranged for a demonstration to potential investors of Tesla's new electrical power system. The results were all Tesla had hoped for. Hence, the former mayor sought immediate financial backing from a Mr. Benjamin to form a company on Tesla's behalf.[51] Sadly, the potential investor showed no interest, and because others who could have helped Tesla were loyal to Edison—their loyalty was based upon their large investments in Edison's DC system—nothing came of the situation. Moreover, Edison brooked no discussion regarding alternating current—one can only surmise that he had a gut feeling that Tesla's more efficient power system would someday rule the day.

It was a seismic disappointment to the burgeoning inventor. The momentary failure continued to shadow him off-and-on for a time. He was devastated, and the incident caused him to find a dark place to hide—where his state of deep depression could find its tea and sympathy. All in all, it was a blow of immeasurable size to him personally—he was gobsmacked. The situation was compounded by learning that a significant sum of money promised to him by the Edison Company in France for all his inventive work, which had improved Edison's direct current power system in immeasurable ways, never materialized. When he appealed to the company's three officials, they each in turn tossed him into what became a vicious circle.[52] This was the first time Tesla was blatantly cheated out of monies due him by Edison, but sadly, it would not be the last. In time, Edison would prove to be conniving, opportunistic, and noticeably jealous of Tesla's abilities, and there was no limit to what he would do in his efforts to destroy Tesla.

In 1907, B. A. Behrend said of men such as Edison: "I can never think of Nikola Tesla without warming up to my subject and condemning the injustice and ingratitude which he has received alike at the hands of the public and of the engineering profession. A child of genius . . . he came to this country [USA] and did great and imperishable work."[53]

Once Tesla finally made his way to America, this adversarial dynamic would soon result in the most intense of scientific rivalries. It would become the historic "War of the Currents."

It was now 1884, and Tesla was urged by Strasbourg's Mr. Bauzin to return to Paris, where he believed the opportunities were better for him. Once back in the French capital, Tesla now found its alternative culture and free spirit too much to cope with at times, for his upbringing was much more old-world and regimented than those of the au courant members of Parisian society.

PARIS, FRANCE, 1884

After months of unchallenging work, Tesla felt the situation in Paris was not to his liking anymore, and he was desperately looking for a way out. Batchelor sensed Tesla was yet again displeased with his present circumstance and suggested that he go to America with him and improve upon Edison's DC system. Batchelor also knew that the Tesla's alternating current invention and his creative mind needed a wider and more receptive audience. And Tesla understood that if the "new physics" he had created was to ever become a true reality, then he also knew that Batchelor's suggestion was his only logical option. In some ways he had outgrown the European culture and was searching for new horizons. His immediate life was filled with the uncertainty of not knowing what was next but feeling the need for change—it was a time for him to move on from Europe. The genius Leonardo da Vinci was confronted with much the same issue when he decided to move to Milan: The fact that he was feeling exhausted and was in a precarious psychological state, filled with fantasies and fears, was reflected in his willingness to leave Florence.[54]

Finally, in June, Tesla set sail on the S.S. *City of Richmond* from Liverpool, after having transferred from the S.S. *Saturnia*, bound for what he believed to be the Promised Land.[55] He excitedly and proudly stood onboard the ship that would take him across the Atlantic Ocean with just four cents in his pocket, a book of his own poems, pages of mathematical calculations, plans for a flying machine, and a mind full of diagrammed inventions—most notably, his alternating current (AC) induction motor and related polyphase system inventions. As the ship pulled away from port, Tesla never looked back, for the Gilded Age of America was on the horizon, and he was dedicated to continuing with the scientific explosion first launched by Sir Isaac Newton in the late seventeenth century.

The newly energized Tesla vowed to himself to be the next sea-change in the world of electrical science. He also fervently believed that his move to the United States of America was a unique opportunity to set things right after his mother's tragic loss of her eldest son, Dane, years before. His brother was the family's first-born male, and in the patriarchal Serbian society such a child was idolized—he was everything. Tesla blamed himself for his brother's accidental death, and much later in life he still suffered from torturous nightmares and hallucinations because of his brother's death.[56] He also saw the move to

America as an opportunity to make his mother proud of her "Niko," and by developing an iron will and an unyielding dedication to hard work, he would give his mother and the world his best. His "personality framework" was now developing at a startling rate.[57]

After surviving a near mutiny during the difficult weeklong ocean crossing to New York City, and having most of his belongings thrown overboard in a mid-ocean melee, Tesla nonetheless arrived with a full mind, boundless optimism, and a letter of recommendation from Charles Batchelor, the man who "discovered" his genius in France, to Thomas Edison. The very to-the-point letter to Edison read as follows: "I know two great men and you are one of them; the other is this young man."[58]

To be sure, at twenty-eight he was already one of the world's great inventors. But not another soul knew it.[59] And so ended Tesla's "tremendous and troubled" journey in Europe to great public prominence and devastating personal pain, as he continued to test the limits.

America was now before him.

· 4 ·

Home in America

Obsessive-Compulsive Disorder

NEW YORK CITY, USA, 1884

The city that never sleeps hit the excited and wide-eyed Nikola Tesla square in the face with a cold reality. Prior to Ellis Island's opening in 1892, immigrants were deposited from fatigued steamships into the Castle Garden (Castle Clinton) Immigration Office, America's first such immigrant processing facility, located at the southern end of Manhattan Island.[1] His first images were of a Dickensian, dystopic city in many ways. Immigrants, mostly poor, crowded the environs of the former fort and were processed with imprecise alacrity. A customs officer even recorded Tesla's place of birth as Sweden, instead of Smiljan, while another ordered him to "Kiss the Bible. Twenty cents!" Another stared him down with a murderous eye. Tesla remembered asking a policeman for directions, who barked back to the startled young man some sort of inane directions. "Is this America?" Tesla asked himself in painful surprise. He then offered, "It is a century behind Europe in civilization."[2]

The lack of wide boulevards and public spaces struck him as primitive, if not backward in some ways. He even said, "In the Arabian Tales I read how genii transported people into a land of dreams to live thru [*sic*] delightful adventures. My case was just the reverse. The genii carried me from a world of dreams into one of realities. What I had left was beautiful, artistic and fascinating in every way; what I saw here was machined, rough and unattractive."[3] Right then and there, he vowed he would change it. He would make it the most modern, fascinating city in the world, and in doing so, he would, in fact, create the modern world! Five years later, after an important trip abroad to the Exposition Universelle of 1889 (World's Fair) in Paris, Tesla became convinced that New York City *"was more than one hundred years AHEAD of Europe."*[4]

~

It was an early June day and an intense sun hovered high in the sky, baking the pulverized, fouled earth beneath his feet. The air was polluted with the stinking dung of draft animals and the accumulating detritus of a burgeoning society. The city's tall buildings, sea of humanity, cacophony of sounds, and scent of opportunity (the nation was transitioning to a full market economy) assaulted Tesla's senses in a way that neither Budapest nor Paris could ever do, and he was ready. There were also the scents of struggle and desperation in the air that reminded him of his bout with cholera back home in Gospić. Nevertheless, because he felt protected by his "letter of introduction" from Charles Batchelor to the celebrated Mr. Edison tucked safely away inside his only black Callahan frock secret pocket . . . he had no fear. He reminded himself of its contents: "I know two great men and you are one of them; the other is this young man."[5] Tesla cracked a broad smile and kept walking, eyes scanning, mind in overdrive—thoughts streaming uncontrollably and continuously.

Although his remaining pockets were stuffed with items he took from home, he carried a well-used portmanteau that held a few pieces of well-worn clothing, his father's favorite book, numerous technical drawings, and other items. But the embroidered bag hanging over his broad shoulders, made by his mother's own hands, was dearest to him. Today it contained a half-eaten overripe apple, some used pencils, and a one-inch-thick, hundred-page writing notebook he found in his father's library—he would fill it with his thoughts, much like a young Isaac Newton did at Cambridge.[6] They were thoughts that would dramatically change the world in the years to come.

Now that he was on the terra firma that was Manhattan, Tesla looked over his shoulder and saw the Statue of Liberty under construction as the last stage of the Brooklyn Bridge was about to be completed. He quickly learned that Edison's first electric generating plant had opened just two years earlier on Pearl Street.[7] It was called the Pearl Street Station and appropriately so; it was located in the city's financial district, where it serviced some 508 customers and kept 10,164 lamps lit.[8] By the summer of 1884, the Edison Electric Illuminating Company was still in a fast-paced growth stage, adding customers at a money-making rate and determined to provide reliable lighting. Edison's company was responsible for lighting such august businesses as the famed New York Stock Exchange, the New Haven Steamboat Company's offices and its pier location, Brown Brothers & Company bank on Wall Street, and even the North British & Mercantile Insurance Company on William Street.[9] It was to Edison's company that he would make his way to in the coming days, but for now he had immediate work to do . . . to find a room to sleep and to secure a job as an electrician, while most electricians of the time were pondering the mysterious medium [electricity] in terms of electric lights and trolley cars, Tesla envisioned it as the means to unite cities and distant villages, to span whole continents and ultimately the world.[10]

As the twenty-eight-year-old Tesla walked briskly uptown, he saw an extreme concentration of wealth expressed by "blocks of palatial brownstones, expensive libraries, fine art galleries, and exquisitely furnished drawing rooms of merchant-princes."[11] With a determined stride, feeling so good on that fine Friday afternoon, he continued on as he pondered his next meal; then, suddenly, he chanced upon a small machine shop. His curiosity piqued, he managed a stealthy peek inside and saw what appeared to be the foreman—he was barking orders to workmen at various machines who appeared overcome with frustration and the heat of the day. But before he could summon up the courage to ask if the foreman needed someone who was a good troubleshooter and had a familiarity with such machines, his thoughts began to race out of control, just like in Budapest. He paused and tried to mentally put his hands around his thoughts in hopes of containing and calming them. Moments later he had them corralled, but before he could ask a question, he noticed the foreman appeared to be trying to repair "an electric machine of European make."[12] The foreman had become exasperated, failing to fix the problem. "He had just given up the task as hopeless and I undertook to put it in order without a thought of compensation. It was not easy, but I finally had it in perfect running condition."[13] In the end, the grateful foreman nevertheless paid him $20 (approx. $525 today). He was now feeling flush and wishing he had come to America five years earlier.[14] One other story has it that Tesla had originally helped design the electrical machine he repaired for the foreman.[15] Tesla had then and there shown that he would always put "principle over profit" . . . a very Serbian quality. But the reward of $20 certainly did make for a very nice meal, in fact, for several very nice meals and a comfortable place to sleep.

With a full stomach, and some much-needed rest, Tesla went the next day in search of Edison's new laboratory—the Edison Machine Works. He found the former ironworks at Goerck Street, situated just a few blocks from Edison's central lighting station at 255–257 Pearl Street—the nation's first. The tall, whip thin Tesla, blue eyes with a splash of silver burning bright, and finely chiseled facial features set off by a moustache and a full head of dark-brown hair was most likely introduced to the thirty-eight-year-old, hard-of-hearing Mr. Edison by Charles Batchelor.[16] Even though Tesla had been working for several of Edison's companies in Europe, he had never met the famous man. But today would be different. Although Tesla described the moment as "a memorable event in my life," there was an instantaneous clash of personalities and cultures that would be front and center in the two men's relationship for the rest of Edison's life—Tesla outlived Edison by nearly a dozen years.[17] Moreover, the clash would eventually give rise to the famous "War of the Currents."

Although essentially broke, save the four cents he brought with him from Gospić and the remainder of the $20 he had earned the day before, Tesla wore an immaculate bowler hat and white gloves, the mark of a true dandy. His frock coat hung on his tall, slim body as if it were bespoke—his aristocratic mien and attire would soon become his personal trademark. He was a young man with a worldly education, his fluency in numerous languages was obvious, and his depth of knowledge on myriad subjects would soon become apparent to anyone who spoke to him for just a few moments.

Juxtaposed to the refined Tesla was the crude Edison. He was short in stature, pudgy and petulant, and his English left something to be desired, inasmuch as each word was infused with chewing tobacco spittle and an upper-Midwestern inflection.[18] He was a trickster, a showman, and a good storyteller.[19] But most of all, he was a "con" artist who raised deception to an art form. Adding to that image of Edison, Tesla detested Edison's carefree, unkempt indifference and said so: "If he had not married a woman of exceptional intelligence, who made it the one object of her life to preserve him, he would have died many years ago from the consequences of sheer neglect."[20]

~

From Edison's unheralded birth, he showed no facility for invention or desire for fame or the ability to make money. He was the spawn of a long line of rebels. His successful grandfather John Edison was a Tory who fought against George Washington and ended up being charged and convicted of treason. The sentence was "hanging," but he fled to Canada before the fateful day. And his father, Samuel Edison, bolted from Canada to Michigan after attempting an unsuccessful coup against the Royal Canadian government.[21] The very same father tied his young son to a whipping post and beat him in public after he had started a fire in a barn that almost burned down the town.[22]

As for Thomas Alva Edison himself, he was born February 1847, in Milan, Ohio. He was the last of seven children, and he answered to the nickname Al. He was curious, a dreamer, a miscreant at times, and not very good at formal school. To put it kindly, his public schooling was short-lived. It was his frustrated teacher G. B. Engle who said of young Al that he was "addled" and claimed he was unteachable.[23] Thankfully for young Al, his doting mother, Nancy, saw what was happening and took to home-schooling her irascible, yet talented and curious son. She gave him sage guidance: she encouraged him to read and love science. Years later, Edison said of his mother, "My mother was the making of me. She understood me. She let me follow my bent."[24]

Edison began inventing in his teens, producing practical products that caught the attention of industrialists and Wall Street investors. Specifically, by his early twenties he had invented his stock ticker, which greatly improved upon the telegraph industry's difficulty with efficiently transmitting data, while his automatic duplex permitted two messages being sent simultaneously on a single wire. His list of practical inventions would continue for decades to come.[25] Some would also include the microphone, an electric pen (forerunner of the mimeograph), and a musical telephone.[26]

~

Curiously, but not surprisingly, Edison's inventive process differed greatly from Tesla's. Although both were tireless workers, Tesla's extensive academic education, training, and high-level mathematical skills gave him a tremendous engineering advantage over Edison, who utilized a fundamental, ponderous approach of "trial and error." In addition, Tesla preferred to work "alone," while Edison was a team player—a collaborator who needed the constant assistance and assurance of others. In a brief moment of reflection immediately after Edison's death on October 18, 1931, the then-famous Tesla said of Edison's technique of invention: "If he had a needle to find in a haystack he would not stop to reason where it was most likely to be, but would proceed at once with the feverish diligence of a bee, to examine straw after straw until he found the object of his search. . . . Trusting himself to his inventor's instinct and practical American sense . . . the truly prodigious amount of his actual accomplishments is little short of a miracle."[27] Tesla went on to add acerbically, "I was almost a sorry witness of such doings, knowing that a little theory and calculation would have saved him ninety percent of his labor."[28]

Tesla was most likely egged-on by Edison during their time together because Edison had great contempt for Tesla's belief in high-level mathematics, and he bragged about his not appreciating scientific theories. Edison defended his practice of an unsophisticated trial-and-error method of invention when he said, "At the time I experimented on the incandescent lamp I did not understand Ohm's law.[29] Moreover, I do not want to understand Ohm's law. It would prevent me from experimenting."[30] Ironically, had Edison understood Ohm's law, he would have realized that his approach to the incandescent lamp was patently wrong.[31] This goes a long way to explaining the myriad trial-and-error attempts Edison required to ultimately come up with a workable incandescent lamp (lightbulb).[32] Even his official biographers, Frank Dyer and T. C. Martin, considered Edison a cut-and-try inventor who eschewed theory, when they stated Edison would "trust nothing to theory, he acquires absolute knowledge" by way of experimentation.[33]

To further illustrate Edison's eschewing of anything very technical or mathematical, later in life he claimed: "I am not a mathematician, but I can get within 10 percent in the higher reaches of the art." He added with greater truth but with still greater arrogance: "I can always hire mathematicians, but they can't hire me."[34] In keeping consistent with his pedestrian views about science, he derided talk about scientific theory, and he even confessed that he knew very little about electricity. When he had to, he relied on the expertise of his various assistants accomplished in science and math to investigate the principles of his inventions—the theoretical foundations were often beyond his limited interest, bragging that he never made it passed basic algebra in school.[35]

As a counterpoint, Tesla expressed himself in a florid, Eastern European, cultured manner with multisyllabic words not often heard on the west side of the Atlantic Ocean at the time. Even his explanation of scientific theories was larded with the patois of someone who truly understood the complexities of scientific principles. It can be said that he exhibited a great aperçu that was both theoretical and mathematical, something Edison not only lacked but rejected in toto. To put a finer point on the differences between the two men, while Edison was a trial-and-error man, Tesla was a man of theory and practice who utilized his considerable education to solve a problem.[36] There were times when Tesla would even employ what many call "methodism": The belief that science's truth-finding power is explained by a special procedure or form of organization, a distinctive way of dealing with empirical evidence, hence, the "scientific method," while Edison oftentimes stumbled around until he hit upon something that made sense to him or to someone in his employ.[37]

That said, the process of *real* innovation is rather ephemeral at times. Often the seed of such innovation is planted in childhood. Just as Tesla was amazed by electricity when he encountered it on his cat's back—as static electricity—Tim Berners-Lee, the English inventor of the World Wide Web, was fascinated from childhood with how the human brain makes random associations. He described his process of innovation this way: "Half-formed ideas, they float around. They come from different places, and the mind has got this wonderful way of somehow just shoveling them around until one day they fit. They may fit not so well, and then we go for a bike ride or something, and it's better."[38]

~

Despite the many contradictions between the two men, within a few weeks Tesla had won Edison's confidence, and he described it this way:

> The S.S. *Oregon*, the fastest passenger steamer at that time, had both of its lighting machines disabled and its sailing was delayed. As the superstructure had been built after their installation it was impossible to remove them from the hold. The predicament was a serious one, and Edison was much annoyed. In the evening I took the necessary instruments with me and

went aboard the vessel where I stayed for the night. The [long-legged Mary-Ann] dynamos were in bad condition, having several short-circuits and breaks, but with the assistance of the crew I succeeded in putting them in good shape. At five o'clock in the morning, when passing along Fifth Avenue on my way to the shop, I met Edison with Batchelor and a few others as they were returning home to retire. "Here is our Parisian running around at night," he said. When I told him that I was coming from the *Oregon* and had repaired both machines, he looked at me in silence and walked away without another word. But when he had gone some distance I heard him remark: "Batchelor, this is a d—n good man," and from that time on I had the full freedom in directing the work. [Later] Edison said to me: "I have had many hardworking assistants but you take the cake." During this period I designed twenty-four different types of standard machines with short cores and of uniform pattern which replaced the old ones. The Manager had promised me fifty thousand dollars on the completion of this task but it turned out to be a practical joke. This gave me a painful shock and I resigned.[39]

Prior to Tesla's resignation, he had been given full run of the machine shop, enjoying good meals, and even showed his advanced billiard skills to Edison's personal secretary Alfred O. Tate, who uttered that "He played a beautiful game. He was not a high scorer but his cushion shots displayed skill equal to that of a professional exponent of the art."[40]

During Tesla's time at the Edison Machine Works, he was never more than a thought away from mentally working out the fundamentals of his AC "electric" motor. To him it was still in the thought experiment stage, and silence was the word of the day. It was probably a good idea considering that while in Paris working for the Société Electrique Edison, none of Edison's men were receptive to his ideas and it was one of the reasons he saw America as his future. There was a brief moment when Tesla was going to broach the subject with Edison on Coney Island, but he was interrupted by a friend of Edison's. From that time on he pledged to himself to never speak about it until it was the right time and the right place.[41]

However, there exists a slightly different version of the story, and probably the more believable of the two, inasmuch as Edison was known as stingy, he was having employee problems, he was feeling anxious and professionally threatened, and he had an immediate need for capital. It is most probable that Edison himself told Tesla that he would give him $50,000 if he fixed the problems and increased the dynamos' power as he had boasted.[42] Edison was struggling to keep his business of DC power distribution profitable. When Tesla came for his money in December 1884, Edison chuckled and said, "You are still a Parisian! When you become a full-fledged American you will appreciate an American joke."[43]

It's not enough that Edison had "cheated" Tesla out of a well-deserved reward for his inventiveness while working for the Edison company in France only months before, but when he arrived in New York City to work for Edison directly . . . it would happen again, and again! Edison was also facing heavy competition for an arc-lighting system that would deliver efficient light to towns, cities, and factories. Major competitors such as Thomson-Houston Electric Company, the Brush Electric Light Company, and the United States Electric Lighting Company were in the midst of doing the same, so Edison was desperate. He filed a patent for an arc lamp in June. He then gave Tesla the basic plan for an arc-lighting system and told him to make it work. In short order, Tesla had developed an entirely new, complete system for which he had expected to be compensated. However, the system was never used and Tesla was never paid.[44] In the years to come the real Edison would reveal himself more and more, demonstrating to anyone who paid the slightest attention that Edison was always for Edison, and that he would do whatever was necessary to advance his *own* position, even if that meant making anyone, particularly Tesla, his victim.

For a fleeting moment Edison tried to woo Tesla back with the promise of a $10 raise of his then meager salary of $18 per week. He even gave Tesla a backhanded compliment when he said, his ideas were "magnificent but utterly impractical," and that he was indeed "the poet of science."[45] Tesla's answer was to simply leave the Edison Machine Works and not look back—that is exactly what he did. But a violent storm was on the horizon, and the two men would soon meet again with each other's future at stake and on different terms.

∼

One has to question Edison's duplicitous actions toward Tesla. Was it that Edison was jealous of the innovative Tesla? Was it that Edison never understood what Tesla was thinking about and was intimidated by the foreigner? Was it that Edison feared Tesla would someday be his competitor and ultimately render him irrelevant? Was it that Edison was more a clever, opportunistic entrepreneur than a true, natural inventor? Or was it that Edison just could not stop the "con"? These are all threshold questions when considering the relationship between Tesla and Edison. And the only answer to any and all of the preceding questions is the same: "Yes."

Tesla made a final entry in the notebook at the Edison Machine Works before he left, and it read as follows: "Good by [*sic*] to the Edison Machine Works!"[46] His employment at the Edison Machine Works in New York City had lasted some six months.

And now it was over. So, no matter how one looks at it, Tesla found nothing humorous about *never* receiving the rewards he was promised by Edison and his agents for his hard work and ingenious problem-solving abilities both in New York and Europe. After all, he was a Serb and Serbs did not joke about such serious matters. Moreover, this series of seemingly innocent jokes, as they were characterized by Edison and his gaggle of worthless jokesters, proved to be a harbinger of things to come for Tesla, because he trusted every man's word. However righteous his naivete, it would not serve him well throughout his long life.

~

Although disappointed, stung by his former hero's laughing in his face, and mocking his accomplishments while making fun of immigrants, Tesla would soldier on.[47] A man of his God-given talents and relentless assiduity could not be held down for too long. He would always attract both entrepreneurs and opportunists—he just had great difficulty distinguishing the difference between the two. He was now hoping his next venture would prove worthy of his talents and help him change the world.

So now he was ready to go! Or was he? Could it be that others had different plans for the unguarded psyche of Tesla?

NEW YORK CITY, USA, 1885

In March, Tesla was still trying desperately to shake off the disappointment he had experienced with Edison, the man he once worshipped. His mind was again operating in overdrive. Thoughts were bombarding him at a relentless rate. He could not process or deal with the onslaught of negativity, but he knew he had to if he was to succeed, for he so wanted to bring his genius to the world for all the right reasons. Yet he began to question himself, thinking that it may never happen. Then he met with well-known patent attorney Lemuel Serrell—formerly in Edison's employ—and his patent artist Raphael Netter. He thought his wants were answered. The two men helped Tesla learn how to deconstruct complex patents into their individualized improvements, and on March 13, 1885, Tesla and his legal advisor applied for his very first patent (no. 335,786), which was a better-functioning arc lamp that produced a uniform light and prevented the annoying flicker of other arc lamps.[48] The patent application process did not stop there for the burgeoning inventor. In May and June, Tesla applied for other patents that addressed the issues of sparking commutators (see chapter 3) and the regulation of the current by means of a new independent circuit coupled with auxiliary brushes. This aspect permitted spent lamps to separate themselves from the circuit until new carbon filaments could be replaced without causing all functioning lamps to go offline.[49]

Tesla's attorney soon put him in touch with one Benjamin Vail, a local council member in Rahway, New Jersey, and his associate Robert Lane. Vail had long flirted with the idea of his town being a leader in utilizing modern technologies. Now was his chance. Joining his investment of one-thousand dollars with another four-thousand dollars from additional investors, the men launched the Tesla Electric Light and Manufacturing Company. At the same time Tesla took up residence in a garden apartment in Lower Manhattan. There children played and often made off with colored glass balls on sticks that Tesla had placed in the garden, which were very fashionable in Europe. Having no choice, he soon replaced them with metal ones, and had his gardener take them inside his apartment each night.[50]

As history continued to repeat itself concerning Tesla's unsuccessful business matters, Vail deceived him when he intimated that he was also interested in Tesla's AC motor, and that's what ultimately brought Tesla into the opportunist's honeytrap of sorts. After nearly a year of working with Paul Noyes from the Gordon Press Works, he finalized the installation of his first and only municipal arc-lighting system that illuminated numerous streets of a town and several factories.[51] Tesla knew that once he had perfected his AC motor, he would be in a position to dominate the lighting industry. But for now, his more efficient and unique approach to Edison's arcane arc-lighting system did produce an unexpected but welcomed result: George Worthington, editor of *Electrical Review*, featured Tesla's recent inventions and his new company in a very large advertisement that showed precise drawings of the "Tesla Arc Lamp" and his improved "dynamo."[52] Needless to say, Tesla was excited and had every reason to believe he was on his way to producing endless inventions of great import. But it was not to be . . . again.

Neither Vail, the company's president, nor Lane, its vice president, cared much for Tesla's dream invention—his AC motor. They thought it was a useless invention and not worth their time—arc-lamp lighting was a big seller.[53] But in time, Vail and Lane realized that the manufacturing of arc-lighting systems was too competitive, so they decided to simply pursue operating the Union County Electric Light and Manufacturing Company for the City of Rahway, New Jersey, and surrounding areas. The duo no longer needed Tesla, but they kept his patents (Tesla could no longer use his own patents) that earlier had been assigned to their joint company bearing Tesla's name.[54] Sadly, they weren't alone in not believing in his AC motor. Tesla had already faced such rejections of his AC motor from those in Europe and even Edison himself—his assault on Tesla's genius would soon make history. Needless to say, the mentally delicate inventor was outraged. After all, he had postponed exploiting his AC system until the Rahway project had proved its worth, thus attracting countless investors—and he knew it would. But, as he so clearly remembered, "Then came the hardest blow I ever received. Through some local influences, I was forced out of the company losing not only all my interest but also my reputation as engineer and inventor."[55] In a duplicitous act that can only be characterized as devasting to Tesla, after the fiasco where he had been "cheated" yet again, he spent the next twelve months unsuccessfully dealing with "terrible heartaches and bitter tears."[56] His lack of finances because of national economic distress and con artists only exacerbated his forlorn condition: "Very often I was compelled to work as a laborer and my high education in various branches of science, mechanics and literature seemed to me like a mockery."[57] He was even mocked for his fine dress and white hands when digging ditches, to which he responded, "I worked harder than anybody. At the end of the day, I had $2."[58]

~

Clearly upset and under tremendous pressure, it is most likely that Tesla experienced normal disappointment at this time, and this period in his life did not represent a major depressive episode. Although Tesla would continue to experience ups and downs, success and failure, as well as mania and depression for the next sixty-two years of his active life as an inventor until his death—his scientific and creative output as measured by the timing of his more than two hundred worldwide patent applications occurred during two specific periods that correlated with times when he was predominantly hypomanic. This is not surprising, since other notable bipolar creative geniuses also did their greatest work during manic episodes. During a likely manic episode, Handel wrote one of the longest compositions of all time, *The Messiah*, with fifty-three movements in three parts in just twenty-four days, with the original 259-page score showing ink literally flying off his pen as he composed. For Tesla, of his one hundred eleven U.S. patents filed over his lifetime, he produced an amazing sixty-four of them during the twelve years between 1886 and 1894—more than five a year or almost one every other month; he had a gap in 1895, likely related to a depressive episode in 1894 and early 1895 (laboratory destroyed in a fire) and then produced another thirty-one patents between 1896 and 1901, with none in 1899 (also likely related to a depressive episode in 1898 and early 1899) for a patent output during these four productive years of almost eight per year. Following 1901, Tesla filed only fourteen more patents in the next forty-three years, and none in the last eight years, times when he was predominantly depressed and had no indications whatsoever of hypomania or mania. His bursts of creative output occurred at times when he was demonstrably manic or hypomanic. Thus, creative output had a good relationship to times of mania and hypomania, with lack of creative output correlating with depression and lack of mania or hypomania.

NEW YORK CITY, USA, 1886

By May, Tesla was determined to bounce back from yet another cheat, but an acute depression took hold of him. He had been there before. He briefly abandoned all hope of succeeding in the world, which until now had defrauded him of his rightful success.[59] His early childhood thoughts and idiosyncrasies were now front and center again. He fought to hold them back. He had to make it all work. There were no options.

Finally summoning up a modicum of intestinal fortitude, he managed to file a patent application for a dynamo commutator; next came another patent application for a "thermomagnetic" motor (Curie-motor) that was based upon the fact that iron magnets lose their magnetism when heated. It was used to convert heat into kinetic energy using the thermomagnetic effect by way of a pyromagnetic generator. He thought the concept was original enough that it would attract support from legitimate investors. Although the invention did not prove profitable, the foreman of the ditch-digging crew he worked on was so impressed with his obvious problem-solving talents that he introduced him to Alfred S. Brown—superintendent of Western Union's New York Metropolitan District—a bona fide electrician and recognized expert in underground telegraph work.[60] It was Brown who then introduced the eager Tesla to Charles Peck, a noted patent lawyer and investor from Englewood, New Jersey. He was also the secretary of the Mutual Union Telegraph Company, which supplied and operated twenty-five thousand dedicated telecommunication lines throughout some twenty states for banks requiring secured data exchanges.[61] Could the team of Brown and Peck (sometimes Peck and Brown) be what Tesla was needing, was waiting for—honest men whose business acumen was unparalleled and who supported and appreciated his rare talents?[62] After all, they were interested in the possibility of generating electricity directly from heat.[63]

Perhaps luck was with Tesla this time. Brown and Peck had the money to invest because Peck had connections to banker John. C. Moore, who in turn had financial links to robber baron J. P. Morgan.[64] Brown in particular was very interested in the advantages of (AC) alternating current over (DC) direct current. Where Edison had failed to see the revolution before him in the name of Nikola Tesla and his AC system, he did not fail to experience what would become the inevitable demise of his DC system of electrification, and hence, dethrone him as the preeminent purveyor of electrical power distribution. However, Brown was prescient enough to see what the future would be with Tesla's AC system of electrification.[65] Together Tesla, Brown, and Peck would form a new company called the Tesla Electric Company in April 1887. Despite the economic recession that would continue into the next year, there was no stopping the troika. Ownership and profits would be divided between Tesla, who would receive one-third; Brown and Peck to divide one-third; and the remaining one-third to be invested in new innovations Tesla created and to pay the costs of patents.[66] In reality, the three shared a single patent for an AC dynamo, while Peck and Tesla split five more patents on commutators, motors, and power transmission, and the remainder of inventions created during this period were placed in the name of the Tesla Electric Company—the first patent was filed on April 30, 1887.[67]

∼

Much to Tesla's surprise, his old friend Anton Szigeti, a talented Hungarian mechanical engineer, had arrived in New York City on May 10, 1887. He was the only individual to be present during Tesla's "Boom" moment in a Budapest park in February 1882, when he envisioned the idea of doing away with the commutator and using a "rotating magnetic field" in his induction motor.[68] Tesla had always felt comfortable around Szigeti and trusted him. The dynamic duo quickly resumed their habits of strolling through parks, with Tesla reciting poems aloud in any number of languages, and the two discussing design ideas.

Now it was time for Tesla to get to work at his new laboratory and offices located at 89 Liberty Street.[69] The Tesla Electric Company was capitalized with $500,000, and Tesla was salaried at $78,000 per year in today's money. Although minimally furnished at its opening, his laboratory was abuzz with activity, so much so, that he had to make a deal with the Globe Stationery & Printing Company on the ground floor to obtain additional power at night after the printing presses were shut-off.[70] With much needed support in place, Tesla set to work on a new prototype of the AC motor he had conceived of years before in Budapest. "The motors I built were exactly as I imagined them," he said years later. "I made no attempt to improve the design, but merely reproduced the pictures as they appeared to my vision and the operation was always as I expected."[71] This aspect of Tesla's rare abilities is at times glossed over, but he often said that he invented, manufactured, and tested his inventions in his mind before bringing them to life in the real world.

The workload at the company was allocated as follows: Tesla as sole designer, Brown as technical expert, and Szigeti as assistant. Soon they were manufacturing the company's first AC "induction" motors. Peck spent the next ten years, as did Brown, silently backing Tesla's work, and he also assisted in patent applications by making trips to investors in several states and dealing with the mountains of paperwork necessary to be granted patents.[72] The legal aspects of patent applications, particularly involving all iterations of the AC induction motor, were the duty of the highly respected law firm of Duncan, Curtis and Page. In particular, the Harvard-educated Parker Page took a great interest in Tesla's AC induction motor.[73]

Tesla produced (invented) machines as rapidly as they could be constructed: three complete systems of alternating-current machinery—for single-phase, two-phase, and three-phase currents—and conducted experiments with four- and six-phase currents. In each of the three principal systems he produced the dynamos for generating the currents, the motors for producing power from them, and transformers for raising and reducing the voltage, as well as a variety of devices for automatically controlling the machinery. He not only produced the three systems but also provided methods by which they could be interconnected, and modifications that allowed a variety of means of using each of the systems. He also calculated the mathematics at the root of these groundbreaking inventions, inventions that would prove be the dominant force in the manufacturing of both electrical power and its distribution as well.[74] Needless to say, it was a complete "turnkey system." And astonishingly, he had not put pen to paper once. He wrote nothing down. It was all recalled from memory, every detail.[75]

After all, the rewards were high, but so were the risks. It was an all-or-nothing scenario that would play out before the whole world: It was AC vs. DC. It was the War of the Currents. The vast riches that were possible were incalculable. By controlling the infrastructure and intellectual property of what was to become the nation's largest industry, wealth was a certainty. Not only was wealth a certainty, but also to the victors would go the power to determine the course of human history: to give birth to a new economy and culture. People's lives would be forever altered, giving them more personal time because workdays would be shorter; elevators and streetcars would allow people to live in expanded urban environments; food could be kept fresh for days; buildings could be kept cool; and AC-electric-powered streetlamps promised to reduce crime rates.[76]

Thus began the process by which Tesla would make a worldwide name for himself, as his scientific reputation rose by the day. He would bring his dream of an AC motor and alternating-current polyphase power system to life. It would soon become the global standard for electric power, and it remains so to this very day! It would take tremendous work, more than most could ever imagine, to accomplish such a sea-change in society, but Tesla was more than up to the task. He worked virtually day and night, with little rest, but with much distress. In this pursuit, his hyperthymic drive without surges of out-of-control mania may have been a blessing. By now he had developed a very strong aversion to germs of all types, and other phobias began to become manifest as his obsessive-compulsive disorder (OCD) morphed into new symptoms the longer he remained in a relatively stable hyperthymic state without highs or lows. Touching others was verboten; he only used freshly laundered napkins, dress collars, and handkerchiefs just once. He washed his hands incessantly and wore cotton gloves most of the time when in public. Intrusive thoughts would at times steal the day, leaving him wondering if he would ever succeed, if he would even live another day. Over time, such oddities would grow, altering his social relationships and perpetuate his schizoid isolation. As he became more financially secure, he would often eat alone in a very ritualistic manner that observers found bizarre and yet entertaining, not from manic psychosis but from odd performances of obsessive compulsiveness like nothing anybody had ever seen before. He lived nearly sixty years of his life in New York City hotels beset by his obsessive-compulsive disorder. It had deeply invaded his life at this point, and it would never leave.

But for now, it was time to sell his "new" physics!

War of the Currents

Mania, Wrath, Anger

NEW YORK CITY, USA, 1887

The challenge of changing the way civilization was powered would seem utterly impossible to most people, but not to Nikola Tesla. Even in the stifling heat of a summer day in his new world, his mind knew what to do. Even after his abrupt resignation from the Edison Machine Works, his mind knew what to do. And even in the face of constant rejection of his AC induction motor, his mind knew what to do, because he was powered by his unrelenting need to change the world and to matter. But at what cost to him?

It was time for "his" new physics, and he was going to deliver it. He had long ago passed the thought-experiment stage of his AC induction motor, for it had been more than five years since the glorious moment when his beautiful, miraculous creation appeared before him . . . complete and whole . . . humming and whirring . . . spinning and shining . . . alternating endlessly. His "rotating magnetic field" would reshape all of humanity, releasing unlimited power and prosperity.

It had been done!

But there was still much to do. To convince the kingmakers of finance, industry, and business, Tesla knew he had to move with the greatest of urgency. He had remembered how he had convinced Peck and Brown to invest in him and his revolutionary inventions. However, even Charles F. Peck, one of Tesla's partners and a distinguished lawyer who hailed from Englewood, New Jersey, was initially among the doubters; he, like most, was a captive of conventional wisdom. Peck was aware of the "failures in the industrial exploitation of alternating currents and was distinctly prejudiced to a point of not caring even to witness some tests."[1] Simply put, it was time for Tesla to convince his own (future) partners, as well as others. The truth was that "the alternating current was but imperfectly understood and had no standing with engineers or electricians and for a long time Tesla talked to deaf ears."[2]

After several meetings with Peck and Brown, Tesla remembered a brilliant demonstration from centuries ago that could prove to be the "convincer" he needed. He called yet another meeting with the two men and asked the question: "Do you know the story of the Egg of Columbus?"[3] Of course they did, or least they said they did. As legend has it, and as Tesla paraphrased to Peck and Brown, Columbus needed the financial support of Isabella, the Queen of Spain. "So the saying goes that at a certain dinner the great explorer asked some scoffers of his project to balance an egg on its end. They tried it in vain. He then took it and cracking the shell slightly by a gentle blow, made it stand upright."[4] Call it fact or fiction, but Columbus was granted an audience by the queen and won her support—she pawned her jewels to finance his famous trip.[5] Then with his audience of two captivated, Tesla segued into his pitch: "Well, what if I could make an egg stand on its end without cracking the shell?" To which one of the men responded, "If you could do this we would admit that you had gone Columbus one better." With that, Tesla responded happily, "And would you be willing to go out of your way as much as Isabella?" The experienced lawyer Peck answered, "We have no crown jewels to pawn, but there are a few ducats in our buckskins and we might help you to an extent."[6]

So now it was showtime: In the dramatic fashion of Columbus, Tesla began with a distinct flair all his own as he set a copper-platted egg—and added several brass balls and pivoted iron discs as a sideshow "for additional effect"—on top of a wooden table; underneath it he had placed a four-coil magnet producing alternating current. With his long arms commanding the moment, as if he were a Renaissance knight wielding a mighty rapier in defense of his honor and his life, Tesla flipped a switch on (delivering a rotating magnetic field). The "egg" started spinning, and as its rotation accelerated, the egg stopped wobbling and amazingly stood on its end, still spinning, thus demonstrating both motion and stasis at the same time—it was the "mysterious" rotating magnetic field at work. The two men had now been convinced in the concept of the rotating magnetic field.[7] They were all-in. And as we now know, the Tesla Electric Company was born in April 1887. He then moved his laboratory to a larger workspace in Manhattan's financial district located at

33–55 South Fifth Street (currently West Broadway), ironically, just down the street from Edison's downtown office.[8]

~

Tesla realized after the success of his egg demonstration that a key ingredient of invention was being able to sell it to investors and the public by virtue of creating the right illusion about his revolutionary designs. He also realized that people invest in projects that capture their imagination and not the mundane. Most people do not have the imagination to think beyond what they see, so if they see something otherworldly, either by way of stories, metaphors, or targeted themes, they then want to see more and to be a part of something extraordinary. Tesla would become a master showman extraordinaire in short order. In fact, once he had achieved great fame and fortune, the most celebrated of the world's glitterati would be held spellbound as he made the mysterious "electricity" his slave before their very eyes in the magician's chamber.

~

Tesla understood that he was now in a position to dramatically influence the course of human history, and to that end he drove himself beyond the expectations of mortal men. Collapsing from overwork was not uncommon for Tesla because he often failed to take breaks for food and rest. And because his history of mental problems was ever present, lurking in the shadows, such days of hard work were a real risk, pushing him to extremes. Either he would drop into the depths of despair, thinking that it was all too much, or he would rise to the highest heights, and invention after invention arose in his mind, asking him to choose.

It was at this time that Thomas Commerford Martin appeared on the scene, seeking out Tesla to write a magazine article on invention. T. C. Martin was immediately consumed by Tesla's presence, his radiant brilliance. As a bona fide electrical engineer in his own right and editor of the respected *Electrical World*, he was influential in the world of electrical sciences. He described the lanky Tesla as having

> eyes that recall all the stories one has read of keenness of vision and phenomenal ability to see through things. He is an omnivorous reader, who never forgets; and he possesses the peculiar facility in languages that enables the educated native of eastern Europe to talk and write in at least half a dozen tongues. A more congenial companion cannot be desired. . . . The conversation, dealing at first with things near at hand and next . . . reaches out and rises to the greater questions of life, and duty, and destiny.[9]

Moreover, as a journalist, T. C. Martin embraced the drama that the unknown inventor was soon to be an electrical titan that would unseat Edison and win the War of the Currents: AC versus DC. T. C. Martin further appreciated and admired the thirty-one-year-old Tesla for his humanistic bent. Tesla had spoken of his deep desire to help the helpless, to unburden humanity, and to make his life matter. He sought to ease the hard labor of the world writ large with his magnificent induction motor and alternating current polyphase system of electrical power and distribution.[10] T. C. Martin also understood that Tesla had no choice but to come to America to seek his fame and fortune—more importantly, to see his revolutionary inventions become reality. He said of Tesla just a few years later:

> It does not follow that such a man shall remain in a confessedly unfavorable environment (Europe). Genius has its own passport and has always been ready to change habitats until the natural one is found. Thus it is, perchance, that while some of our artists are impelled to set up their easels in Paris or Rome, many Europeans of mark in the fields of science and research are no less apt to adopt our nationality, of free choice. They are, indeed, Americans born in exile, and seek this country instinctively as their home, needing in reality no papers of naturalization. It was thus that we welcomed Agassiz, Ericsson, and Graham Bell. In like manner Nikola Tesla, the young Servian [Serbian] inventor with whose work a new age in electricity is beginning, now dwells among us in New York.[11]

The mustachioed, dapper T. C. Martin now knew what he must do. He would soon come to be a major player in Tesla's life and a champion of the lifelong naïf. The man himself was English by birth, born less than two weeks after Tesla. His father worked for the famous Lord Kelvin, and he was associated with Thomas Edison from 1877 to 1879. Thereafter, he sought the security of editorial work, and he subsequently became a founding member of the prestigious American Institute of Electrical Engineers (since 1963 it has been known as the Institute of Electrical and Electronics Engineers, IEEE)—a place where Tesla would soon make a name for himself. In fact, he was perfectly positioned to help Tesla, for he understood the ins-and-outs of the glitzy futuristic field of electrical engineering, and how the vicious rivalries that existed took no prisoners.[12] He believed Tesla held in his hands the epochal nature of a device called "the" AC motor and polyphase system, all of which would lead to deserved fame and fortune.[13] The moment T. C. Martin left the Liberty Street laboratory—after a personal demonstration of Tesla's AC motors—he set out to create a major public relations campaign to bring Nikola Tesla to the world.

~

To begin with, T. C. Martin assisted Peck and Brown in arranging for the celebrated Professor William Anthony, of Cornell University, to come to the Liberty Street laboratory to test the efficacy of Tesla's new AC motors. To keep the level of secrecy at its highest, Tesla and Professor Anthony met several times in secret.[14] After the meetings and demonstrations (in both Manchester, Connecticut, and New York City) were concluded, Professor Anthony was very excited with the test results (Tesla's AC motors were 50–60 percent more efficient than DC models). More specifically, Professor Anthony was so enamored of Tesla's inventions, he said in a letter to Dugald C. Jackson, a teacher of electrical engineering at the University of Wisconsin and a recipient of the Edison Medal: "I [have] seen a system of alternating current motors in N.Y. that promised great things. I was called as an expert and was shown the machines under the pledge of secrecy as applications were still in the Patent Office. . . . In all this you understand there is no commutator. The armatures have no connection with anything outside."[15] Recalling the moment when Tesla's Professor Poeschl at the Polytechnic School in Graz publicly admonished him for stating affirmatively that "a commutator was not needed in an electric motor," Tesla was now once again vindicated, but there was still much to do to "make the sale." He now had to really "sell" it; he had to go on a bona fide publicity tour; he had to use all his powers to get it done. He was not happy, but he knew there was no avoiding the obvious. Internal tensions fought to take control again, but he was still able to hold them in check for the moment.

Now with T. C. Martin's assistance, and with Peck, Brown, and Professor Anthony's imprimatur, and after much difficulty because of his reluctance to speak about his unique creation, Tesla was now convinced he needed to conduct a lecture and demonstration of his motor before the American Institute of Electrical Engineers, which had been formed the year he immigrated to America.[16] He would do so some two weeks after being granted a major cache of his forty fundamental patents for his new AC motors on May 1, 1888.[17] The invitation was certain recognition that he had indeed arrived. Now the members were to meet him and his AC motors in person . . . and he was expected to perform.

~

Tesla's paper was cobbled together in haste—he had been struggling for many days with physical and mental exhaustion and some unidentified illness from overwork. He was experiencing mixed episodes of mania and depression, most easily recognizable to him because of his traumatic experience in Budapest only a few years before when a nervous breakdown forced him to bed, not knowing if he would ever survive. His "evil twins of genius and mania" had once again inhabited his every thought. It had all happened the night before he traveled on a chilly Tuesday evening up to Columbia College at Madison Avenue and Forty-Seventh Street, where the august body, already known as the American Institute of Electrical Engineers, was meeting.

The May 16, 1888, meeting began with many congratulatory remarks praising T. C. Martin's successful term as president. At that point, the room, now abuzz with childlike anticipation, breathlessly awaited the reclusive yet naturally dynamic Tesla. Word of his new AC motors preceded him, and now the hour was "ripe." With the mien of an eastern European aristocrat who draped himself in a dark swallowtail coat, Tesla spoke in faultless, yet accented English, first thanking many in attendance, including his most vociferous supporters, Thomas C. Martin and Professor William Anthony, as well as Franklin Pope, editor of a highly regarded electrical publication. He then went on in his high-pitched voice to politely apologize for what he believed would be a rather incomplete presentation. His face exhibited the look of an exhausted genius still pushing to discover new laws of nature and to make electricity work at the command of the common man. He more specifically expressed regret by saying, "The notice was rather short, and I have not been able to treat the subject so extensively as I could have desired, my health not being in the best condition at present. I ask your kind indulgence, and I shall be very much gratified if the little I have done meets your approval."[18] Before he began to demonstrate his AC motors, he reminded himself of his patent lawyers' admonition: do not give too many details. As such, Tesla was careful not to describe in his paper all his work, hence, it dealt in reality with the Tesla Rotating Magnetic Field, motors with closed conductors, synchronizing motors, and rotating field transformers.[19]

The remarkable scene had been set and ready to unfold before the renowned body of electrical engineers. Tesla's tall image appeared extramundane in the dim light of the grand room, for he was an imposing figure, no doubt, and his mind was burning red-hot. He held his breath for a beat to raise the level of anticipation, as his piercing-blue eyes scanned the immediate area before him. His lecture was titled: "A New System of Alternate Current Motors and Transformers." It presented the theory and practical application of alternating current to power engineering; thus AC became, and still is, the foundation upon which has been built the entire electrical system of the world today.[20]

After his unneeded apology, he paused again for another beat and tried to push to the back of his mind his belief that he preferred self-imposed seclusion over socializing, and he saw no need to give speeches or write articles.[21] Then, with everyone's attention focused on him, the aristocratically styled Tesla, standing rod-straight, began his lecture by starting and stopping his miraculous motors and showing drawings and diagrams to prove his points. Simply put, with the flip of a switch a new era of technology began.[22]

As magic was happening before everyone's eyes, Tesla said,

> The subject which I have the pleasure of bringing to your notice is a novel system of electrical distribution and transmission of power by means of alternate currents, affording peculiar advantages, particularly in the way of motors, which I am confident will at once establish the superior adaptability of these currents to the transmission of power and will show that many results heretofore unattainable can be reached by their use; results which are very much desired in the practical operation of such systems, and which cannot be accomplished by means of continuous currents.[23]

Tesla made it quite clear that his motor had rendered commutators and brushes unnecessary complications. He had invented an electric motor that had never been seen before, an electric power producer that operated in a system specifically designed for it by him.[24]

Tesla peppered his talk with insights that served to keep those in attendance in a state of rapt awe. His words poured out of him in a manner of pure contemporaneous speech—no notes needed—because his photographic, eidetic, and didactic memories were in fine form. He had made his point and made it unequivocally. So now it was time to open the meeting to discussion. T. C. Martin stood before those in attendance and said, "Prof. Anthony I believe, is here, and as he has given this subject some attention I think he might very properly supplement Mr. Tesla's paper with some remarks."[25]

Professor Anthony stood and responded by saying, "I confess that on first seeing the motors the action seemed to me an exceedingly remarkable one."[26] The professor once again reiterated his delight with the Tesla polyphase system and touted its efficiency. Then came Elihu Thomson, the English-born inventor and electrical engineer, who had also been working on an AC motor since 1884. He was a vocal proponent of commutators and used them in his AC motors. Interestingly, he had given a presentation paper before the AIEE in June 1887, unaware of Tesla's work, but he was now faced with the Serb's claims of having developed a "practical" AC motor. He politely warned Tesla that he [Tesla] was not the only engineer developing AC motors and that he [Tesla] could expect competition in the U.S. Patent Office and in the marketplace.[27] Moreover, he attempted to diminish Tesla's efforts by commenting in a very condescending manner, saying, "I have certainly been very much interested in the description given by Mr. Tesla of his new and admirable little motor."[28]

Inasmuch as Thomson did not know Tesla, he must have thought his admonition and belittling comment would cause Tesla to back off and fade away. But Tesla stood strong, just as he had with Professor Jacob Poeschl and other cynics, skeptics, and nonbelievers. While he did tip his hat to Thomson, recognizing his preeminence in the world of electrical engineers, he did respond in kind, saying that he had built a motor such as Thomson's, but logic and his research dictated that the best AC motor would be one without a commutator.[29]

Moreover, Tesla's polyphase motor had several obvious advantages over Thomson's concept. Notably, it was less expensive to manufacture since its insulation and windings were simpler, and it did not require costly brushes or commutators—unnecessary complications. It was also less costly to operate because it needed fewer moving parts that could wear out quickly. It was even more robust and could be scaled up or down from smaller to greater sizes and from lesser to higher voltages. Ironically for Thomson, the U.S. Patent Office confirmed Tesla's primacy and rejected Thomson's patent application, arguing Thomson's "teaser current" method was actually based on Tesla's invention.[30]

The evening ended with Francis R. Upton, the AIEE's vice president and treasurer of the Edison Lamp Company, who presided over the proceedings at the time, having said, "I believe that this motor—Mr. Tesla can correct me if I am not right—is the first good alternating current motor that has been put before the public anywhere—is that not so, Mr. Tesla?" Upton closed the extraordinary meeting with the announcement that Tesla had invited everyone to see his magical motors spin and whirr at his Liberty Street laboratory.[31]

Tesla's AIEE lecture proved to be the centerpiece of the promotional campaign orchestrated by Peck and Brown (with assistance from T. C. Martin), and it set him on his way to becoming the embodiment of the electrical age's zeitgeist. Until that moment, Tesla's obscurity—having chosen anonymity by not joining any professional organizations such as the American Institute of Electrical Engineering, the Electrical Club of New York, or the National Electric Light Association as well as not meeting many people in the electrical engineering community—was an existential problem.[32] As an enigmatic Serb, he could seem aloof to lesser mortals, and he truly enjoyed the habits of a recluse.[33] T. C. Martin had said of Tesla prior to the groundbreaking lecture, "He stood very much alone, [as] the majority [of the electricians] were entirely unfamiliar with [the motor's] value."[34] In effect, Tesla proved to be a mysterious man of many powers, plagued by internal thoughts that would challenge him throughout his life.

As Tesla took his seat, those who had witnessed the remarkable sea-change that he had just introduced to the world became uneasy about their futures and resentful. After all, having mistakenly thought Tesla was a mere young upstart, they quickly realized he had just created a new religion in the electrical sciences, and had surpassed their wildest dreams, thus rendering them irrelevant. T. C. Martin looked on, knowing he had just become Tesla's willing prophet.[35] To be sure, it would not be that simple. Edison had already begun his anti-Tesla public relations campaign. He saw Tesla's AC motor and polyphase system as "unsafe and unfit for human habitations."[36] It would not be the last time Edison would attack Tesla, as he desperately fought to keep his place in the community of scientists practicing the art and science of electric power and distribution—for the purpose of making money. In fact, things would get much worse for Tesla, because Edison and his many hirelings went on an all-out, full-frontal assault on Tesla and his inventions. Edison's history of cheating Tesla was well chronicled, and his history of being primarily driven by money was known by most, so things were about to get mercilessly worse for the trusting Tesla. Edison had declared war!

The startling success of the AIEE lecture was to have the effect of adding fuel to the fire of the AC versus DC debate, which was still in its kindling-wood stage. But soon it would grow into a global conflagration and be known as the "War of the Currents." It can easily be said, and it should be said that Westinghouse was an important component of Tesla's success in winning the War of the Currents. But it must not be said, and it should never be said that the War of the Currents was between Edison and Westinghouse. The War of the Currents was between Tesla and Edison, for although Westinghouse was a loyal ally in Tesla's war, it was Tesla's obligation to defend himself and "his" patents, and Edison had drawn first blood long ago. Even the documentation in books and newspapers has always been incorrect. Tesla's name deserves to come first because he won the war between his superior AC current and Edison's inferior DC current—and his winning was inevitable.

Tesla, to his surprise, discovered himself to be a natural and brilliant lecturer; and his lecture soon became a landmark in the field of electrical engineering.[37] He had carried the day and dazzled them all. In fact, Tesla's ideas seized the imagination of the electrical engineering community, and his lecture at the AIEE became an instant classic, reprinted by electrical and other technical journals across the discipline. Yet one must ask the question: In such a moment, what must it have been like to be Nikola Tesla? He triumphed that night, but the importance of the War of the Currents must be examined through the lens of Tesla's mental health.

Now began the practical phase of the process. With Tesla's "polyphase electric motor" proclaimed as "an advance in the art," the stage was now set for Peck and Brown to offer Tesla's patents to the highest bidder.[38]

~

Enter Civil War veteran, inventor, and successful businessman George Westinghouse Jr., who hailed from the Westinghausen family of Westphalia, Germany. Westinghouse invented several much-needed devices after the Civil War, and he grew to be a supporter of Tesla's work. Always trying to be honest yet firm, he was a most astute businessman who made it a point to never act like a "robber baron." In 1865, he received his first patent for a rotary engine. He went on to invent several devices integral to the railroad industry, most notably the "air brake," which he perfected in 1869. It very soon became the gold standard for railroad car safety, and the system was quickly accepted in both the United States and Europe.[39]

Westinghouse was a man of substantial physical bearing—a solid round-chested man some six feet tall—sporting a well-groomed, walrus-style mustache and muttonchops, as well as a full mane of chestnut-colored hair. He was a charming man, but when matters turned to money, he was most serious—although he was known to enjoy the occasional joke. Much like Tesla, he was influenced by his father and other family members who possessed an inventive bent. His eyes were kind but often appeared to be aimed intensely at anyone in his sight. His dark-suited dress and habit of carrying an umbrella only increased his formidable appearance.[40] At times individuals did not know what to make of him, but Tesla did.

In a tribute to Westinghouse after his passing on March 12, 1914, Tesla extolled his virtues:

> The first impressions are those to which we cling most in later life. I like to think of George Westinghouse as he appeared to me in 1888, when I saw him for the first time. The tremendous potential energy of the man had only in part taken kinetic form, but even to a superficial observer the latent force was manifest. A powerful frame, well proportioned, with every joint in working order, an eye as clear as a crystal, a quick and springy step—he presented a rare example of health and strength. . . . Not one word which would have been objectionable, not a gesture which might have offended—one could imagine him as moving in the atmosphere of a court, so perfect was his bearing in manner and speech. And yet no fiercer adversary than Westinghouse could have been found when he was aroused. . . . His equipment was such as to make him easily win a position of captain among captains, leader among leaders. His was a wonderful career filled with remarkable achievements. He gave to the world a number of valuable inventions and improvements, created new industries, advanced the mechanical

and electrical arts and improved in many ways the conditions of modem life. He was a great pioneer and builder whose work was of far-reaching effect on his time and whose name will live long in the memory of men.[41]

〜

But make no mistake about it, as a highly successful entrepreneur and titan of industry—having survived and thrived in the cutthroat world of railroad businesses and robber barons—Westinghouse's first thought was to get his piece of the burgeoning electric power industry. Initially, it mattered not to him whether it was AC current or DC current—would it make money was his only concern. However, after he read Tesla's AIEE lecture, his interest was piqued, and he became concerned that his own company was not in the forefront of developing AC patents. He immediately dispatched a legal emissary to Turin, Italy, to secure control of Galileo Ferraris's induction motor. He also acquired the rights to a transformer developed by Frenchman Lucien Gaulard and Englishman John Dixon Gibbs.[42]

Westinghouse quickly understood the incalculable value of the Tesla patents, thus, he needed to act with all due dispatch. It was time. On May 21, he sent Henry M. Byllesby, vice president of Westinghouse Electric, directly to Tesla's laboratory, along with Thomas B. Kerr, general counsel, to New York City. Charles Peck, a clever lawyer and the primary backer of the Tesla Electric Company, asked Tesla to demonstrate his polyphase motors to the two men.[43] Byllesby reported back in a querulous tone to his boss: "Mr. Tesla struck me as being a straight-forward, enthusiastic sort of a party, but his description was not of a nature which I was enabled, entirely, to comprehend."[44] That said, Byllesby was thoroughly impressed with the demonstration when he commented to Westinghouse, "The motors, as far as I can judge from the examination which I was enabled to make, are a success."[45]

Now Peck was about to demonstrate his legal skills to Tesla because dealing with Westinghouse regarding the selling of the Tesla patents was not going to be easy. After all, Westinghouse was anything but a simple mark. And although he was at that point a step-or-two behind Edison in building major lighting projects—Edison had begun in 1882, while Westinghouse began in 1886—his business had expanded very rapidly. In just three years, Westinghouse Electric generators were powering 350,000 incandescent lightbulbs.[46]

So it was time to make *the deal*.

〜

Peck understood that he had to play the game. In essence, he had to try to high-roll Westinghouse. He came straight out with a well-disguised but blatant bluff. The moment Byllesby and Kerr voiced an interest in buying the Tesla patents on behalf of Westinghouse, Peck answered their interest with the statement that he had already been in talks with a San Francisco capitalist (a Mr. Butterworth . . . headed a syndicate) who had tendered an offer of $200,000 plus a royalty of $2.50 per horsepower for each motor installed.[47] Gamesmanship had been enjoined. Byllesby told Westinghouse of the outrageous terms, and he added, "and so I told them . . . I told them that there was no possibility of our considering the matter seriously. . . . In order to avoid giving the impression that the matter was one which excited my curiosity I made my visit short."[48] Peck in turn told Byllesby that he must match or surpass the offer by Friday. Things did not look promising for the Westinghouse deal makers, but they did manage to obtain a six-week option for the sum of $5,000.[49] By now, Byllesby and Kerr had recommended to Westinghouse that he should purchase the Tesla patents because they knew the motors were a success. Not wanting to be bested by Peck and seem too eager, Westinghouse decided to send his best inventors—two of the Westinghouse boys—Oliver Shallenberger and William Stanley Jr., to scrutinize Tesla's actual work, perhaps then making Peck believe he was not in the strongest position after all.[50]

With the deadly blizzard of March 1888 behind the people of New York City, on June 12 Shallenberger made his trip to the Tesla's Liberty Street laboratory to do just that. The demonstration involved the unexpected usage of motors operating on four wires. Shallenberger instantly understood Tesla to be far ahead of his own work on AC motors with regard to the rotating magnetic field. Shallenberger quickly made his way back to the Westinghouse in Pittsburgh and advised him not to hesitate in the purchasing of the Tesla patents.[51] But Westinghouse's investigation of Tesla's patents did not end there. Less than two weeks later, Stanley made his way to Tesla's laboratory for a look-see of his own. He left the laboratory with the word from Peck that the patents were about to be sold.

Westinghouse was coming to the realization that if he controlled the patents for polyphase motors, he would have more to offer his customers. Also, within his own company Shallenberger was working on his own AC project that was proving troublesome. Westinghouse had already purchased the rights to other versions of alternating current motors that were in development, so getting his hands on the best-of-the-bunch made perfect sense. The opportunity was there for him to win the day with purchase of the Tesla patents and at the same time eliminate the threat that others would control the massive future that alternating current promised. As Shallenberger admitted, "Their motor is the best thing of the kind I have seen. I believe it more efficient than DC motors [of Edison]. I also believe [the patent] belongs to them [Tesla]." At that point Westinghouse was determined to buy the Tesla AC patents and their corollaries. He expressed his need to own the Tesla patents and avoid potential lawsuits when he said, "If the Tesla patents are broad enough to control the alternating motor business, then the Westinghouse Electric Company cannot afford to have others own the patents."[52] Moreover, he would hire Tesla to oversee the development and manufacturing of his motors and to secure his own position as the dominant producer of AC electric power.[53]

Westinghouse was now feeling the squeeze from Peck, but he knew it was time to move on the patents or face the hard fact that someone else would have the incredibly valuable patents to use against him as business offerings to customers. So the final deal played out in dollars and cents as follows: On July 7, 1888, Peck and Brown agreed to sell the Tesla patents to Westinghouse for the then princely sum of $25,000 in cash, $50,000 in notes, and a royalty of $2.50 per horsepower produced for each motor. Tesla and his partners were also guaranteed that the royalties would scale up from $5,000 the first year, $10,000 in the second year, and $15,000 in subsequent years. Over the seventeen-year life of the patents the troika of Tesla, Peck, and Brown stood to make at least $315,000 ($8,720,000 in 2021).[54] Moreover, Tesla unwillingly acquiesced and arranged to move to Pittsburgh to assist in the mass production of his motors, converters, etcetera, at a monthly salary of $2,000 per month— equivalent of $600,000 per year in present dollars—at the Westinghouse Electric Company (Westinghouse Electric and Manufacturing Company).[55]

Tesla was not comfortable with the move to Pittsburgh, despite the large salary, but he knew he had no choice. Yet again he found himself working for another inventor since his arrival in America. Although his position was different from his lowly job in the employ of the Edison business empire, and his position at Westinghouse was one of great importance, his personal dilemmas would get the better of him.[56]

Westinghouse thought the royalty per horsepower was ridiculously high, but he understood that there was no other way to operate a motor but by alternating current, and "if it is the applicable to streetcar work, we can unquestionably easily get from users of the apparatus whatever tax is put upon it by the inventors."[57] In essence, a historic deal was made, and Westinghouse once again demonstrated his ability to get what he wanted. If others—customers— had to pay, then so be it. As for the troika, Tesla demonstrated his generosity and once again showed that he would always put "principle over profit" by giving Peck and Brown five-ninths of the proceeds from the deal, thus keeping four-ninths for himself. It was a tip-of-the-hat to his partners—an all too gracious gesture.[58]

As an example of Tesla always guarding his peculiar "principle" regarding compensation, it is certainly possible that he received no direct salary for his time in Pittsburgh. He held to the belief that because he had devoted himself to scientific laboratory research and would, hence, never accept fees or compensations for such professional services. That said, he had been paid for his patents and was receiving royalties from them. Moreover, from time to time he would make other contributions to Westinghouse's company, and in return he would receive either cash or company stock.[59]

At this point Westinghouse was already in a pitched battle with Edison over the shortcomings of their respective lighting systems. With Tesla's innovative patents in hand, a full-scale industrial war erupted.[60] It was inevitable, for the future of industrial development in the United States was on the table. Moreover, in the Gilded Age of unbridled capitalism, the question of who would control the winning technology was the true concern of both Westinghouse and Edison.[61] Hence, going to war over such technology was the obvious result. But it must be said that there would be no war worth fighting without Tesla's genius. Period.

PITTSBURGH, PENNSYLVANIA, 1888

In the last days of July, Tesla relinquished his New York apartment and escaped the intense heat of Manhattan on a ferryboat across the windy Hudson River. Once on land, he boarded the Pennsylvania Railroad for the arduous ten-hour journey to Westinghouse's corporate headquarters in Pittsburgh.[62] Once in there, he would stay for a short while at Westinghouse's mansion, known as Solitude. The white-brick estate, which included ten acres of finely manicured grounds, served as a brief respite for Tesla—he dined and engaged in conversation with Westinghouse, his wife Marguerite, and notables that lived nearby. But soon he sought the quietness of a hotel room. Several different hotels, including the Metropolitan, the Duquesne, and the Anderson, served his purpose. Hotels would become his usual domiciles for the rest of his days.[63] It was a lifestyle that served his need to be alone. Yet by the summer of 1888, he was already achieving great fame as a result of his AIEE lecture. How would he cope?

By the summer of 1889, Tesla was more than anxious to return to New York City. He had been dealing with the malignant envy of other Westinghouse employees who were jealous of the generous deal he had made with their boss. The situation was exacerbated by the fact that some of the Westinghouse Boys had convinced themselves that they had developed a better AC motor, and others simply could not brook what they saw as a "pompous foreigner."[64] In addition, Tesla found himself thoroughly bored with practical engineering and the bureaucracy of operating a business. He came to America to change the world for the better, and being holed up in southwest Pennsylvania was not going to work for him. He despised the quotidian repetitive work, and he thought those who questioned him were contemptuous. Then there was the issue of Westinghouse engineers using 133-cycle current on Tesla's motors, which was not workable. Early on, Tesla had concluded that the 60-cycle current was ideal for his motors, and he designed them so. Within a decade, 60-cycle current became the standard for power production in the United States and remains so today.[65] Once again, there were those at Westinghouse who were not enamored of Tesla, his genius, and the fact that all the Westinghouse AC central lighting systems had to be retrofitted to the Tesla standard—but they knew he was right. Moreover, they felt threatened by Tesla's inventiveness, and in turn they made life at the factory difficult for him. The last straw was small but proved most significant. Tesla had informed Westinghouse engineers not to use graphite bearings because they would overheat, but they went against his wishes.[66] They overheated, and so did Tesla.

By August 1889, Tesla's time in Pittsburgh had taken its toll on him. He felt that since he had created the "perfect" AC motor, it was the charge of others to work out the final details. He was ready to move on to new territories in the vast, limitless landscape of electrical science.[67] Years later he said of his one year working in Pittsburgh,

> I experienced so great a longing for resuming my interrupted investigations that, notwithstanding a very tempting proposition by him [Westinghouse], I left for New York to take up my laboratory work. But owing to pressing demands by several foreign scientific societies I made a trip to Europe where I lectured before the Institution of Electrical Engineers and Royal Institution of London and the Societe de Physique in Paris. After this and a brief visit to my home in Yugoslavia I returned to this country in 1892 eager to devote myself to the subject of predilection on my thoughts: the study of the universe.[68]

PARIS, FRANCE, 1889

Ironically, Paris was his first stop on his return to Manhattan—he traveled there as a bona fide member of the delegation of the American Institute of Electrical Engineers to attend the Exposition Universelle.[69] Once there, he visited the recently completed Eiffel Tower, which featured the American Otis Brothers' passenger lifts. He also attended the Congrès internationale des électriciens and even lectured before the august body. In addition, he heard the lectures by many of the world's foremost scientists, which included a talk on vibrating diaphragms by Vilhelm Bjerknes—a Norwegian physicist who had conducted experiments with Heinrich Hertz on electromagnetic waves.[70] It is most probable that Bjerknes introduced the now famous Tesla to the discovery of electromagnetic waves by Heinrich Hertz—it was a lecture that he found to be most stimulating after a year of perilous boredom in Pittsburgh.[71]

After a host of exhilarating but exhausting days in Paris, it was truly time for Tesla to make his way back to New York City. He was excited by the virgin field of electromagnetic waves and the possibilities it held, as well as other electrical wonders yet to be discovered. A decade later he would remark fondly: "The journey is not finished yet, and the wanderer is well-nigh exhausted. He longs for more sweet berries, and anxiously asks, 'Did anyone pass this road before?'"[72]

Inasmuch as Pittsburgh proved to be a great disappointment and a distraction to the thirty-three-year-old "electrical" phenomenon, Tesla buoyed himself with the thought of journeying once again down a road of his own. But first there were personal matters that needed attending to, for he was secretly lingering between euphoria and depression. He needed to discover a delicate equipoise between the realms of idealism and reality and between the present and the future. He felt his experiences in Pittsburgh had forced him to the present; hence, he became unhappy and uncreative.[73]

He was stymied while in Pittsburgh, and declared so, "I was dependent and could not work."[74] But, he later recalled, "When I became free of that situation ideas and inventions rushed through my brain like Niagara."[75]

Meanwhile, the War of the Currents was ginning up.

· *6* ·

New Creations

Unhappy, Uncertain

NEW YORK CITY, USA, 1889

*A*s we have learned, before returning to New York City in late summer, after having gladly left Westinghouse Electric in Pittsburgh, Tesla, physically tired and mentally deflated from his experience in the steel town, took a brief detour through Paris to attend some scientific discussions and demonstrations that were being held at the Exposition Universelle. His earnings while in Westinghouse's employ allowed him the much-needed time in Paris.[1] He needed time to take stock of where he was; and he needed time to access what was his next move to parry the constant, horrendous attacks by his seemingly forever nemesis, Thomas Edison. Tesla was in a black mood, and he didn't like it. Nevertheless, there was a diversionary event that served to give the intense man a moment of reprieve from his scientific pursuits and war with himself and Edison.

It was a short-lived experience in the City of Light that all too often goes without serious consideration by Tesla's previous chroniclers. Although it was a fleeting moment of unexpected delight and a brief distraction from his troublesome days in Pennsylvania that took place while he was dining out with French engineer and physicist André Blondel, it was nonetheless significant. It was yet another of the unusual facets of his diamond-in-the-rough character to the point that is instructive as we discover what made Nikola Tesla, Tesla—the architect of the modern world.

∽

As previous biographers have told the story, it was the famous soubrette, the stunningly dressed French actress Sarah Bernhardt—the "divine Sarah"—who passed by Tesla's table on that occasion and dropped her lace handkerchief.

Tesla quickly leaned down and retrieved the handkerchief, and as he returned it to the actress, it was said that their eyes met with a burning intensity.[2] However, another version tells of the dashing Tesla returning the lace handkerchief without even looking at the beautiful actress as he bowed and said, "Mademoiselle, your handkerchief."[3] Yet still another version, possibly apocryphal, has Tesla stating it was a scarf and not a handkerchief, which he kept the rest of his life. The certainty of whether the incident took place in 1889 or rather when Tesla worked for Edison in France years before is also still an uncertainty.[4] Nonetheless, one can look at the event as something very instructive as we seek to understand the mechanics of Tesla's complex mind. His relationship to and with women has often been a thorny issue for biographers, hence, they tend to stay away from it, choosing to spend their labors exclusively on what he did rather than how he did what he did, given who he was.

As a respected scientific journal of the time reported, Tesla might have been "invulnerable to Cupid's shafts," for the most part, but Sarah Bernhardt was perhaps an exception.[5] Tesla gives us greater specificity as to his intentions during a published interview:

> I have planned to devote my whole life to work and for that reason I am denied the love and companionship of a good woman, and more too. . . . I believe that a writer or musician should marry. They gain inspiration that leads to finer achievement. . . . But an inventor has so intense a nature, with so much in it of wild, passionate quality that, in giving himself to a woman, he would give up everything, and so take everything from his chosen field. It is a pity, too; sometimes we feel so lonely [despite] . . . that sublime moment when you see the labor of weeks fructify in a successful experiment that proves your theories.[6]

Having had an early, problematic history with wine, woman, and song, so to speak, during his college years, Tesla consciously forged a path of asceticism in his adult life because he knew his purpose was of a higher order. It could be said that his womanizing while in college in Graz may have contributed greatly to his never having completed his studies.[7] Even a professional palm reader had pontificated on Tesla's hand, stating that it indicated "a flirtatious streak and hypersensitivity."[8] Such theories are supported by his appreciation of the activities of numerous women who showed a real interest in his well-being: Mrs. Clarence Mackay, Mrs. Jordan L. Mott, and the beauty Lady Ribblesdale, the former Mrs. John J. Astor, were such women, along with Miss Anne Morgan and Miss Marguerite Merington.[9]

Throughout Tesla's life, friends and the media alike were forever trying to get him married, but his resistance to matrimony never flagged. Kenneth Swezey, a young journalist, science writer, and confidante of Tesla's remarked

years later that "Tesla's only marriage has been to his work and to the world
. . . believing that the most enduring have come from childless men."[10] Tesla
consciously chose celibacy and, vanity aside, his focus can be best described as
"lifting the burdens from the shoulder of mankind," which he never wavered
from. It was his raison d'être. It was his "life force" to be sure.

~

Upon his return to New York, Tesla was fired by ideas for inventions pouring
forth. It often left him exhausted beyond description, but it was better than the
intellectual stagnancy he had experienced during his "lost year" in Pittsburgh.
And because of his need to work out his new inventions, as he always had,
inside his mind first, he was perpetually pushing himself. His antennae were always up, ever seeking to discover new laws of nature that others simply could
not detect. But there was still an unhappiness, a mental darkness chasing him,
much like Churchill's "black dog."[11] His mental baggage continued to prove
very problematic, and at times it was constraining to the point of approaching debilitation. Then what? Would Tesla be able to control if not overcome
his black dog? Would he prevail in his war with the fraud Edison? Would he
survive the war with himself?

To be frank, he was stalked by the ever-present image of a rancorous,
vindictive, cutthroat Thomas Edison who had waged an unbridled war against
him—the War of the Currents produced everything from newspaper images
and articles of Edison attacking Tesla to verbal attacks in public aimed at him.
As a consequence, just as the Republic of Venice had built its Venetian Arsenal of naval power during the late Middle Ages to the high Renaissance as its
defense against warring nations who sought its treasures, so too did Tesla focus
on building his arsenal of inventions, discoveries, and patents at his Grand
Street laboratory, for he too was at war! Edison had declared it, and as far as
Edison was concerned, it was too be fought with the gloves off. Edison was
smart enough to understand the ramifications of a loss: it would be humilia-
tion for him personally and economic devastation for him and his company.[12]
But he was not without a plan, a plan that would put Tesla dead-center in
the bull's-eye. And as the Venetians led their naval forces to the Holy Lands
during the Crusades to save Eastern Christianity, so too did Edison martial his
forces and alliances to save himself and destroy Tesla.

~

In the remaining few decades of the nineteenth century "electricity," that
most mysterious of energies, was viewed by most people as nuclear power
was once viewed years later: dangerous! Admittedly, electricity, like nuclear
power, had its benefits, yet most people knew no better. And because of that

lack of a fundamental understanding regarding electricity, it was very easy to cause dread among the general public. Edison claimed high voltages produced by alternating current were deadly to living beings. So Edison, whose real talent was in the realm of marketing, made the decision not to compete with Tesla's alternating current on its merits, but rather to attack by raising the specter of the dangers of Tesla's product in terms of public safety.[13]

Really, what were Edison's options? He knew his system was inferior from its transport limitations—requiring a direct current power plant about every mile—to its high cost of construction.[14] So, before he let Tesla get too far ahead, the Wizard of Menlo Park—a sobriquet that followed from his first laboratory facility in New Jersey—recruited Samuel Insull, his personal secretary and a visionary in his own right, and his publicity man Harold P. Brown to orchestrate a massive propaganda campaign to frighten the world into believing that alternating current was both dangerous and evil, while direct current was seemingly safe and benign.[15] It would take some convincing, because Tesla's demonstrations of his AC were undeniably successful and proved its supremacy as the obvious choice to power the world.

After Edison arrogantly rejected an invitation by Westinghouse to visit his factory in Pittsburgh for the purpose of calling a truce of sorts, Edison officially answered Westinghouse by saying, "My laboratory work consumes the whole of my time." To further make known his dislike of the invitation, Edison metaphorically kicked sand in Westinghouse's face by saying, "Direct current [is] like a river flowing peacefully to sea, while alternating current [is] like a torrent rushing violently over a precipice."[16] He doubled down on his full-frontal assault on alternating current for any purpose by stating, curiously in a literary publication of the time, the following: "The electric-lighting company with which I am connected purchased some time ago the patents for a complete alternating system, and my protest against this action can be found upon its minute-book. Up to the present I have succeeded in inducing them not to offer this system to the public, nor will they ever do so with my consent. My personal desire would be to prohibit entirely the use of alternating currents. They are as unnecessary as they are dangerous."[17] Edison's animus toward Tesla and his outright fear of alternating current was patently obvious. He went on to say, "I can therefore see no justification for the introduction of a system which has no element of permanency and every element of danger to life and property."[18]

∼

Throughout his lifetime, Tesla faced obstacles, had setbacks and was stung by disappointments; he also experienced "eureka" moments of joy from his discoveries, and jubilance from seeing his inventions "lift the burdens from the shoulders of mankind." It is likely that many of these event-driven fluctuations in mood were normal, as they are experienced in everyone. Thus, it can sometimes be difficult to distinguish from the record Tesla's reactions of disappointment from his major depressive episodes. Taking to bed for a prolonged period of time likely indicates a major depressive episode, and Tesla's own complaining in his autobiography and such complaints reported by others also likely indicate a major depressive episode (comments such as broken in spirit; poor health; exhausted; spent, having no more to give, and peculiar excessive sleeping spells). Indeed, these types of comments by patients or those who are around them are the very indicators of mood state used by modern psychiatrists when taking face-to-face patient histories. Such observations are thus also the tools used here to make an approximation of Tesla's mental status, as though he or those who knew him were providing this information directly to a psychiatrist.

Some of his likely major depressive episodes appeared to be spontaneous, but most appeared to have been triggered by "psychosocial stressors" that would impact anyone (such as death of a loved one, being cheated or betrayed, being turned down when seeking funding, or failure of any sort). However, in Tesla such stressors seem often to have caused him to fall beyond mere discouragement into much more disabling depressive episodes associated with poor physical health and absence of productivity such as inventing and filing patents. More complicated yet is when Tesla seems to have experienced mixed episodes of both mania and depression, such as the feeling of being worn out, psychologically spent, haggard, overworked while at the same time being manic, overactive, spouting new theories, designing and developing new inventions, and applying for patents.

In 1889, Tesla was in the midst of his most productive years and spent a good deal of time during these years between 1886 and 1901 mostly in a manic/hypomanic state when almost all of his inventions and patents were completed. During this period, he did have some bouts of mixed features of mania combined with lower levels of depression, and even a few periods of major depression as well. However, after 1901, and for the remaining forty some years of his life, there are no more convincing indications in the record of experiencing mania, but to some extent he lost his charisma and ability to raise funds. Tesla likely spent most of these years at the end of his life in a prolonged state of moderate bipolar depression. This course of bipolar disorder in Tesla is not unlike what happens to many patients with this condition: namely, more mania when young and then more depression when old.

∽

As a consequence, the fall of 1890 saw Edison's agitprop campaign against alternating current, which included an eighty-four-page denunciation of al-

New Creations   97

ternating current—"A Warning," bound in red—that attacked Westinghouse (and Tesla) by alleging, in excruciating detail, the manifold dangers of alternating current. It even included an appendix listing the names of individuals supposedly killed by it.[19] This manifesto also pleaded with all electricians "to unite in a war of extermination against cheapness in applied electricity, whenever they see it involves inefficiency and danger."[20] Westinghouse, not a man to be cowed nor trifled with, was nevertheless taken aback by Edison's response and his "method of attack which has been more unmanly, discreditable and untruthful than any competition which has ever come to my knowledge."[21] He fired back, producing his own brochure that went after Edison's very own safety record. It showed that while none of Westinghouse's central stations sustained any problems due to fires, "of the 125 central stations of the leading direct current company [Edison's] there are numerous cases of fire, in three of which cases the central station itself was entirely destroyed, with the most recent example being the Boston station, as well as the complete destruction of a large theater at Philadelphia."[22]

While Tesla was drawn to the mysteries of electricity as a young boy, most people were afraid of the unknown and Edison preyed upon their fear. So what was Edison's propaganda campaign? and how far would he go to maintain his monopoly on electric power? What tricks would he play on the public? What lies would he propagate in his relentless effort to destroy Nikola Tesla's life and preserve his? And was their no limit to what Edison would do to maintain his stranglehold on electric power?

To better understand how easily the general public—even scientists of the time were flummoxed by the mysteries of electricity—at the dawn of the twentieth century could be "swayed" into believing Edison's lies, we need only to step back a half century in time to gain the perspective needed.

Sir John F. W. Herschel, the nineteenth-century polymath of the Romantic era in Europe, accurately suggested that electricity and electromagnetism still possessed secrets yet to be discovered, and their investigation would undoubtedly become the prominent science of the new age.[23] He further confirmed his belief in their mysteries when he said: "To electricity the views of the physical enquirer now turn from almost every quarter, as to one of those universal powers which Nature seems to employ in her most important and secret operations. This wonderful agent, which we see in intense activity in lightning, and in a feebler and more diffused form traversing the upper regions of the atmosphere in the northern lights, is present, probably in immense abundance, in every form of matter which surrounds us, but becomes sensible only when disturbed by excitements of peculiar kinds. The most effectual of these is friction, which we have already observed to be a powerful source of heat. Everybody is familiar with the crackling sparks which fly from a cat's back when stroked."[24] Tesla, as we have seen, was first alerted to the mysterious wonders of electricity when he rubbed the back of his cat Mačak and saw bolts of "static electricity" jump off of the cat's back.[25]

∿

While in Europe, Tesla had continued to pay the rent on his laboratory so that his longtime assistant Anton (Anthony) Szigeti could continue to test devices of Tesla's own making.[26] At the same time, with funds from his first patent sales to Westinghouse in the bank, he instructed both Peck and Brown to arrange for a fresh, larger workspace. Upon his welcomed return, a new, second laboratory—his Arsenal—was soon setup within a five-story factory building at 175 Grand Street, the corner of Grand and Lafayette. It was here that he began serious work on high-frequency apparatuses, and wireless transmission, and he also had solid notions on the connection between electromagnetic radiation and light. In particular, he looked to the German Heinrich Hertz's findings for inspiration.[27]

In 1887, Hertz reported his important research detecting the electromagnetic waves that Scottish physicist James Clerk Maxwell had predicted in his theoretical work on electricity and magnetism.[28] It was the work of both Maxwell and Hertz that instigated a curiosity that would lead Tesla to experiment with various aspects of electromagnetism as well as development of his oscillating transformer.

Although Tesla always insisted on working essentially "solo" from conception to the end product, rarely if ever writing anything down—it had all been worked out in his mind beforehand—he now had the financial wherewithal to expand his supporting workforce. Eager to increase his creative output and to save time, he hired several specialists. They included a glassblower of German descent named David Hiergesell, as well as two mechanics, one named Charles Leonhardt and the other F. W. Clark, who had previous experience at the Brown & Sharpe Works—highly respected machine tool builders. Tesla rounded out his specialist workforce by employing Paul Noyes from the Gordon Press Works, who had helped him with the arc-lighting system in Rahway, New Jersey.[29]

But while Tesla felt renewed to be back in New York City after what he saw as an absolute waste of his intellectual powers in Pittsburgh, there was an undercurrent of mental unrest beneath the surface. It was difficult for him at times to face others, especially when his highly sensitive emotional state flared up. As an example that repeated itself throughout his long life . . . he once again took up residence at a hotel. Now a wealthy man, free from the economic bondage of an earlier time, he chose the Astor House, built by John Jacob Astor as New York City's first luxury hotel. It soon became the most celebrated hotel in America, as it played host to many of the world's glitterati, as well as robber barons and famous racounteurs. Tesla quickly became the center of attraction. He was the famous young scientist who was changing the world before their very eyes, and he would remain so for the entirety of his creative years. The War of the Currents only added to his growing celebrity, but were it his choice, he would have none of it.

That said, he still relied on his longtime confidante Szigeti, who Tesla would often say was "a man who had a considerable amount of ingenuity and intelligence. . . . He was not exactly a theoretical man, as myself, but he could understand every idea fully."[30] The two's friendship grew over the years, and Tesla called Szigeti "a very intimate friend of mine and I treated him as well as I possibly could."[31]

Tesla was never comfortable to be in the employ of another, despite being a well-paid technical consultant. Upon his return, he continued his work with the Tesla Electric Company, wherein during the spring months of 1890 he filed three subsequent patents on AC motors and assigned them to this company. However, it is interesting to note that these would be the last patents held by the company. All future patents would be held in Tesla's name only. A final period was placed upon the business relationship with his partners Peck and Brown when Peck fell ill and died that summer. For the next few years Tesla would occasionally query Brown for his business advice, although he did not have the business acumen that Peck had, which had helped in the early success of the AC motor.[32]

At this point the War of the Currents had become a series of high-stakes battles for electric power supremacy that played out between Edison, his minions, and the inferior Direct Current that he promoted versus Nikola Tesla, Westinghouse, and the superior alternating current that Tesla had invented—it would prove to be his force-multiplier in his war with Edison. By the late 1880s, the winner was still not clear. After all, Edison had considerable funding, including that of financier J. P. Morgan, and hence, he had the momentary advantage that he would press at every turn. That said, Tesla and Westinghouse certainly had the technical advantage with their superior electrical power system. But there was another challenge that entered the War of the Currents from the sidelines: Electrical power distribution was facing the force of natural gas companies for lighting and heating, because the money to be made was astronomical.[33] Add to that challenge another that appeared on the horizon when it was announced that English engineer Charles Parson was in the process of inventing a steam engine "that could produce electricity far more efficiently than traditional piston-fired engines."[34] To which Tesla and Westinghouse answered all comers with a new metering device that allowed the more efficient selling of alternating current to end users. The duo also added Tesla's alternating current transformer into the mix, which could step up the alternating current to high voltages, thus making possible the transmission of electrical power over very long distances with greater efficiency and reliability.[35]

Because of the rapid proliferation of electric power that was taking place across the United States and Europe during the late 1880s and 1890s, the immediate "need" for a well-functioning infrastructure was never more critical.[36] It was radio engineer Harald Friis of Bell Telephone Laboratories (Bell Labs) who, in the mid-twentieth century, spoke on the necessity for new inventions: They "always originated because of a definite need."[37] Enter Tesla's unique ability to invent on the fly, which had always served him well. From his inventions of the sound adjuster, voice amplifier (loudspeaker), and numerous other devices that would streamline voice transmission—none of which he even bothered to patent—while working at the Budapest Central Telegraph exchange prior to his move to the United States, he had demonstrated early on his extraordinary abilities time and again. And now he was doing the same again by fine-tuning his AC system for mass-production and electrical power distribution, which would be the foundation of the required infrastructure as AC electrification of the world became a reality.

By virtue of his genius and inexhaustible work ethic as well as his extraordinary ability to actually build what he invented, he had rendered himself indispensable as the world was becoming electrified.[38] But there was a cost, a mounting cost to his fragile psyche. Edison's war against his former employee was of great disappointment to Tesla, for he had always believed that people would generally do the right thing, as he had always done. His belief in "principle over profit" was embedded in his DNA, and any attempt to challenge his core belief was an intolerable affront to him. He had willingly sought to give the fruit of his genius for the betterment of mankind and Edison was doing everything he could to see to it that it would never happen.

Tesla valued implementation of ideas over theory; he also valued discovery and invention over commercialization. His adversary was ignorance, not Edison, even though Edison's adversary was Tesla. Tesla's goal was to conceptualize how energy and communication could be brought to civilization, not simply to make a profit off it. He was both flummoxed by those who did not share his values and condescending toward them.

~

It should now be self-evident that Tesla was fighting a very different war on many fronts and levels including his war against himself and how he would make his way in the world. He was still relatively unknown as a scientist outside the world he inhabited, as he set about uncovering truths regarding alternating current and how to make it work for humanity. He was not fighting Edison for fortune. He was not fighting Edison for fame. But he was baffled at the idea that the best science would not win, as his nemesis wantonly eschewed truth and merit, even to the point of demonstrating his inhumanity.

∼

Tesla would eventually triumph in the War of the Currents and defeat Edison. However, his ability to have his future ideas prevail would fade, especially when he lost his hypomanic infectious charm and the de facto salesmanship of his mania. Despite continuing to have endless ideas that would eventually be validated after his death, Tesla was unable to understand the importance of commercialization in bringing his ideas to fruition, the power of the profit incentive, and how "clever invents; genius steals."

The creative Tesla had the ideas, but many inventions that would transform society were stolen from him by entrepreneurs during his lifetime and after his death. Tesla likely was confused by this, especially when funding dried up for his later ideas, and he was likely disappointed by this way of the world. He never accepted the critical value of commercialization. However, this attitude did not seem to harm the implementation of his early ideas during the productive manic phase of his career when his charm did lead to sufficient financing to get some ideas implemented. But once Tesla lost his charm and his mania in later years, his condescending attitude to financing and commercialization led to many of his ideas not being realized, with many of his patents and even his life expiring before the ideas were commercialized. It also led to his own personal financial impoverishment during the latter years of his life when he was no longer manic, no longer productive in fundraising, patent filings, converting his ideas into practice, or even getting others to exploit his ideas. One can only wonder how much sooner the internet would have been developed, exploration of Mars would have unfolded, and his other transformational ideas would have been implemented if only he had understood the value of commercial exploitation of science, or indeed, if only he had remained hypomanic for many more years.

∼

As his Pittsburgh experience was fading in the distance, he was in a state of serious turmoil, trying to advance the frontier of knowledge, while his ignorant adversary was engaging him in a war to make (DC) direct current triumph, irrationally, over (AC) alternating current. It was simply about fame and fortune for Edison; it was not about what worked the best. To Tesla, it was akin to playing chess in the corner of a room where somebody keeps batting a tennis ball at you expecting you to play their game. But he was engaged in a very different war, and to him, it was no game!

~

The year 1888 was fast becoming a macabre period in the War of the Currents. Brown had been manufacturing electric chairs and selling them to prisons at the princely sum of $1,600. He also offered himself as a bona fide executioner. He, an electrical engineer in his own right, was troubled by the dozens of deaths, some friends, caused by AC over a few years. While understanding that DC also had its share of victims, he nonetheless was convinced that AC had to go.[39]

It was known that Brown had previous experiences with the electrocution of animals when he held a demonstration of such that summer in Professor Chandler's lecture room at Columbia College–School of Mines. One shocking example was reported by the *New York Times* on July 31, 1988.[40] Brown claimed before an unsuspecting audience of electricians that he had no financial ties to anyone or any company and that he simply believed direct current to be safe and alternating current deadly—he lied. He had exchanged numerous conspiratorial letters with the duplicitous Edison.[41] He then proceeded to jolt an innocent dog with 1,410 volts of direct current with no deadly results, then he gladly admitted that he had repeatedly killed numerous dogs with between 500 to 800 volts of alternating current. But the murderous demonstration was not over. He then presented a Newfoundland canine mix weighing about seventy-six pounds to the astonished audience. It was muzzled, shaking uncontrollably, and cruelly restrained inside a wire cage. He then shocked the horribly terrified dog with dosages of direct current ranging from 300 to 1,100 volts. The dog yelped, struggled mightily, breaking its muzzle, but it survived the brutal trial of torture and terror. At that point Brown was to have said, "He will be less trouble when we try the alternating current." Finally, in short order the dog succumbed to just 330 volts of alternating current. But before Brown could continue, an agent from the American Society for the Prevention of Cruelty to Animals stopped any further demonstrations.

Another report by a journalist who was present revealed that . . . "many of the spectators left the room unable to endure the revolting exhibition." Before things quieted in the death chamber, some fifty animals, including cats, calves, and a horse met their untimely deaths by shocks of alternating current.[42] The day ended with Brown insisting that alternating current should only be used at the "dog pound, slaughter house and the State Prison."[43] Brown proved to be the perfect proxy for Edison's dastardly deeds.

Seemingly, Edison's desperation to destroy Tesla had reached its most reprehensible and inexcusable level, but had it? After all, he had had his chance to join with Tesla, as the young inventor, while under his employ, was working with his alternating current and induction motor creations—something that Edison could never fully comprehend on a technical level. So he sought out and welcomed engineer Harold P. Brown (known as an activist against alternating current) and Samuel Insull into his lair of lies. Edison's commercialization of direct current for the transmission of electricity forced him to play it out by ending Tesla's future in every way possible.[44] It was then that the propaganda campaign took a very deadly turn. With assistance from Edison Electric—where Edison had employed teams of scientists to invent and reinvent the inventions of others—the two hirelings staged the systematic electrocutions of animals from dogs and elephants to horses and other helpless creatures at Edison's famous Menlo Park laboratory.[45] It was Edison at his very worst. But it still didn't stop there.

The dreadful demonstrations were shots across the bow to Tesla and to those who supported Tesla's alternating current, and the eventual result was legislation that forced the installation of alternating current underground after hot wires flowing with its current had killed several dozen people. Direct current also had claimed myriad victims, but it was the misuse of alternating current that allowed Brown to make his specious claims. Believing that alternating current was the enemy of the people, Edison planned yet another attack on Tesla that can only be described as both diabolical in its purpose and Medieval in its execution.

AUBURN, UPSTATE NEW YORK, 1890

The absolutely horrific day had finally arrived—August 5, 1890, 9:30 p.m. At this point Edison had reached down as low as any human could possibly go to achieve his goals: to protect his fame, to protect his fortune, and to destroy Nikola Tesla. One William Kemmler, warehoused in Auburn Prison, in upstate New York, was to be the unwitting victim of Edison's continued crusade against Tesla's alternating current. He was to die a horrendous, painfully gruesome death by means of the misuse of alternating current at the hands of an incompetent executioner whose charge it was to *kill* Mr. Kemmler, at the conniving behest of one Thomas Alva Edison.

The world was not prepared for what was about to happen, but Edison certainly was. He had even managed to lull the public into looking at the horrors of electrocution by alternating current as being "Westinghoused"—a maniacal euphemism if there ever was one.[46] Edison also released a promotional brochure to homeowners which said, "Don't let your house get Westinghoused."[47] To further make his point, Edison used his deep connections to state and federal politicians, as well as high-ranking employees in New York State when he recommended to the state's legislature that electrocution by alternating current offered a more "humane" method of capital punishment than the conventional method of hanging until dead—which had in recent years become a problem, what with too many botched hangings.[48] As such, the Medico-Legal Society of New York was established to study the topic of human execution by means of AC, and Brown signed-on as its chief spokesman.[49]

As these serial acts of demonstrating the presumed deadliness of AC above three-hundred volts by Brown continued unabated, Westinghouse was not without his voice. He stated that "a large number of persons can be produced who have received a one-thousand-volt shock from alternating currents without injury." He went on to add, "We have no hesitation in charging that the objects of these experiments [are] not in the interest of science or safety."[50] And just when one couldn't imagine anything getting more bizarre or desperate, the arrogant Brown publicly challenged Westinghouse to an "electrical duel," stating that the great industrial titan cared not for anything but only for money, while countless individuals died by "the death-dealing alternating current." He further challenged Westinghouse to defend his AC "and take through his body the alternating current while I [Brown] take through mine the continuous current [DC]."[51] Westinghouse's response . . . nothing.

The predaceous plot thickened when Brown, in May 1889, purchased three of Westinghouse's alternating current generators, never saying why. A year later it was learned that Auburn State, Sing Sing, and Clinton prisons would use such a device as the only means of executing those sentenced to death.[52] Edison, still claiming no association with Brown and his idea for execution, had as far back as the early 1880s thought about killing humans with electricity. In fact, in December of 1887, he wrote Dr. Alfred P. Southwick, a dentist from Buffalo, New York, who was a member of New York Governor David B. Hill's committee, stressing that he, Dr. Southwick, could expect his support in the use of "electricity," despite his personal dislike for capital punishment. Edison added with emphasis the following:

> The best appliance in this connection is, to my mind, the one which will perform its work in the shortest space of time, and inflict the least amount of suffering upon its victim. This, I believe, can be accomplished by the use

of electricity, and the most suitable apparatus for the purpose is that class of dynamoelectric machinery which employs intermittent currents. The most effective of these are known as "alternating machines," manufactured principally in this country by Geo. Westinghouse. . . . The passage of the current from these machines through the human body even by the slightest contacts, produces instantaneous death.[53]

In the end, it was Edison's support of electrocution as a certain and reliable means for the "extinguishing" of life that was the tipping point in the committee's deliberations, thus directing the governor to make his deadly decision: death by alternating current electrocution![54]

So the witching hour was fast approaching, as William Kemmler had exhausted all of his legal appeals, thus giving Edison the very moment he was waiting for when the deadly nature of alternating current would be demonstrated in dramatic fashion.[55] Kemmler was a thirty-year-old alcoholic and convict, who had bludgeoned his mistress (sometimes referred to as his common-law wife or the wife of another man) Matilda "Tillie" Ziegler with an axe some twenty-six times to the skull and other body parts until dead, who was now about to violently meet his maker.[56] And he was also about to become the first "test case" for exacting the death penalty by way of AC electrocution.[57] But it was far more complex than a simple test case; it was Edison once again wanting to preserve his fame and fortune, thus making the most of his war against Tesla, rather than wanting a civilized way to kill convicted felons sentenced to death. Many would argue there was nothing civilized about executing a human by any means, let alone electrocution. Needless to say, those in attendance that fateful day were horrified. The descriptions of what they saw stretched the imagination to its breaking point.

The death march was short, just half a hallway down from the condemned man's basement cell, as warden Charles Durston introduced his charge to the some two-dozen witnesses—including two reporters—that were present to chronicle in their own ways what was to take place. One attendee said of the condemned man that he was a vegetable-peddler turned murderer and that he was a "spruce looking, broad-shouldered little man."[58] As if that mattered. The *New York Times* described the act of first-degree murder this way: "William Kemmler was a vegetable peddler in the slums of Buffalo, New York. An alcoholic, on March 29, 1888, he was recovering from a drinking binge the night before when he became enraged with his girlfriend [elsewhere referred to as his common-law wife] Tillie Ziegler. He accused her of stealing from him and preparing to run away with a friend of his. When the argument reached a peak, Kemmler calmly went to the barn, grabbed a hatchet, and returned to the house. He struck Tillie repeatedly, killing her. He then went to a neighbor's house and confessed by saying, 'I killed her. I had to do it. I meant to. I killed her and I'll take the rope for it.'"[59]

As the warden read the death warrant to Kemmler, Kemmler appeared to be calm. He called out to those present and let them know that he was "going to a good place, and I am ready to go."[60] He even stated to a jailer earlier at breakfast: "Thy say I am afraid to die, but they will find that I ain't." Without delay, Warden Durston and his assistant secured Kemmler in the wooden chair, attached by eleven straps, as the convicted man supposedly said to the warden, "My God, Warden, can't you keep cool? Take your time."[61] He even asked the warden to tighten the straps so that everything would be right.[62] He then calmly tested each one until he was satisfied.[63]

As the final preparations continued, which included a metal cap with an electrode attached to his shaved head, those present must have found it startlingly difficult to hear such words being said. And when one considers that every individual in attendance, handpicked by the warden, was a witness to history—that is, the first legal electrocution of a person by the state—it must have been an indescribable and heart-rendering scene at times.[64]

All was set. The warden rapped twice on the wall of the control next to him, giving the predetermined signal to prison electrician E. F. Davis to "flip-the-switch." The deadly alternating current, said to be about one-thousand volts, suddenly poured out of the massive dynamo into William Kemmler's body, thus rendering him dead by electrocution![65] But was he!? Yes, his body had stiffened as his fingers clutched the chair, all the while fighting to escape the inevitable. Suddenly one index finger turned inward and punctured his palm, letting loose a stream of crimson blood. He said nothing. The attending physician Dr. E. C. Spitzka looked at his stopwatch, which indicated some ten seconds had passed. It was over. He said, "Stop!" By the time the power was cut another seven seconds had transpired. The doctor said, "He's dead." Kemmler's body now appeared to be in a state of total relaxation with only the eleven straps holding him somewhat upright in the "electric chair."[66]

It had been done and none was happier than Dr. Southwick, who had waited some ten years for this moment to prove his ignominious invention worthy. He had gleefully confirmed the death when one witness suddenly pointed toward Kemmler. Blood was pulsating out of the "dead" man's hand. The doctors in attendance looked incredulously at each other, then one yelled out, "Great God!" Another said, "He breathes!" Dr. Spitzka ordered the alternating current to be turned on again, shouting, "This man is alive!"[67] Because the dynamo had been shut down, it took a few minutes before it could produce the necessary voltage again. Once ready, the warden gave the same signal and the voltage coursed again through Kemmler's body, causing it to arch and contort in all directions, lurching and seemingly struggling to break free. Dr. Spitzka insisted that the current continue until, as all in the room saw, Kemmler's head began to smoke. The acrid scent of burning flesh permeated the death chamber, as electrodes set his shirt and the back of his vest afire. Foam erupted from his mouth. Dark, purplish spots were now visible on his skin as the command to "cut the current" rang out.[68]

No one knows for certain how long the current was on the second time, but the official report indicated that William Kemmler was electrocuted to death using two-thousand volts of alternating current for seventy seconds. And what remained of Kemmler was a "smoking lump of mutilated body."[69] It was most certainly a bungled execution, carried out inhumanely.

Although the barbarous spectacle turned public opinion against Brown and Edison for a period of time and newspapers across the land likened it to something only seen in "the darkest chambers of the Inquisition of the 16th Century," there was something else besides the eventual ushering in of executions by electrocution.[70] Something that spoke to Nikola Tesla in the deepest and most dark corners of his mind.

The execution itself was horrific. No one can say otherwise. And there would be more and more and more. Edison understood this. "Just when objectivity matters most, scientists—great scientists, perhaps, above all—are apt to draw on their deepest rhetorical and political resources to skew the course of inquiry to favor their own needs."[71] And Edison did just that. But not Tesla, for he could not understand why all scientists did not feel the same. After all, he had given the world "his" alternating current for the betterment of all, not for punishment and death. And now people looked at him and Westinghouse and alternating current and wondered. He had always believed what many in the scientific community believed then and now: the inviolate code of intellectual conduct against which scientists' reasoning is periodically compared—allowing science to "self-correct"—had been violated by Edison.[72] Edison had demonstrated in the cruelest way that nothing was beneath him in his quest to win! He was convinced that Tesla's alternating current and everything that flowed from Tesla's genius would undermine what he saw as the "scientific paradigm."[73] And after all, to Edison the scientific paradigm was sacrosanct, and no one, not even Nikola Tesla, was going to change it. But just as Charles Darwin's theory of evolution supplanted the theory of special creation (God having created every species), so too would Tesla's alternating current poly-phase system supplant Edison's direct current, all for the advancement of science and mankind.[74]

Was Tesla unhappy about the execution? Yes, he was. Was he uncertain about his future? Yes, he was. Would he cower in fear? Never.

II

THE WAR CONTINUES

· 7 ·

Searching for the New

Charisma, Showman, Mania Sells

NEW YORK CITY, USA, 1890

$\mathcal{I}$n the aftermath of the infamous execution by electrocution event per-formed with the imprimatur of the New York State legislature and its prison system, Edison's propaganda campaign against Tesla continued unabated at a steep emotional cost to the young inventor, for he had by now amassed several portmanteaus of emotional baggage. It was not enough that the very same leg-islature had legalized such a horrendous method of execution some two years earlier, but now it was deemed de rigueur in the very state Tesla called home.[1] After all, as a child, the killing of a mayfly was more than he could bear; to now have his discovery of alternating current used for killing a human being was far beyond the pale.

Just think of it. Three hours after William Kemmler took his last breath, the autopsy revealed in the most gruesome terms that he had been "roasted." His cooked corpse, sans several internal organs, was then secretly interred in darkness in the prison courtyard. For good measure, a large quantity of quick-lime was added in the grave to ensure that no trace of "the body" would ever be found.[2] Could there be anything worse to contemplate for Tesla? At this point in his life, no. But the overriding thought that would not leave him was why others (Edison et al.) could not seek what he was seeking: the right direc-tion for the science of electricity and ultimately for the betterment of human-ity. As Karl Popper, one of the twentieth century's preeminent philosophers of science, observed decades ago: Individual scientists . . . can be famously bull-headed, overly enamored of pet theories, dismissive of new science, and heedless of their [personal] fallibility.[3] Was Popper thinking about Edison?

~

Some years ago, upon the discovery of his whirling magnetic field, Tesla had proclaimed to his then doubting friend and confidante Anton Szigeti, "No more will men be slaves to hard tasks. My motor will set them free, it will do the work of the world."[4] This profound statement by Tesla is the quintessence of who he was, the man who gave it all to "lift the burdens from the shoulders of mankind" by inventing the modern world. Should this not be the goal of all science, Tesla questioned time and again?

Even George Westinghouse had difficulty processing the electrocution. He emphatically called it "a brutal affair. They could have done it better with an axe. My predictions have been verified. The public will lay the blame where it belongs and it will not be us. I regard the manner of the killing as a complete vindicator of all our claims."[5] And Tesla himself, still pained over the matter and the electric chair some forty years later, declared it "an apparatus monstrously unsuitable, for the poor wretches are not dispatched in a merciful manner but literally roasted alive. . . . An individual under such conditions, while wholly bereft of the consciousness of the lapse of time, retains a keen sense of pain, and a minute of agony is equivalent to that through all eternity."[6]

NEW YORK CITY, USA, 1891

By the end of 1890, the wealth of the United States of America had soared beyond anyone's imagination, and the commercial promise was being fulfilled in grand style. In the postbellum period, the national fortunes exceeded the total wealth of all the aristocrats and merchants of Great Britain, Germany, and Russia combined. By now Tesla himself was enjoying his hard-earned financial success, which translated into a Tesla who was the epitome of European sartorial elegance, from his bespoke suits, kid-leather gloves, and silk handkerchiefs (used only for one week each) to the finest shoes, spats, and topcoats. He even set up residence at the Astor House, the nation's most sophisticated hotel, and dined nightly in the grand dining room at Delmonico's, New York City's celebrated restaurant, amid imported dinnerware, the aromas of fine French cuisine, the bouquet of the premium burgundy wines, and its famous and infamous clientele.[7] But despite Tesla's new lifestyle, his constant passion remained: to realize his dreams of electrifying the world with his alternating current. His envied lifestyle did encounter a challenge for a brief period, however, when without warning, word sailed across the Atlantic Ocean from London that Baring Brothers & Company, the world's largest banking house, was on the verge of utter collapse.

The financial freefall caused a panic, and Westinghouse's creditors were forced to call in the many loans extended to him while his business was expanding at an exponential rate. As a financial restructuring of the West-inghouse Company was underway to bring it out of receivership, investors (moneymen) demanded that George Westinghouse terminate Tesla's lucrative contract, which weighed heavily on the balance sheet of the company. This contract (of which no hard copy exists) with Tesla amounted to the payment of $2.50 per horsepower generated by each installed generator—which were accruing to Tesla in the millions of dollars.[8] Moreover, Westinghouse's pro-pensity for funding young innovators and buying other businesses, spending staggering sums on litigation, and the ongoing feud with Edison proved to be a deadly financial cocktail.[9] Westinghouse was in a veritable "no win" situa-tion. What made matters worse was that he knew Tesla to be the greatest of inventors and a man of unquestionable character. Moreover, the Tesla patents were his own personal pot at the end of the rainbow. All these factors caused him to initially balk. But he knew the hard reality. He had to confront Tesla and lay out the details.

The meeting—which does not appear in Westinghouse's official biog-raphy—between the scientific giant and the manufacturing giant occurred in Tesla's fourth-floor laboratory on Fifth Avenue where Westinghouse had acquired the patents four years before.[10] The magician's chamber was mys-teriously lit by proto-fluorescent lamps of his own invention. They did not require filaments, just gases, and produced light by reacting in a highly charged electrical atmosphere that required no electrical wires.[11] One must admit, even a twenty-first-century denizen of a major American city would be overcome with awe entering the magician's chamber and seeing the laboratory lit by wireless lightbulbs not plugged into any external power.

By 1891, Tesla was somewhat aware of the financial hardship Westing-house was experiencing in his business, so he had some expectation that things would change to a degree; hence, the sight of the full-bodied George West-inghouse, attired in the dark formal wear of the day, let Tesla know something was amiss. Westinghouse, never a man who cared much for preambles, set to bluntly explaining to Tesla what he was facing.

Tesla described to his first biographer the decisive event as follows:

> "Your decision," said the Pittsburgh magnate, "determines the fate of the Westinghouse Company."
>
> "Suppose I should refuse to give up my contract; what would you do then?"
>
> "In that event you would have to deal with the bankers, for I would no longer have any power in the situation," Westinghouse replied.

"And if I give up the contract you will save your company and retain control so you can proceed with your plans to give my polyphase system to the world?"

"I believe your polyphase system is the greatest discovery in the field of electricity," Westinghouse explained. "It was my efforts to give it to the world that brought on the present difficulty, but I intend to continue, no matter what happens, to proceed with my original plans to put the country on an alternating current basis."

"Mr. Westinghouse," said Tesla, drawing himself up to his full height of six feet two inches and beaming down on the Pittsburgh magnate who was himself a big man, "you have been my friend, you believed in me when others had no faith; you were brave enough to go ahead and pay me . . . when others lacked courage; you supported me when even your own engineers lacked vision to see the big things ahead that you and I saw; you have stood by me as a friend. The benefits that will come to civilization from my polyphase system mean more to me than the money involved. Mr. Westinghouse, you will save your company so that you can develop my inventions. Here is your contract and here is my contract—I will tear both of them to pieces and you will no longer have any troubles from my royalties. Is that sufficient?"[12]

And so the story goes, for Tesla, in a grand flourish, did indeed tear up the contracts and with it went millions upon millions of dollars due him then and in the future.

The magnanimity of the moment cannot be overstated. Although Tesla had a loyalty to Westinghouse, he had a higher loyalty to himself and his ever-constant belief that he would always put "principle over profit." As he had stated myriad times throughout his long life, it was not about money; it was never about money; it would never be about money. This was his modus operandi all his living days, and it is what separated Tesla from Edison, and all others, including Westinghouse, for they were all driven by profit and only profit!

∼

Trying desperately to put the negativity behind him, Tesla was eager to expand his horizons, and he began numerous experiments with a concerted effort. He sought to increase the frequencies from his oscillating transformer and other phenomena. He also developed a spherical carbon button lamp that illuminated when positioned at one end of a wire attached to one terminal of his oscillating transformer (later called the Tesla coil) the high-frequency, high-voltage current illuminated his carbon button lamp to an incandescent brilliance. When compared to Edison's incandescent light bulb, Tesla's carbon button lamp amplified light twenty times as much for the same amount of current consumed. It was a high-frequency carbon button lamp, a "directed ray" if you will, that Tesla described as a pencil-thin line of light—the precursor to the laser beam, able to vaporize substances from zirconia to diamonds and at various distances. Even hard metals were not beyond the reach of his directed ray, noting that it could heat and mold such substances and thus revolutionizing the field of metallurgy.[13] Tesla recognized his carbon button lamp's commercial value inasmuch as his lamp required only one connection to the power source, thus reducing the wiring needed for electric life by half.[14] Following the same business strategy taught to him by his former lawyer Peck—patent, promote, sell—he immediately sought the necessary patents to protect his invention.[15]

But as Tesla's work was progressing, he was blindsided, as if body-punched by a prizefighter, when Szigeti, his closest friend of nearly ten years, decided to leave him and develop inventions of his own, one of which was an advanced compass for steering ships. After some five to six months away, Szigeti returned, only to hear from Tesla, who uncharacteristically lashed out and told him that his invention had already been invented by Sir William Thomson (later Lord Kelvin). This news delivered by an angered Tesla caused Szigeti to leave a second time in 1891, presumably to South America, never to return.[16] But then, before Tesla could attempt to salvage the friendship, he learned Szigeti had suddenly passed away at such a young age. Feeling abandoned, Tesla immediately wrote his family—specifically Uncle Pajo—which was something he rarely did, and told them, "I feel alienated and it is difficult [to live the American way of life]."[17] The two men shared many great times together as friends as well as employer and employee, both in Budapest and during his early years in New York City. However, Tesla never viewed their relationship as simply employer to employee, but much more. He saw Szigeti as "the only one who had supported me during my first attempts and whom I loved dearly because of his virtues and respect." And some two decades later he expressed what he felt was a great loss when he said, "I would have much desired to see him, because I would have wanted him."[18]

It is at this point in Tesla's tremendous and troubled life that some biographers had seized upon the previous sentence as if it were a testament to his (presumed) homosexuality. There is no denying his closeness to Anton Szigeti; after all, as we know, he was the only person present at the "creation."[19] Moreover, it is the custom of many Europeans, both male and female, particularly those of an Eastern European persuasion, to show a companionable closeness for each other that is rarely seen in the United States. Add to that, there is no bona fide evidence to prove or disprove the claim of these biographers other than the fact that he was a germophobe and did not like to make physical contact with any person of either sex.[20] So, to suggest that Tesla was homosexual is a reach at best. There were other times in his life where this issue arose with no solid evidence, just what others surmised. His relationships with some men can be understood as deep and platonic, because he was engaged in scientific inquiry and experimentation that exceeded what most could comprehend.

∼

With the War of the Currents taking its toll and the sudden death of Szigeti, as well as the death of one of his greatest supporters and promoters, the lawyer Charles Peck in the summer of 1890, Tesla felt challenged.[21] This confluence of personal events, coupled with financial losses due to the Westinghouse Electrical Manufacturing Company restructuring, proved to be an inflection point for the beleaguered inventor. He needed to salve his emotional wounds, so he did the only thing he knew how to do better than anyone else—invent. At the same time, he began sending home sizeable amounts of money to his mother, sisters, and extended family members. This philanthropy continued throughout the 1890s and beyond.[22] It was not common for Tesla to communicate with his family back home, but he knew that helping them financially, now that he was a man-of-significant-means, at least showed some level of concern.

∼

Tesla continued his search for new horizons in the science of electricity, venturing deeper into the mysteries of high-frequency energy. Still in its nascent stage, he knew electricity was the future, that his alternating current would light the way, and that he would be among its greatest explorers.[23] It was at this point that T. C. Martin and Professor William Anthony, two of Tesla's most ardent devotees, knew that Tesla was indeed the man to bet on. T. C. Martin urged him to make public and publish his research and then present his results in another celebrated lecture.[24]

While furiously inventing and then patenting, promoting, and selling, Tesla simultaneously wrote articles for several magazines and decided, somewhat reluctantly, to give another lecture. After all, it was his first AIEE lecture that had catapulted him to international fame and made all comers, including Edison, sit up and take serious notice. He was still in one pitched battle after another with Edison regarding what alternating current could do—that is, send power over great distances in a safe manner, which direct current could not do. In fact, Tesla took to writing about his progress (intellectual thrusts) in the February 1891 issue of *Electrical Engineer*. Edison parried by writing several acerbic verbal attacks in numerous trade journals.[25] At the same time, yet another Tesla antagonist, one Elihu Thomson, whom Tesla had easily humiliated during his first AIEE lecture in 1888, was nipping at his heels. Thomson had complained that he did not always observe the same effects with high-frequency currents as Tesla had. Reason: His experiments were under 10,000 cycles. In turn, the two fought it out, both throwing verbal punches at each other in electrical journals—most notably the *Electrical Engineer*—throughout March and April 1891.[26] Additionally, Thomson admits in one of his articles directed at Tesla that part of his dislike for the young superstar inventor was "elicited . . . by Mr. Tesla having on a former occasion misunderstood my motives."[27]

Seemingly, the stage had been set, or rather the ring for the main bout. What was about to happen that night would become historic before the sun rose the next morning. One could not measure just how historic, how impactful Tesla's lecture would become to the world of science and to him personally. But once again, at what cost to Tesla's delicate psyche?

In the spring of 1890, Thomson had given a keynote address at the AIEE wherein he discussed his work with AC phenomena. What better place for a rematch of sorts with Thomson and, at the same time, indirectly with Edison than Columbia College, the very place where he had introduced to the world his rotating-magnetic field motor in May 1888. The two seconds in Tesla's corner were T. C. Martin and William Anthony (president); both men were in positions of importance in the American Institute of Electrical Engineers (AIEE) that year, so they set a date for the lecture on "high-frequency phenomena." Tesla's lecture would once again set the scientific world afire, and it was to be delivered during the AIEE's three-day symposium on May 20, 1891, in the lecture hall of Theodore W. Dwight at the college, located between Park and Madison Avenues; the public was invited.[28] In addition, Professors Francis B. Crocker and Michael Pupin served as hosts of the newly formed Columbia School of mines, and they were pleased that Tesla could bring such great attention to them and their school.[29] But before he committed to the second Columbia college lecture, Tesla did what he did the last time he spoke before the august body of electrical engineers. He knew he could not be too protective of his genius and his work, so he set about applying for several patents for high-frequency incandescent lighting, and he filed additional applications for legal protection in Britain, France, Germany, and Italy.[30]

∼

Tesla's wizardry for the night's lecture and demonstration required copious amounts of electrical power, so he simply installed his high-frequency alternator in the college's workshop area and supplied it with power by an electric motor of his own design. When necessary, he could easily adjust the power output by adjusting a switch he had attached to his demonstration table on-stage. This allowed him to adjust the speed of the motor and thus have control over the electrical frequency produced by his alternator.[31]

Journalists from numerous publications flooded the lecture hall, as did a myriad of the foremost electrical scientists of the day—many of whom had seen Tesla's first groundbreaking Columbia College lecture in 1888—and curious members of the public, comprised of investors, aspiring inventors, and the like. One journalist in particular, Joseph Wetzler, M.E., reported on the event for *Electrical World*. Interestingly, he had sensed that the miraculous was about to take place and, believing that his publication was too narrow in its scope in both interest and circulation, he shopped the story around until the highly respected *Harper's Weekly* decided to give it the maximum treatment— believing it deserved an exciting, full-page account.[32]

It would be nearly a decade before Tesla would begin to commercialize his "World-System" of wireless transmission that comprised the Tesla Transformer (Tesla coil), the Magnifying Transmitter, the Tesla Wireless System, the Art of Individualization, and the terrestrial Stationary Waves.[33] Although he had been visualizing and creating these miracles of electricity for years, he knew everything had its time. And it was time to make his next move in the field of electrical science that he now dominated.

With the Serb inventor's physical bearing and mien drawing everyone's eyes to him, at center stage, Tesla began. It is worth remembering Joseph Wetzler's firsthand account of just what Tesla accomplished with his second Columbia College lecture:

> At one bound he placed himself abreast of such men as Edison, Brush, Elihu Thomson, and Alexander Graham Bell. Yet only four or five years ago, after a period of struggle in France, this stripling from the dim mountain border-land of Austro-Hungary landed on our shores, entirely unknown, and poor in everything save genius and training, and courage inherited from many a [Serb] chieftain who shed his blood in the ceaseless warfare with the unspeakable Turk. [His earliest inventions], dealing as they do with the difficult problems of the utilization of alternating currents for motive power, are in themselves fundamental and far-reaching; but Mr. Tesla has now utterly eclipsed them by his experiments and methods for obtaining the electric light electrostatically. In a word, his lecture at Columbia showed two things very clearly. It showed not only that he had gone far beyond the two distinguished European Scientists Dr. Lodge and Professor Hertz in the grasp of the electro-magnetic theory of light, but that he had actually made apparatuses by which electrostatic waves or "thrusts" would give light for ordinary every-day uses.[34]

Simply put, Tesla not only described but demonstrated in the most dramatic fashion the many manifestations of his genius at work. The lecture hall was lit up again by Tesla, by gas-filled tube lights, some phosphorescent and still others made with uranium glass formed into the names of his favorite scientists and poets.[35] No one could deny the fact that Tesla, unlike Edison or any of his other contemporaries, was a true man of letters in every way as he paid homage to poets. He now appeared, center stage on a riser, showing "the exhausted tubes, which looked like a luminous sword held by an archangel representing justice."[36] Those were the excited words of one the many reporters in attendance. Even the famous who attended the lecture, which included Professor Anthony, Elmer Sperry, Alfred S. Brown, William Stanley, Elihu Thomson (another Tesla nemesis), Francis Upton, and others said they would never forget what they experienced that May night in a Columbia College lecture hall. And Robert Millikan, a graduate student at the college at the time and a future Nobel Prize winner for his work in cosmic rays expressed what he experienced at the Tesla lecture when he said, "I have done no small fraction of my research work with the aid of the principles I learned that night."[37]

Now, standing with his tall body draped in a bespoke swallowtail coat, his long arms began to wave back and forth, as if he were summoning the scientific spirits, and he proceeded to yet again hold up his carbon button lamps that were brilliantly lit . . . but had no physical connection to power! There were no wires, no heat whenever he passed the lamps and gas-filled tubes between the electrostatic field generated by charged zinc plates on either side of the stage—thus putting the lie to Edison's unfounded charge that AC was deadly.[38] His high-frequency coil stood sentinel nearby, giving no indication to the audience that it was the secret source of the power, oftentimes producing 250,000 volts, and Tesla said he had gone as high as 100,000,000 volts to date.

Tesla understood that there was an organized opposition to what he was doing, and it was led by Edison and carried out by his hirelings, other inventors, investors, and anyone else who sought profit from his work. As a true visionary, he was a direct, existential threat to the status quo represented by Edison and many of the scientists in attendance that night. In one fell swoop, his lecture rendered Edison's proudest achievement (the incandescent lightbulb) simply technologically passé and it needed to be replaced without delay.[39]

As he continued to work his magic before the adoring crowd of onlookers at Columbia College, one can only imagine what he was thinking. But happily, we actually do know what he was usually thinking whenever he gave such spectacular lectures and demonstrations because he wrote about it in his autobiography. As an example, here are his thoughts in his own words:

> In investigating the behaviour of high frequency currents I had satisfied myself that an electric field of sufficient intensity could be produced in a room to light up electrodeless vacuum tubes. Accordingly, a transformer was built to test the theory and the first trial proved a marvelous success.

It is difficult to appreciate what those strange phenomena meant at that time. We crave for new sensations but soon become indifferent to them. The wonders of yesterday are today common occurrences. When my tubes were first publicly exhibited they were viewed with amazement impossible to describe. From all parts of the world I received urgent invitations and numerous honors and other flattering inducements were offered to me, which I declined.[40]

Suddenly, much to the amazement of the audience, many of them well educated, hundreds of thousands of volts of high-frequency currents produced by one of Tesla's oscillating units were coursing through his upright body. With the lecture hall seemingly silent, save for the hum and crackling sounds of electricity, the now surreal figure of Tesla was seen holding up in one hand one of his "carbon-button" lamps. Energy now contained in his body caused gas molecules in the glass lamp to bombard a small button (amount) of carborundum until it glowed to incandescence, thus producing a light some twenty times more luminous than any other lamp then in use, including the lamp produced by Thomas Alva Edison.[41]

Tesla, comfortable in the environment he now controlled, completed his nearly three-hour tour de force by giving voice to his thoughts about the future of electricity. He said,

> We are whirling through endless space with inconceivable speed. All around us everything is spinning, everything is moving, everywhere is energy. There must be some way of availing ourselves of this energy directly. Then, with the light obtained through the medium, with the power derived from it, with every form of energy obtained without effort, from stores forever inexhaustible, humanity will advance with great strides. The mere contemplation of these magnificent possibilities expands our minds, strengthens our hopes and fills our hearts with supreme delight.[42]

Tesla then stood and courteously received the massive show of approval as hundreds stood applauding and shouting bravo in unison. And still others remained seated, silent, and entranced by what they had just seen and experienced.

Tesla's high-frequency device that he simply referred to as his oscillating transformer soon would achieve its own world fame, and today we call it the "Tesla coil." Tesla knew it was another quantum leap forward in the science of electricity. "This apparatus," he said, "is in the production of electrical vibrations as revolutionary as gunpowder was in warfare."[43]

Tesla closed out his very productive year of 1891 by fielding endless requests for money or investments from all manner of relatives, some of whom he had never heard of. Still others asked for jobs in the laboratory of the great inventor. But Tesla had one issue that was front and center in his mind: obtaining U.S. citizenship! He was granted what he considered the highest of honors in late July 1891. His happiness and depth of gratitude in becoming a citizen of the United States of America was more precious to him—and he said so throughout the remainder of his life—than "my orders, diplomas, degrees, gold medals, and other distinctions . . . packed away in old trunks."[44] Some years later he was asked if he considered himself a good American. He looked with disbelief at the man and answered: "I a good American?" he said. "I was a good American before I ever saw this country. I had studied its government; I had met some of its people. I admired America. I was at heart an American before I thought of coming here to live. What opportunities this country offers a man. Its people are a thousand years, yes, a thousand years, ahead of the people of any other nation of the world."[45] He always kept his citizenship papers secreted away in a safe for all his days.

That said, let it be known that Nikola Tesla never deserted his pride and delight in being born and raised a Serb. He considered his most important title to be the "Grand Officer of the Order of St. Sava," which would be conferred upon him by King Alexander I Obrenoviç in Belgrade (present-day Serbia) only months later during his successful progress across the European continent.[46] Moreover, the celebrated Serbian poet Jovan Jovanović Zmaj composed a poem in Tesla's honor entitled "Pozdrav Nikoli Tesli" ("Greetings Nikola Tesla") that was read aloud to the throngs of well-wishers in attendance at the heroic son's return home.

Now it seemed to Tesla that after his second triumphant AIEE lecture—surely a command performance—the world lay before him and his grand visions. A fawning Europe of 1892 eagerly awaited his arrival.

· 8 ·

Tour de Force

Triumph, Tragedy

LONDON, ENGLAND, 1892

$\mathcal{I}$n the months immediately after Tesla's Columbia lecture, he was deluged with requests ranging from demands on his time for discussions with other scientists and all manner of opportunists to everything from dinners, honorifical presentations and awards, and other engagements with investors and celebrities who simply wanted to be in his presence. Even though Tesla did his best to keep focused on his work with high-frequency experiments, events in Europe caused him to ready himself for a trip back to the Continent, both to deliver lectures on his inventions and discoveries and to defend them before the many pretenders to the throne.[1] The enthusiasm for Nikola Tesla's coming to Europe and speaking before scientists, journalists, sycophants, and courtiers was palpable. The *Electrical World*, one of several celebrated technical publications of the time, published in December of 1891 an article that read:

> During the past six months Mr. Tesla has been steadily at work developing the beautiful principles that he enunciated in his striking lecture before the American Institute of Electrical Engineers. In his skillful hands the experiments have extended far beyond their merely theoretical importance in the direction of important practical applications. Part of the fruits of his industry has already appeared in his method patents on incandescent lighting, another patent on condensers and divers[e] applications are now in the Patent Office. Meanwhile, the transmission of power has not been forgotten, and some interesting improvements in that line may soon be expected, especially as the use of the methods embodied in Mr. Tesla's early patents has recently made such a sensation abroad. About the middle of January Mr. Tesla expects to sail for Europe, to deliver, at the urgent request of his English friends, a lecture on his high frequency researches and to look

after his foreign interests generally. The trip will be a flying one of only a couple of months, as the pressure of work at home forbids a long absence. On his return some of the latest commercial developments of his work may be expected to appear as promptly as circumstances will permit. The revolutionary character of his methods will make any extensive application of them of unique interest.[2]

It is noteworthy that in February of 1891, several months before his trip to Europe, Tesla applied for the first of a troika of critical patents establishing his primacy regarding the conversion and distribution of energy.

These constant requests caused him great angst, but after having declined invitations as best he could for as long as he could, he knew it was an exercise in futility. He nonetheless tried to put himself in the correct frame of mind because the inevitable was before him. He knew that London had long been a center of science as much as it was of business, and he was going there to do both. After all, the famed Royal Society (of London for Improving Natural Knowledge, est. 1660) called London home, and its membership included many of science's greatest practitioners, as did the Royal Institution, before whom he was also about to lecture.[3]

Regarding the onslaught of requests for his valuable time, he later wrote:

If my memory serves me right, it was November of 1890, that I performed a laboratory experiment [see chapter 7] which was one of the most extraordinary and spectacular ever recorded in the annals of science. But in 1892 the demands became so irresistible and I went to London where I delivered a lecture before the Institution of Electrical Engineers. It had been my intention to leave immediately for Paris in compliance with a similar obligation, but Sir James Dewar [1842–1923, inventor of the Dewar flask or thermos] insisted on my appearing before the Royal Institution. I was a man of firm resolve but succumbed easily to the forceful arguments of the great Scotchman. He pushed me into a chair and poured out a half glass of a wonderful brown liquid which sparkled in all sorts of iridescent colors and tasted like nectar. "Now," said he, "you are sitting in Faraday's chair and you are enjoying whiskey he used to drink." In both aspects it was an enviable experience. The next evening I gave a demonstration before that Institution, at the termination of which Lord Rayleigh [Nobel Laureate in physics] addressed the audience and his generous words gave me the first start in these endeavors. I fled from London and later from Paris to escape favors showered upon me, and journeyed to my home where I passed through a most painful ordeal and illness. Upon regaining my health I began to formulate plans for the resumption of work in America. Up to that time I never realized that I possessed any particular gift of discovery but Lord Rayleigh, whom I always considered as an ideal man of science, he said so and if that was the case I felt that I should concentrate on some big idea.[4]

Was Tesla's reference to returning home that of the home of his birth—then in the Austro-Hungarian Empire (formerly the Austrian Empire)? Also, was his reference to a "most painful ordeal and illness" a euphemism for suffering another case of bipolar disorder in the form of a great depression caused by the pressures placed upon him because of his fame and fortune and success as a scientist? Perhaps he had a premonition that his mother's death was near?

It is worth noting that his hard work and fame and pressures were often blamed for his depression, but hard work and pressures also accompanied his mania and he thrived on it. So, maybe Tesla and others are trying to find reasons for depression, rather than just accepting that often depression has a mind of its own when it comes and when it goes. It is entirely possible that he did not have a major depressive episode until his mother died and that triggered not only grief, but a major depressive episode that was more severe than expected grief reaction.

~

As the challenges in Europe to his considerable accomplishments began to gather steam—as all too often happens to many of those who dare greatly—even electrical journals in Great Britain had on occasion asked questions as to whether Ferraris had developed an electric motor utilizing Tesla's rotating magnetic field or not—Tesla's patents show him to be the first. Then there was the German F. A. Haselwander, who made a specious claim of developing a three-phase induction motor that generated ten horsepower in the summer of 1887; however, he did not produce a functioning motor until after Tesla had demonstrated his rotating magnetic field to great success and filed the appropriate patents before any action by Haselwander.[5]

The false claims multiplied at the Electrotechnical Exhibition in Frankfurt, Germany, when during the rather balmy summer of 1891, he was confronted with a situation that he found most troubling. It was there that other pretenders to the throne sought to assert primacy as the inventor of three-phase AC. Once it had been decided that the electrical-power motors would use AC instead of DC technology, Michael von Dolivo-Dobrowolsky, a Russian by birth, of the German firm Allegemeine Elektrizitäts-Gesellschaft was commissioned to build AC motors, along with Englishman Charles E. L. Brown of the Swiss firm of Maschinenfabrik Oerlikon for the Imperial German Postal System to transmit power 110 miles from Lauffen to Frankfurt.[6] The argument can be made that Dolivo-Dobrowolsky conveniently sidestepped the issue of the Tesla patents by calling his system *drehstrom*, which in German means "three-phase rotary current."[7] Needless to say, Tesla was perturbed and disturbed when he learned that Dolivo-Dobrowolsky and Brown were given credit by journalists for the concept of a three-phase AC electrical current. Oddly though, Brown found it within himself, in a moment of honesty, to do the right thing and admit that Tesla's work made his and Dolivo-Dobrowolsky's work possible. But as is most often the case, infringers were all around, despite Tesla's patent protections in numerous countries throughout Europe, including England and Germany.[8]

With the great success of the Dolivo-Dobrowolsky and Brown configuration for the power plant, the perceived War of the Currents with regard to Europe was seemingly put to rest, with the engineers employing Tesla's AC system fundamentals, even following his suggestion that "oil" should be used as an insulator. More specifically, the power produced at the Frankfurt site was so great that a thousand incandescent lamps lit a large advertising sign.[9]

The stage was set. Tesla was to deliver a much-anticipated lecture before the Institution of Electrical Engineers on February 3, 1892, in London. The IEE's normal lecture hall at the Institution of Civil Engineers held only about four hundred people. Expecting a much larger crowd, it was prudently decided to move the event to the Royal Institution, which could accommodate some eight hundred people. However, the Royal Institution, which gladly offered its facility for the lecture, did ask a favor in return. It wanted its own extremely erudite members to benefit from Tesla's revolutionary genius, so a second lecture was scheduled. When Tesla heard of the request, he balked, as we read in his description above, but James Dewar, the Royal Institution's Fullerian Professor of Chemistry, managed to convince him that it would benefit him greatly to give an additional lecture. He would be lecturing on the very same stage where English scientist Michael Faraday (1791–1867) had set forth the fundamentals of electromagnetic induction in the 1830s.[10]

So now it was showtime. The electrical wizard appeared onstage and scanned the onlookers, who were assembled in an amphitheater-style setting. The scientific greats in attendance included Joseph Swan, J. J. Thompson, Sir James Dewar, Sir William Preece, Sir Oliver Lodge, and Lord Kelvin, but one particular gentleman caught Tesla's discerning eye. It was Sir William Crookes (1832–1919), the celebrated chemist who was known for conducting some of the first experiments with radiation.[11] Tesla admitted that while in college he had read a small book (paper) entitled "On Radiant Matter" penned by Crookes that had instigated his interest in electricity. Moreover, he added that he then began to enter the "same vague world that Professor Crookes so aptly explored," before him.[12]

It can easily be surmised that Tesla must have been at the very least momentarily intimidated by the gathering of such brilliant scientists who had convened in standing-room-only conditions to hear him speak and demonstrate his inventions. But he did not disappoint. Tesla's lecture was aptly named "Experiments with Alternate Currents of High Potential and High Frequency." He was to demonstrate lights lit wirelessly; tubes so sensitive that they could pick up an electronic impulse across the Atlantic Ocean; and even a motor that was able to be powered by a connection to the ground and a basic antenna. Tesla, his emotions firing white-hot, was in his domain. The audience, now held spellbound by his mercurial genius—many thinking him a Prometheus personified to be sure—followed his every word:

> It is quite possible that such "no-wire" motors, as they might be called, could be operated by conduction through the rarefied air at considerable distances. Alternate currents, especially of high frequencies, pass with astonishing freedom through even slightly rarefied gases. The upper strata of the air are rarefied. To reach a number of miles out into space requires the overcoming of difficulties of merely mechanical nature. There is no doubt that with the enormous potentials obtainable by the use of high frequencies and oil insulation luminous discharges might pass through many miles of rarefied air, and that, by thus directing energy of many hundreds or thousands of horse-power, motors or lamps might be operated at considerable distances from stationary sources.[13]

It was closing in on an honest two hours of lecturing before Tesla seemingly took a breath. He had stated to the audience that nearly all his experiments were new and not a mere easy repetition of his successful lectures in America.[14] During his tour de force, he asked the rhetorical question: "Is there, I ask, can there be, a more interesting study than that of alternating current?"[15] The question was asked after a portion of his demonstration was completed. So, in order for the reader to feel a fuller contextual experience, let us eavesdrop on Tesla as he is delivering a part of his groundbreaking lecture:

> We may take at random, if you choose any of the many experiments which may be performed with alternating currents; a few of which only, and by no means the most striking, form the subject of this evening's demonstration; they are all equally interesting, equally inciting to thought.
> Here is a simple glass tube from which the air has been partially exhausted. I take hold of it; I bring my body in contact with a wire conveying alternating currents of high potential, and the tube in my hand is brilliantly lighted. In whatever position I may put it, wherever I move it in space, as far as I can reach, its soft, pleasing light persists with undiminished brightness.

Here is an exhausted bulb suspended from a single wire. Standing on an insulated support, I grasp it, and a platinum button mounted in it is brought to vivid incandescence.

Here, attached to a leading wire, is another bulb, which, as I touch its metallic socket, is filled with magnificent colors of phosphorescent light.

Here still another, which by my fingers' touch casts a shadow—the Crookes shadow—of the stem inside of it.

Here, again, insulated as I stand on this platform, I bring my body in contact with one of the terminals of the secondary of this induction coil with the end of a wire many miles long and you see streams of light break forth from its distant end, which is set in violent vibration.

Here, once more, I attach these two plates of wire gauze to the terminals of the coil; I set them a distance apart, and I set the coil to work. You may see a small spark pass between the plates. I insert a thick plate of one of the best dielectrics between them, and instead of rendering altogether impossible, as we are used to expect, I aid the passage of the discharge, which, as I insert the plate, merely changes in appearance and assumes the form of luminous streams.[16]

A cablegram sent by an observer on February 3, 1892, to one of the most respected technical journals of the day stated, in part, the following:

After a short introductory as to the theory which led him up to his new line of work, involving the application of enormously high potentials and frequencies far exceeding those heretofore employed, Mr. Tesla began his experimental demonstrations with the exhibition of various forms of discharges. He showed first an induction coil operated by condenser discharge. With this coil he produced various remarkable discharge phenomena. Among others he imitated the spark of a Holtz machine and showed a brush discharge covering an area of something like 4 or 5 square feet, and also a luminous wire outlining the name of William Thomson. Mr. Tesla then took up some very novel and exceedingly interesting phenomena discovered by him some time ago consisting of a brush rotating in an exhausted globe. He showed that the brush was affected electrostatically, magnetically (even by the earth's magnetism), and that the study of these phenomena might lead to important discoveries as to the nature of electrostatic or electromagnetic fields.[17]

One cannot underestimate the power of Tesla's delivery and the importance of his work. It was charisma on command. His "magnetic" persona caused some to say that he spoke as if he were a sorcerer, an alchemist, a conjurer, but it was all real. Even the great Sir Isaac Newton was at times said to have exhibited such characteristics. He knew how to draw out knowledge from all possibilities, making his method at times seem extremely diverse.[18] It was even suggested that Tesla had the knowledge and capability to animate that which was inanimate.

> One reason, perhaps, why this branch of science is being so rapidly developed is to be found in the interest which is attached to its experimental study. We wind a simple ring of iron with coils; we establish the connections to the generator, and with wonder and delight we note the effects of strange forces which we bring into play, which allow us to transform, to transmit and direct energy at will. We arrange the circuits properly, and we see the mass of iron and wires behave as though it were endowed with life, spinning a heavy armature, through invisible connections, with great speed and power with the energy possibly conveyed from a great distance.[19]

Tesla went on to add,

> We observe how the energy of an alternating current traversing the wire manifests itself not so much in the wire as in the surrounding space in the most surprising manner, taking the forms of heat, light, mechanical energy, and, most surprising of all, even chemical affinity. All these observations fascinate us, and fill us with an intense desire to know more about the nature of these phenomena. Each day we go to our work in the hope of discovering, in the hope that some one, no matter who, may find a solution of one of the pending great problems, and each succeeding day we return to our task with renewed ardor; and even if we are unsuccessful, our work has not been in vain, for in these strivings, in these efforts, we have found hours of untold pleasure, and we have directed our energies to the benefit of mankind.[20]

Once again, we are reminded that the central theme of Tesla's life is best demonstrated by the words to "lift the burdens from the shoulders of mankind." In them he found his purpose—he was most altruistic when in a manic state and feeling grandiose and capable of all things. Hence, it is clear that he was manic during his lecture.

Tesla's words and demonstrations continued at a torrid pace as hour three approached. It was one spellbinding demonstration after another. As his lecture journeyed deeper into his work and deeper into the cold London night, even the most sophisticated in the audience were emotionally fired and warmed by his original genius. A report in a highly respected New York electrical journal on March 19, 1892, led with the headline: "Mr. Tesla before the Royal Institution, London." The writer described the event thus:

> Mr. Tesla is the lion in electrical circles without doubt and deserves every honor given him. The lecture given by Mr. Tesla on Wednesday last week

before the Institution of Electrical Engineers, and repeated on Thursday before the Royal Institution, will live long in the imagination of every person in the brilliant scientific audiences that heard him, opening, as he did, to many of them, for the first time, apparently limitless possibilities in the applications and control of electricity. Seldom has there been such a gathering of all the foremost electrical authorities of the day, on the tiptoe of expectation to witness the experiments, details of some of which had already been given to us from the other side of the Atlantic, but of which no written account could convey the true significance and beauty.[21]

By now Tesla had learned that in some situations the "lecture" was the lingua franca of science. This was one of those situations. After all, there was no internet to spread the word; there were no banner ads to advertise one's inventions, discoveries, and products to potential investors and buyers. There were few ways to get the word out and tell the world what he was doing and how it could benefit them (AC over DC) other than newspapers and personal lectures and demonstrations. As a consequence, he had to become a brilliant showman. He had to make drama part of his delivery, and he had to engage the audiences' imaginations as well as their minds and emotions.[22]

There was no doubt that "spectacle" had to be the set-piece of such a performance. He became a demiurgic force once he stepped onstage and presented himself to the audience with his grand physical movements in his trademark swallowtail coat, top hat, and rubber boots, setting off the technological devices that surrounded him. By now Tesla knew just how to reach the crescendo in his performance, saving the best for last. Many times his presentation would end with tens of thousands of volts of wild electricity coursing through his body as lightbulbs were turned on and blue-colored sparks of lightning-like electricity shot from his long fingertips.[23] He was alive with electricity in every way.

Mentally weary and physically worn, Tesla nonetheless achieved a triumph in London in myriad ways. He had successfully reimagined the world and made it real before many of the most brilliant minds Britain had to offer. He had indeed become the *imagineer* of his time and was the consummate outlier of science. To do so, he had rejected the received wisdom of the day and understood, much like Newton and Einstein, that beauty lies within nature; it is the light that lights the path to truth. Moreover, he also had the awareness of the ancients, who understood such beauty as it was expressed in *prisca sapientia* (the ancient knowledge).[24]

After the second lecture, the world-renowned physicist Lord Rayleigh broke with the Institution's convention of not commenting at the conclusion of a lecture when he said, "Mr. Tesla has not worked blindly or at random, but has been guided by the proper use of a scientific imagination. Without the use of such a guide we can scarcely hope to do anything of real service. I do not think there is anything I need add; it does not require any great capacity to see that Mr. Tesla has the genius of a discoverer, and we may look forward to a long career of discovery for him."[25] At that moment in time the London press was consumed with the young man with the magic, the wizard who wowed them all.[26] The lectures produced several more invitations by fellow scientists to meet, dine, and discuss matters of science, and so he accepted.

After the lecture, Tesla was invited to Sir William Crookes's laboratory and as a lagniappe to his host, he built a "Tesla coil" and taught his fellow scientist how build such coils on his own. As was his style, Tesla persisted in playing in the laboratory until it was time for a respite. Afterward there was a sumptuous dinner, and then the two relaxed and conversed intently on topics that ran the gamut from their own scientific work to wireless communication, theology, spiritualism, metaphysics, and even Tesla's homeland.[27]

When the conversation began to delve deeper into research on supernatural phenomena, Tesla balked, never having had any real belief in the afterlife. But when Sir Crookes presented him with tranches of documentation and the support of others whom Tesla admired, he began to entertain the idea, He did have a desire for immortality, but he had always felt that it was his work that would provide it.[28] Suddenly, his mind began to spiral off in various directions. He found conversation difficult, and Sir Crookes looked at him with the concern of a parent. Tesla had left behind old-world convictions of superstitions to make his way in the New World, and now he was feeling like he never left home. His view of the world was once again challenged, and he didn't like it.

Admittedly, Tesla was under an onerous pressure, what with his rigorous travel schedule and now having to process what Sir Crookes had told him. Sir Crookes could sense that he was in great need of rest. He later wrote in a letter to Tesla, "I hope you will get away to the mountains of your native land as soon as you can. You are suffering from over work, and if you do not take care of yourself you will break down."[29] Tesla thought it sage advice, but he knew that his time was not his own. The excitement, the lectures, and the dinners in London had to come to an end because Paris was calling. He had previous experiences in the City of Light—good and bad. With trepidation in the forefront of his mind, he wanted to return to New York City, but he had made a commitment, and he was always a man of his word.

Perhaps Sir Crookes noted in his own scientific analysis the paradox of Tesla moving from mania to mania with mixed features of depression—a harbinger for Tesla that he was about to slip into a deep depression. He seemed to pick up on Tesla's melancholy and the fact that his output was no longer seemingly effortless, despite continuing prodigious output of performance energy at the lecture.

PARIS, FRANCE, 1892

Impressionist painters, some famous, some trying to be, and some nothing more than poseurs, had planted their easels across the city, dabbing paint on reused canvasses in an effort to reveal what their eyes had seen, hoping others would like them enough to buy them. Cafés lined each and every promenade and the grand boulevards swarmed with tourists drunk on the city's splendor. Paris, like Florence, often casts a spell upon those who come to experience its beauty. In fact, many call it Stendhal's syndrome, while others refer to it as the Paris Syndrome.[30] It is a psychological condition wherein individuals are overcome by the beauty of a magnificent city or a great work of art, or even by an elegant mathematical formula, theory, or machine.

Leaving England's nasty mid–February weather, Tesla undertook an uneventful crossing of the English Channel. He immediately made his way to the Hotel de la Paix and refuge from the demands that were taking their toll—his February 19 lecture before a joint gathering of the Société de Physique and the Société International des Electriciens was fast approaching, and he felt anxious and burdened. Nonetheless, he thought it prudent to seek out one of the pioneers in the fields of electrophysiology and diathermy; hence, he met with Jacques-Arsène d'Arsonval (1851–1940), a French physician and physicist. Recalling the meeting, he wrote: "When . . . Dr. d'Arsonval declared that he had made the same discovery [concerning the physical effects caused by sending extremely high frequency through the body], a heated controversy relative to priority was started. The French, eager to honor their countryman, made him a member of the Academy, ignoring entirely my earlier publication. Resolved to take steps for vindicating my claim, I . . . met [with] Dr. d'Arsonval. His personal charm disarmed me completely and I abandoned my intention, content to rest on the record. It shows that my disclosure antedated his and also that he used my apparatus in his demonstrations." The final judgment is left to posterity. Tesla writes,

> Since the beginning, the growth of the new art [of electrotherapy] . . . and industry has been phenomenal, some manufacturers turning out daily hundreds of sets. Many millions are now in use throughout the world. The currents furnished by them have proved an ideal tonic for the human

nerve system. They promote heart action and digestion, induce healthful sleep, rid the skin of destructive exudations and cure colds and fever by the warmth they create. They vivify atrophied or paralyzed parts of the body, allay all kinds off suffering and save annually thousands of lives. Leaders in the profession have assured me that I have done more for humanity by this medical treatment than all my other discoveries and inventions. Be that as it may, I feel certain that the MECHANICAL THERAPY, which I am about to give to the world, will be of incomparably greater benefit.[31]

Once again, in this essay Tesla reaffirms his belief that it is his moral duty to do whatever he can to benefit his fellow man. Moreover, his work with "mechanical therapy" has proven to be important and beneficial.

~

As was expected, Tesla's lecture and its attendant demonstrations on the night of February 19 in Paris were simply the rage. However, as reported, Tesla did repeat some of his earlier experiments, thus indicating that the lecture tour was taking its toll on him physically and mentally. However, the French electrician Édouard Hospitalier reported that Tesla presented experiments of "a character of novelty." He added, "Mr. Tesla, in his memorable experiments, has shown us that, on periodically varying, with very great frequency, an electrostatic field, it is possible to place apparatus of great simplicity therein, such as tubes of rarified gases, which collect a portion of such energy and render it luminous. To him the light of the future resides in incandescence of solids, gases, and phosphorescent bodies. . . . The young scientist is convinced of this as a precursor, and almost as a prophet. He introduces so much warmth and sincerity into his explanations and experiments that faith wins us, and, despite ourselves, we believe that we are witnesses of the dawn of a nearby revolution in the present processes of illumination."[32] A U.S. journal reported that Tesla's lecture in Paris established his scientific reputation in a single stride.[33] Another journalist said of Tesla's performance that night that "Tesla's eyes glowed as he spoke of his work. Leaning forward, peering almost each moment into the eyes of his [audience] to make sure that his meaning has been understood, he proved a talker from whose train of reasoning there was no escape while a man was under his influence."[34]

Nonetheless, there were always the doubters, as most great inventors will tell you. Yes, Edison's head popped up again, like a grifter looking for his next mark, as he continued to peddle his inferior DC system and feared that Tesla would capitalize on his AC system in Europe—the War of the Currents was still problematic for Tesla. And then there were others who scoffed at Tesla's claims insisting that his motor could never work and even one brazen American insisted that he had invented an alternating-current system.[35] Such negativity only served to exacerbate an already difficult time for Tesla, brought about by the lecture tour, self-doubt at times, and the curious thoughts that continued to infect his psyche. As a counter to such negativity, Tesla met with several dignitaries and principals of companies both in France and Germany, with the purpose of producing income from his foreign patents. It was during this time that the physical stress and mental exhaustion was about to pour over into a boiling cauldron of emotions.

In the midst of all the triumph in London and Paris, Tesla was soon to experience his greatest tragedy. He realized his issues were mounting, and in his autobiography he gives us insight into his feelings after his Paris lecture. As he wrote,

> I found it too hard to break away from the laboratory, and several months elapsed during which I had succeeded in reviving all the impressions of my past life up to the spring of 1892. In the next picture that came out of the mist of oblivion, I saw myself at the Hotel de la Paix in Paris just coming to from one of my peculiar sleeping spells, which had been caused by prolonged exertion of the brain. Imagine the pain and distress I felt when it flashed upon my mind that a dispatch was handed to me at that very moment bearing the sad news that my mother was dying. I remembered how I made the long journey home without an hour of rest and how she passed away after weeks of agony! It was especially remarkable that during all this period of partially obliterated memory I was fully alive to everything touching on the subject of my research. I could recall the smallest details and the least significant observations in my experiments and even recite pages of text and complex mathematical formulae.[36]

What can we take away from Tesla's comments? Was his amnesia about his previous life a defense mechanism? Perhaps, but despite his great loss—his mother was the wellspring of his inventing abilities—he was still able to focus on his life's work and its purpose.

Tesla spent much of his time in these years in a manic/hypomanic state in which, like most manics, he had a diminished need for sleep. When mania goes away, there is not only the need for normal sleep, but often hypersomnia, with lots more than normal sleep, and that sleep being non-refreshing and non-restorative, which must have seemed quite peculiar to him, since he was definitely depressed.

GOSPIĆ, AUSTRIAN-HUNGARIAN EMPIRE, 1892

The news of his mother's imminent passing hit Tesla like a thunderbolt of lightning, reminiscent of the night of his birth—he was confronted with life eventually ending in death, despite his belief that his work would grant him "immortality." Upon returning to Gospić, after nearly a decade away, and seeing that little had changed save "electric" street lamps, he was met by his three sisters, all of whom had married Serbian Orthodox priests. His Uncle Petar, the local bishop, was also there to comfort him, for he was in a dreadful state of mind. He was immediately led by his family members to his mother's bedside. Looking at his teary-eyed son, she summoned up the motherly words, "You've arrived, Nidzo, my pride."[37]

Although not given to outward emotions, Tesla cried out at the sight of his dear mother, his greatest inspiration, as she was in the last throes of life— her body frail; her face wan with the fading of light that death brings to all. Tesla tells us he was in a clairvoyant-dream state controlled by the emotions of the moment and spiritualism as his mother expelled her last breath.[38]

> Ever since I was told by some of the greatest men of the time, leaders in science whose names are immortal, that I am possesst [*sic*] of an unusual mind, I bent all my thinking faculties on the solution of great problems regardless of sacrifice. For many years I endeavored to solve the enigma of death, and watched eagerly for every kind of spiritual indication. But only once in the course of my existence have I had an experience which momentarily impressed me as supernatural. It was at the time of my mother's death. I had become completely exhausted by pain and long vigilance, and one night was carried to a building about two blocks from our home. As I lay helpless there, I thought that if my mother died while I was away from her bedside she would surely give me a sign. Two or three months before I was in London in company with my . . . friend, Sir William Crookes, when spiritualism was discussed, and I was under the full sway of these thoughts. I might not have paid attention to other men, but was susceptible to his arguments as it was his epochal work on radiant matter, which I had read as a student, that made me embrace the electrical career. I reflected that the conditions for a look into the beyond were most favorable, for my mother was a woman of genius and particularly excelling in the powers of intuition. During the whole night every fiber in my brain was strained in expectancy, but nothing happened until early in the morning, when I fell in a sleep, or perhaps a swoon, and saw a cloud carrying angelic figures of marvelous beauty, one of whom gazed upon me lovingly and gradually assumed the features of my mother. The appearance slowly floated across the room and vanished, and I was awakened by an indescribably sweet song of many voices. In that instant a certitude, which no words can express, came upon me that my mother had just died. And that was true I was unable to understand the tremendous weight of the painful knowledge I received in advance, and wrote a letter to Sir William Crookes while still under the domination of these impressions and in poor bodily health. When I recovered I sought for a long time the external cause of this strange manifestation

and, to my great relief, I succeeded after many months of fruitless effort. I had seen the painting of a celebrated artist, representing allegorically one of the seasons in the form of a cloud with a group of angels which seemed to actually float in the air, and this had struck me forcefully. It was exactly the same that appeared in my dream, with the exception of my mother's likeness. The music came from the choir in the church nearby at the early mass of Easter morning, explaining everything satisfactorily in conformity with scientific facts.[39]

These types of compelling thoughts, almost experienced as visions and hallucinations, are a common part of a normal grief reaction. In Tesla at the time, he was transitioning from mania to a mixed manic state while lecturing in London, from anticipation of problems with his mother to a grief reaction triggering a full depressive episode with sleep episodes.

$\sim$

April 19 marked the day Djouka (Đuka—Georgina) Mandić Tesla died at the age of seventy.[40] It was Easter Sunday. She was laid to rest next to her beloved husband, Milutin, in the Jasikovac Cemetery in Divoselo. Because both the Tesla and Mandić families had a long history of involvement in the Serbian Orthodox Christian Church, the funeral service was officiated by no less than six priests. Tesla arranged for and paid for two very tall white obelisk headstones erected over both his mother and father.[41] The weathered headstones still stand sentinel today in the cemetery, engraved Serbian Orthodox crosses occupying the top two-thirds of each one.

Upon reflection, Tesla wrote very poignantly to a relative, "I don't have to tell you that I am very sad and holding myself in restraint. I was afraid of this event a while ago, but the blow was heavy." Even years later the thought of his mother's death haunted him, just as his brother Dane's death had since childhood.[42] Tesla's ability of precognition sadly showed itself again years later when he had thoughts of his sister Angelina's imminent death. He hastily sent a telegram home, only to receive a return telegram confirming her death. During his life, such experiences of impending doom were not uncommon.[43]

After the funeral, feeling emotionally wounded, Tesla spent the next four to six weeks in Gospić reconnecting with family and friends. He journeyed to Plaski to visit his sister Marića, then to Varazdin to see Uncle Pajo and then on to Zagreb to lecture at the university. He took a side trip back to Budapest to consult with Ganz & Company on their 1,000-horsepower generator. In May, he even made his way down to Belgrade where he received honorariums and the like. There Tesla was feted by the king and was read to by Serbia's greatest poet, honoring Tesla's accomplishments. Tesla remarked to the adoring crowd of well-wishers: "If I were to be sufficiently fortunate to bring about at least some of my ideas it would be for the benefit of all humanity." He concluded, "If these hopes become one day a reality, my greatest joy would spring from the fact that this would be the work of a Serb."[44] Once again, in the time of his darkest hour, Tesla was always thinking of how he could lift the burdens from the shoulders of mankind—he never wavered on this noble desire.

〜

Once Tesla had finally decided it was time to go home to his laboratory and hotel in New York City, he made yet another side trip to Germany, meeting with German scientific luminaries such as Heinrich Rudolph Hertz (1857–1894), with whom he had a major disagreement regarding the fact that he had rendered Hertz's mathematical results as well as his device both obsolete and primitive at best. Hertz flinched at Tesla's comments, yet ironically, today's wireless frequencies are incorrectly referred to as Hertzian waves, when in truth they are Teslian waves—generated by high-frequency continuous-wave oscillators of Tesla's design.[45] It is no wonder, then, that Hertz made no mention in his diary of his meeting with Tesla.[46]

During his return voyage home to New York City, Tesla strode upon the ship's deck, as if gliding effortlessly, while thinking, pondering, calculating. Then in a beat, he recalled a hike he had taken in the mountains of his homeland while on his European trip. A thunderstorm had hit with great intensity, but thankfully he was able to find shelter before the deluge. He noticed that the reason he had time to seek cover was because the rain did not begin until he saw a streak of lightning that seemed to puncture the heavens, thus releasing the biblical downpour. As he described in his autobiography:

> This observation set me thinking. It was manifest that the two phenomena were closely related, as cause and effect, and a little reflection led me to the conclusion that the electrical energy involved in the precipitation of the water was inconsiderable, the function of lightning being much like that of a sensitive trigger.
>
> Here was a stupendous possibility of achievement. If we could produce electric effects of the required quality, this whole planet and the conditions of existence on it could be transformed. The sun raises the water of the oceans and winds drive it to distant regions where it remains in a state of most delicate balance. If it were in our power to upset it when and wherever desired, this mighty life-sustaining stream could be at will controlled. We could irrigate arid deserts, create lakes and rivers and provide motive power in unlimited amounts. This would be the most efficient way of harnessing the sun to the uses of man. The consummation [of this theory] depended on our ability to develop electric forces of the order of those in nature. It seemed a hopeless undertaking, but I made up my mind to try it and immediately on my return to the United States, in the Summer of

1892, work was begun which was to me all the more attractive, because a means of the same kind was necessary for the successful transmission of energy without wires.[47]

In time, Tesla's trip to Europe to deliver lectures and license his patents; to visit his dying mother and be present at her funeral; and to discuss topics of great import with fellow scientists, would reveal itself to be profoundly beneficial to him, because it gave him a greater understanding and clarity regarding his work. In August of 1892, Tesla steamed his way back across the Atlantic Ocean to New York City from Hamburg, Germany, on the luxurious *August Victoria*, as he had in 1884. This time he was rich and famous, although it did not assuage his feelings of emotional desperation, loss, and exhaustion, both mental and physical. But he was home and ready to continue his lifelong pursuit of changing the world for the better.

· 9 ·

The Show Must Go On

The Glory of AC

NEW YORK CITY, NEW YORK, 1893

The dawn of the twentieth century was about to break over the horizon and New York City was the place to be. Opportunity was there for the taking if hard work and an entrepreneurial spirit were in you. Nikola Tesla returned home to New York in late August. Although he was physically battered and mentally spent from his lectures and the unexpected loss of his dear mother, he was still young, vibrancy was in his bones, and he was determined to continue his quest to expand upon his new inventions and discoveries. To do so after his long, successful progress in Europe required a fresh start, so he began with a change of residence.

Although he could have easily afforded his very own home, after three years at the Astor House, he made the move to the very tony, eleven-story Gerlach Hotel (The Gerlach Family Hotel—nos. 49, 51, 53, and 55) on Twenty-Seventh Street, just between Sixth Avenue and Broadway was his choice.[1] This was yet another example of Tesla's need to be alone, where he could guarantee refuge from the masses and where his personal needs would be taken care of by others. The hotel suited his style, and no one can deny the magnificence of the structure, situated just south of Washington Square. It was built in the Queen Anne style at a cost of $1 million in 1888. Its first three stories are covered in rough-cut stone and the additional eight stories reveal redbrick walls. Bowed bays added to its elegance and allowed for the capturing of gentle breezes during the hot summer months, while bas-reliefs were placed about the structure to give it an old-world appeal. It was ahead of its time, having elevators, and featuring elegant dining rooms and a very European feel. It was also attractive to Tesla because the entire building was fireproof—a godsend for the era. Unlike most city apartments of the time,

138

Tesla's personal quarters were more of a French flat layout—the Gerlach offered both European and American plans as well as permanent and transient accommodations for guests. Here he would experiment with some of the first "radio "transmissions to and from his downtown laboratory. An added amenity was that it was located just blocks from the trendy Madison Square Garden galleria of shops, theaters, and restaurants that catered to the public.[2] And in a tip of the hat to honor Nikola Tesla, the hotel is presently called the Radio Wave Building—as of 1977, a prominent brass plaque recognizes the genius of the great inventor and discoverer who lived there for many years.

Now eager to continue his important work with high-frequency inventions and feeling the need to constantly improve upon his AC polyphase motors and power system, Tesla had moved his laboratory from Grand Street to a larger workspace on the fourth floor of an unremarkable building located at 33–35 South Fifth Avenue (today LaGuardia Place). The first few floors held a pipe-cutting factory and a dry-cleaning business that produced noxious fumes and sights Tesla's hypersensitivities found offensive. However, he always smiled to everyone he passed. Once on the fourth floor, he unlocked the door of his new laboratory and entered his inner-sanctum.[3]

In the laboratory Tesla found privacy, despite the worldwide attention he received everywhere he went. It was the private retreat of a modern wizard, where he spent most of his days attired in a dark lab coat. On hot days, a white dress shirt, sleeves rolled up, would serve the purpose—but ever immaculate. As one excited reporter wrote after an interview with Tesla: "While awaiting my opportunity in an anteroom, I caught glimpses through the adjoining office and library of the mystic laboratory itself, which, as I ascertained later, opened into an immense machine room [some twenty-five feet square]."[4] The reporter had essentially stalked Tesla, much like a big game hunter, for an interview until he had succeeded in getting one. Outside the building, the streets were afire with all manner of human activity—the noises and chaos of life that Tesla tried to avoid at all costs.

At this point, as Tesla's fame was in its ascendency, he became more secretive, more reclusive, more emotionally charged, and it showed. The intrepid reporter added upon first encountering the very secretive Tesla, "I may candidly state that I was shocked the first time I saw Nikola Tesla, as he suddenly appeared before me and sank into a chair seemingly in a state of utter dejection. Tall, straight, gaunt, and sinewy of frame like a true Slav [Serb], with clear blue eyes and small, mobile mouth fringed with a boyish mustache, he looked younger than his thirty-seven years. But what arrested my attention chiefly at the moment was the pallid, drawn and haggard appearance of the face. While scanning it closely I plainly read a tale of overwork and of tremendous mental strain that must soon reach the limits of human endurance."

Tesla stated emphatically, but politely, "I would like to talk with you, my dear sir, but I feel far from well to-day. I am completely worn out, in fact, and yet I cannot stop my work. These experiments of mine are so fascinating, that I can hardly tear myself away from them to eat, and when I try to sleep I think about them constantly. I expect I shall go on until I break down altogether."[5]

These comments about breaking down from Tesla are the quintessential description of a mixed state. He was still manic and driven and creative and fascinated, but driving himself to the point of exhaustion and wearing out was likely a forewarning of the major depressive episode he was about to experience later that year of 1893 and in 1894, and this is supported by the fact that during this era of a flurry of patents every year, he suddenly filed no patents in 1895, most likely from the lack of productivity earlier in the year when mixed mania and depression led the way to a pure major depressive episode.

Needless to say, the wide-eyed reporter was waiting to see with his own eyes the famous inventor demonstrate what all the fuss was about since his arrival in New York in 1884. As the kindly Tesla was expected to do and did—ever the showman—he set about moving through the room overflowing with all manner of machines from dynamos spinning and Tesla coils firing off long streaks of blue-white lightning to cables surrounding the two like snakes in a sci-fi thriller, shouting orders to employees, whereupon "exquisitely beautiful luminous signs and devices of mystic origin began to flash about me with startling frequency." The reporter observed: "What impressed me most of all, perhaps, was the simple but cheerful fact that I remained unscathed while electrical bombardments were taking place on every side. Curiously though, the polyphase currents of high frequency and high potential, of say 200,000 volts, have, as Mr. Tesla has demonstrated repeatedly on the platform, no harmful effect whatever on the human body, although a like energy exerted in individual currents would prove instantaneously fatal."[6] The reporter was simply overwhelmed by just a few of Tesla's many electrical miracles.

It is imperative to understand the importance of this reporter's interview and observations. Although Tesla had just returned from his European travels and was readjusting to life in New York City, he took the time to indulge the reporter because he knew his life had purpose. The reporter went on to write, "Mr. Tesla makes no boasts, but is willing to abide his time. Throughout the interview I was constantly impressed with the man's loftiness of purpose, innate modesty and utter indifference to public applause. 'I should much prefer not to be written about at all,' he remarked; 'but if it must be done, I trust you will take due pains to quote me correctly.'"[7]

The reporter further noted: "Mr. Tesla speaks our language with the idiomatic range and choice diction of a native who is also a scholar and a trained speaker, the guttural accent of the Slav, of course, being slightly noticeable. He told me he felt equally at home in six languages, not to mention the same number of dialects." "Before I made a regretful farewell to this kindly wizard of Washington Square," the reporter concluded, "he confided to me that he was engaged on several secret experiments of most abundant promise, but their nature cannot be hinted at here. However, I have Mr. Tesla's permission to say that some day he proposes to transmit electric vibrations through the earth; in other words, that it will be possible to send a message from an ocean steamer to a city, however distant, without the use of any wire."[8]

In 1893, predating the earliest attempts in Hertz wave telegraphy by some three years, Tesla first explained his unique wireless system and took out the supporting patents on a number of innovative devices that were until then but poorly understood, if at all. Even the whole of the electrical world derided these patents as bogus. But nonetheless, large wireless companies had to pay him homage in the form of real money, because his "fool" patents were recognized to be fundamental. He actually antedated every important wireless invention up until that time.[9]

Today, as was the case nearly every day, Tesla left the confines of his swank hotel to journey by foot some thirty blocks down to an area in the center of Greenwich Village to his new laboratory, said to be immaculate in every way—nothing out of place, dust-free, and orderly—just as Tesla was.[10] He had already risen hours earlier, eaten a simple breakfast, performed a few physical exercises, dressed in his signature black swallowtail coat and pressed-collar shirt, and readied himself for the day. As he walked, he passed the newly completed Madison Square Garden arena designed by the famed architect Stanford White, who would years later figure as a friend in Tesla's life, as would Augustus Saint-Gaudens, the celebrated sculptor, designer of one of history's most beautiful gold coins—the Saint-Gaudens $20 gold piece (double-eagle), sought after by numismatists the world over. He also encountered gilded carriages on stylish Fifth Avenue and came upon horse-drawn carts transporting goods to market, as all manner of street merchants and mendicants plied their trades. By now the juxtapositions in his life became obvious, for his daily journey to work began at the front door of a fancy hotel and ended in a village full of old-brick buildings that predated the Civil War.[11]

After having hired several workers and a secretary to care for tranches of correspondence during his first weeks back in the city, he had two things in mind. First, his primary focus, was to advance the progress he had already made in fluorescent lighting and wireless transmission of power; and, second, he set about to begin in earnest by dictating an essay on his recent experiments with Hertzian frequencies and their relationship to the surrounding medium.[12] He found it necessary to tweak his oscillators, after which he devised an experiment: "Assume that a source of alternating currentss [*sic*] be connected with one of its terminals to earth (conveniently to the water mains) and with the other to a body of large surface."[13] He then documented electrical vibrations at various points within the city after which he concluded with, "I think that beyond doubt it is possible to operate electrical devices in a city through the ground or pipe system by resonance from an electrical oscillator located at a central point. But the practical solution of this problem would be of incomparably smaller benefit to man than the realization of the scheme of transmitting intelligence, or perhaps power, to any distance through the earth or environing medium. If this is at all possible, distance does not mean anything. Proper apparatus must first be produced by means of which the problem can be attacked and I have devoted much thought to this subject. I am firmly convinced that it can be done and hope that we shall live to see it done."[14]

As his laboratory quickly became a beehive of activity, by the end of September, Tesla was receiving many guests, none more important to him at the moment than George Westinghouse, who made an appearance with Swiss electrical engineer Albert Schmid to discuss the future of Tesla's AC polyphase system. The War of the Currents was still an existential threat to both Tesla and Westinghouse and decisions had to made—it was all on the line. Ironically, each man had his reasons: Westinghouse needed a way to raise his fading profile among the public and his competitors due to publicized financial problems, while at the same time to feed his great dislike of General Electric's chief, Charles Coffin, even if he had to lowball his bid to make certain that Edison and Coffin would lose the contract to light the Chicago World's Fair. While focused on winning, and he was convinced he would, Westinghouse was not focused on Tesla's induction motors or polyphase system, nor did he plan on having such motors displayed at the international fair. No, he was more concerned with designing the power equipment and lamps that would light the months-long event.[15] He had to get the job done, and if there was another, less expensive and less complicated way, that is what he would do; it was all business to him. Tesla was not pleased with Westinghouse's present view of his polyphase system—knowledge and analysis had proven many times over that his inventions were essential to a successful polyphase system.[16]

In the spring of 1892, fair officials, led by Colonel George R. Davis, director general of the fair, sent out word that bids were being entertained from those interested in the City of Chicago's World's Fair site. Financier J. P. Morgan was eager to make an offer because he saw mountains of money if he controlled electric power, so he set upon the mission to acquire the Edison Company, the Thompson-Houston Company, and others who would get in his way as he formed what would be known as the new General Electric Company. In fact, GE bid more than $1 million—the lion's share going to the cost of copper wire that DC electricity required—for the contract. As a counter to Morgan's offer, Westinghouse's AC system (using Tesla patents) drew the redline by proposing that his system could light the exposition for half the price. Moreover, the Westinghouse system would power both electric motors and lights simultaneously. And much to Edison's disgust, the AC system saved the fair hundreds of thousands of dollars just in copper wire costs alone.[17] The copper-wire problem associated with Edison's DC system was one of numerous reasons why Tesla's AC system was superior in terms of costs, manufacturing, power generation, and reliability.

Thankfully, in May of 1892, Westinghouse had finally won the much-coveted contract to power and light the upcoming World's Columbian Exposition at a cost of $399,000. Edison's outrageous bid proved to be the result of his cockiness and the fact that he believed he would win his incandescent-lamp suit, thus giving him a monopoly on the manufacturing of such lamps.[18] When Westinghouse slashed the cost of each light (lamp) to just six dollars per unit, it was game over, even though C. F. Locksteadt, a local contractor's bid of $335,000 undercut all other bids, it was believed he could never complete the project.[19] Frankly, Westinghouse had risked it all; he had put his company at the door of bankruptcy court, what with the country in a recession and numerous other companies fighting their own patent wars before judges across the land.[20]

Was this the death-blow to Edison that Tesla and Westinghouse had hoped for?

CHICAGO, ILLINOIS, 1893

The grand world exposition was slated to be held in Chicago from May 1 to October 30, 1893, to commemorate the four-hundredth anniversary of Columbus's discovery of America—a year late.[21] The securing of such an important contract proved to be another victory in the War of the Currents. With knowledge of Westinghouse's success against Edison and Morgan, Tesla was excited, for he and Westinghouse would now have their best opportunity to slay the dragon—Edison. As we now know, in the end Westinghouse was convinced to see the error of his ways and began to realize that Tesla's polyphase system was the answer. Although Westinghouse's motivation was different than Tesla's, the end game was the same: winning the War of the Currents.

With the lucrative contract in hand, Westinghouse asked for Tesla's assistance in building his Chicago World's Fair exhibit to produce all the power and light for the fairgrounds. In return, Tesla was promised an exhibit of his own to appear under the Westinghouse name. There were seemingly not enough months to accomplish the gargantuan task, but it had to be done. At times Tesla made myriad trips to Pittsburgh during this chaotic time to direct the building of dynamos and at other times he would receive several of Westinghouse's engineers in his New York laboratory to provide technical assistance.[22]

While Tesla toiled away on the world's fair project, his reasons were quite different from his partner Westinghouse's. As the discoverer of the "rotating magnetic field" and the inventor of his induction motor and the subsequent alternating-current polyphase power system, his war, as we recall was only partly with Edison; more accurately, it was a war within himself. He knew his polyphase system was superior to Edison's in every way, and it still is to this very day. Knowing that his mental demons had pursued him since childhood, a victory over the minacious Edison would go a long way to ameliorating what troubled him . . . at least for the moment. He hoped that the crossing-of-swords with Edison was about to come to an end.

Admittedly, it might seem that Tesla had "unknowingly" fired the first shot in the War of the Currents when he gave his groundbreaking lectures on his alternating current polyphase system in 1891 at New York's Columbia College. However, the final battlefield could very well be the international fair, where Tesla and Westinghouse would demonstrate the superiority of his alternating current (over Edison's direct current system). Thus, winning the contract could very well put the issue of AC versus DC to rest for all time.

Morgan and Edison were both enraged by the outcome, but to quote the old saw: "The show must go on." In retaliation, GE forbade Westinghouse from using its "one-piece Edison lamp" technology (federal court upheld the patent), which caused Westinghouse to invent a workaround solution—he and his team invented a new stopper lamp.[23] Only days after the Edison patent was upheld, Westinghouse sent out a circular that he was prepared to supply new lamps that did not infringe on the Edison patent. With the possible infringement issue put to bed, Westinghouse knew he would complete the power and lighting project on time.[24]

The Chicago World's Fair (officially the World's Columbian Exposition) was a massive undertaking that comprised six hundred acres of swampland turned into a Disneyland-like wonderland by a legion of engineers and laborers. When completed, it ultimately drew a total of 27,300,000 visitors—100,000 on the very first day. Visitors were treated to countless exhibitors and amusements. The promoters of the grand show turned a profit of $2.25 million. An endless-series of events tantalized the eyes and energized the spirit in what became known as the "White City." Local architect Daniel H. Burnham and other planners were inspired in their design by the city of Venice and its many waterways. As an interesting sidebar: On October 9, 1893, designated as Chicago Day, the exposition set a world record for attendance at an outdoor event, drawing 751,026 people.

The fairgrounds were home to gleaming, neoclassical buildings—forming the Court of Honor—at every turn. The Manufacturers & Liberal Arts Pavilion, then the largest building in the world, stretched a third of a mile long and nearly three football fields wide. The building consumed over thirty acres and boasted a seating capacity of seventy-five thousand visitors. Then there was the Electricity Pavilion (Hall), itself rather impressive in size and the center of the most important exhibits at the fair. These exhibits featured a pantheon of scientific companies and their products, but what was most surprising to all was that in the Electricity Hall there was erected a forty-five-foot monument that proclaimed to the world who was the "true inventor" of alternating current. The proclamation read: Westinghouse Electric & Manufacturing Co. Tesla Polyphase System. The moment it was revealed to all in attendance in Machinery Hall, Westinghouse gave the word and the switch was flipped, giving electric life to 250,000 Sawyer-Mann stopper lamps, which resulted in more than three times the electrical power being utilized than in the city of Chicago itself.[25]

Entertainment was the focus of the fair, with Buffalo Bill's Wild West show performing no less than 318 times. Edison's latest photographs were displayed, as was the kinetoscope—a device for showing a loop of pictures that replicated motion. American Bell Telephone wowed the onlookers with the first long-distance call from New York City to Chicago. The first American-built automobile (Duryea Brothers) and Ferris Wheel (Chicago Wheel) were on display, as was the revolutionary zipper. But nothing was to equal the visual impact and spectacle created on opening evening, when President Grover Cleveland pushed the button that lit one hundred thousand incandescent lamps, as if they were a new sun rising in the east. With flags of every nation fluttering in the breeze, a full orchestra played Handel's *Hallelujah Chorus*. And as if that were not enough, electronically operated fountains shot water to the heavens and cannons roared across the fairgrounds. This signaled that Chicago had become the true City of Light (inspiration for the Emerald City in *The Wizard of Oz*), and all credit was due Nikola Tesla and his partner George Westinghouse and the 12,000-horsepower Tesla AC alternators (60-c/sec, single-phase, wired for two-phase voltage) housed in the Hall of Machinery that supplied "all" the power and light to the fair for its entire six-month run. The most enduring moment of the fair, and its most consequential, was to be found in the Great Hall of Electricity. It was there that most visitors first came into contact with Tesla's AC polyphase system of power generation and distribution. The result: From that day onward 80 percent of all the electrical devices purchased in the United States were powered by alternating current.[26]

Thankfully, Tesla was rightly given his own time to shine before all when his display area was illuminated by his latest invention . . . neon light. He also demonstrated, with the precision and sprezzatura he was known for, the principle of his discovery of the magnificent "rotating magnetic field," just as when he recreated his Egg of Columbus demonstration that the reader is familiar with. The wonders included several of his earliest attempts at AC power generation, which included induction motors, generators, and phosphorescent signs fashioned in the name of celebrated electrical engineers. The electrical magician also wowed the audiences by illuminating vacuum tubes without wires; otherworldly neon-lights displaying "Westinghouse" and "Welcome Electricians"; and ear-piercing sounds from high-frequency discharges between two insulated plates that radiated bursting rods of light reminiscent of the night of his birth.[27] The wonders never ceased, as he had all eyes on him in the magician's chamber, but Tesla's grand finale was beyond incredible. With the overflowing crowd in the palm of his hands, Tesla stood rod-straight on a stage, proud and confident. While wearing rubber-soled shoes, he gave the order to switch on the dynamo. Without warning, the dynamo sent two million volts of electricity coursing through his body, which produced a halo of electric flames that framed his physical presence.[28]

Awestruck doesn't begin to describe what the attendees experienced that day, but when they witnessed the millions of volts of electricity pass through Tesla's body, leaving him unharmed, they were speechless, as one would be today. He explained the phenomenon this way:

> The amount of energy which may thus be passed into the body of a person depends on the frequency and potential of the currents, and by making both of these very great, a vast amount of energy may be passed into the body without causing any discomfort except perhaps in the arm, which is traversed by a true conduction current. The reason why no pain in the body is felt, and no injurious effect noted, is that everywhere, if a current

be imagined to flow through the body, the direction of its flow would be at right angles to the surface; hence the body of the experimenter offers an enormous section to the current, and the density is very small, with the exception of the arm perhaps, where the density may be considerable. But if only a small fraction of that energy would be applied in such a way that a current would traverse the body in the same manner as a low frequency current, a shock would be received which might be fatal. A direct or low-frequency alternating current is fatal I think, principally because its distribution through the body is not uniform, as it must divide itself in minute streamlets of great density, whereby some organs are vitally injured. That such a process occurs I have not the least doubt, though no evidence might apparently exist or be found upon examination. The surest to injure and destroy life is a continuous current, but the most painful is an alternating current of very low frequency. The expression of these views, which are the result of long-continued experiment and observation, both with steady and varying currents, is elicited by the interest which is at present taken in this subject and by the manifestly erroneous ideas which are daily propounded in journals on this subject.[29]

The denouement of the fantastic story that was the Chicago World's Fair—recognized by most to be the greatest of all world's fairs—was that Tesla had vanquished Edison. The message had been sent across the globe: The War of the Currents had been won, and Tesla had won it, because the future of power generation and distribution was his AC polyphase system and his alone. But was there still another war to fight for Tesla?

At this point in Tesla's upward rise, it is important to understand that he never deviated from his only purpose in life, to make man's lot easier. To date, he had given several groundbreaking lectures that were in and of themselves watershed moments in the science of electricity. He continued such efforts because his life was purpose-driven, and because it had become obvious to all that the war with Edison was over. However, Tesla knew that there was still much work to be done if his system of electrical power was to succeed as a commercial entity on all fronts. To that end, on the evening of August 25, 1892, during the great fair, he delivered another simply astonishing lecture before the august body of some one thousand electrical engineers—who were collectively known as the International Electrical Congress—in the Agricultural Assembly Hall. It was a standing-room-only event, with throngs of curious visitors seeking admission to see the man introduced by the Congress's president, Dr. Elisha Gray (an inventor who knew the disappointment of not getting credit for a major invention: the telephone) this way: "I give you the Wizard of Physics." As reported in the *Chicago Tribune*, one can feel the moment:

> Those who know him knew it would be utterly impossible to predict what the subject of his lecture would be, for Tesla is a man of surprises. He uses no notes, prepares nothing in advance, in fact is originality personified. People crowded about the doors and clamored for admittance. Ten dollars was offered for a single seat, and offered in vain. Only members of the Electrical Congress, with their wives, were admitted, and not even they unless they were provided with credentials. On the little platform at the south end of the hall stood the curious machines, small cylinders of heavy steel mounted on steel pedestals on insulated wooden bases. A table at the right was piled up with curious mechanical appliances while the floor was strewn with wires, rubber tubes, tools, and what not. From the wall projected a two-inch pipe with a valve and a gauge. At one side were electric switches, rheostats, and wires. Such men as Silvanus Thompson, Prof Carhart, and Prof. W. H. Preece gazed in wonder and confessed they could not guess what the curious apparatus was. They lumped off the whole lot under the generic term of "Tesla's animals."[30]

Then it happened. Tesla appeared before the crowd looking gaunt and seemingly physically spent; his eyes, tired as he lowered his gaze, were oddly spirited with a liveliness of a man excited by the possibilities of his work. He stood tall in an immaculate four-button coat of brownish gray before the adoring crowd. A friend who had dined with him a week prior, reported that Tesla could barely be heard across the table. It was said he was exhausted from his incessant need to invent and approaching a state of dissolution, yet he would not yield.[31]

Tesla managed to summon up his magic, as if Merlin were looking over his broad shoulder giving him an assist. He demonstrated newly designed mechanical and electrical oscillators as well as steam generators and a continuous-wave transmitter. Then there was a clock that appeared to be electrical in nature and kept perfect time. He went on to show new applications of his "alternating current" in the form of various AC motors.[32] He even demonstrated a "Pocket Edition of a Dynamo." It was attached to an oscillator that one could easily carry inside of one's hat, which would develop one-half horsepower, and the scientists and electricians cheered again—he had revolutionized dynamo building, for he had anticipated the difficulties attending rotating inertia.[33]

Simply put, despite his mental state of mind and the fact that he was physically exhausted, Tesla once again did not disappoint the audience of experts who clamored for more.

NIAGARA FALLS, NEW YORK, 1893

With the resounding victory in Chicago now behind them, Tesla and Westinghouse looked forward to their greatest challenge to date: winning another big contract (with Tesla's patents), this time the vast hydroelectric opportunities that awaited anyone who could harness the power of Niagara Falls. Engineers and dreamers alike had for decades pondered what seemed to be the impossible—taking the endless flowing cataract of water that plunged off of three drops totaling 167 feet and transmitting the power that could be generated and then sent over long distances from the source. It seemed an unattainable fantasy, but not to Tesla. Remember that, as a young boy, Tesla told his uncle that he would one day go to America and harness the power of Niagara Falls, having been inspired by the concept of the waterwheel when he played with his boyhood friends during a spring snowmelt. Today two massive bronze statues of Nikola Tesla are situated on the U.S. and Canadian sides of the falls in honor of his creation: the hydroelectric power plant (utilizing his turbines) at Niagara Falls, New York. But how did this engineering wonder of the world come to be?

It began in earnest after demonstrations at the Chicago World's Fair proved the efficacy, reliability, and technical advantages of Tesla's AC polyphase system. It was now the obvious choice to generate and transmit electric power over long distances. At that point, Westinghouse, because of this success, wanted more. He wanted the contract to manufacture the equipment necessary to harness the endless, constantly renewable, evenly flowing water of Niagara Falls (the Niagara River is the conduit for the drainage of the Great Lakes) and turn it into electrical power. Early on, Westinghouse understood in the clearest possible terms that "energy," hence power, in all its forms, was the keystone—the fundamental resource—of American industry and the country's growing economy. And that fact is as true today as it was in 1893, as countries across the globe are converting to electric power from industry to automobiles and more at a staggering rate.

Tesla and Westinghouse knew that harnessing Niagara's power could not be done in any efficient, reliable way without Tesla's alternating current polyphase system, and that Niagara Falls was ideally located near industrial centers and trade routes in both the United States and Canada. Moreover, by 1890 at least one-fifth of all Americans resided within four hundred miles of the falls.[34] However, they faced myriad engineering difficulties as well as the protestations of conservationists who objected to the project. In addition, the falls actually had too much water in relation to the land available to allow for rope or belt transmission from waterwheels. Any effort to utilize the full force and amount of Niagara's flow would thus require astronomical capital costs.[35]

Enter the rapacious financial entrepreneur and banker John Pierpont Morgan (1837–1913). Morgan's name was synonymous with capital, and he would do whatever it took to make money. So let us step back for a moment and consider an example. General Electric was, ironically, the brainchild of Sam Insull, none other than Edison's closest confidant and general operating officer. He hesitantly advised his boss to merge with J. P. Morgan. His motivation: to help Edison save face by protecting his legacy to some degree, because it was obvious to all that Edison's DC power system could not win the War of the Currents. It was just that simple. Also, to Edison's great embarrassment, J. P. Morgan, the principal partner in the Cataract Construction Company, had already rejected his efforts to provide the wiring and power transmission of electricity to New York City.[36]

In the end, Edison came to the painful conclusion that his time at the top of the electrical sciences mountain was over, and J. P. Morgan had had a hand in toppling him. Yes, Edison continued as a director of the newly formed General Electric Company; yes, he received millions in stock certificates; and yes, the company would finance his aimless tinkering in his many laboratories, but his name was off the door. Edison's best days were behind him, although some of his famous friends, such as Henry Ford, tried in vain to preserve his fading reputation.[37] Tesla had won, again! But did he win what he wanted? Was it the mental salve he so desperately needed? We shall see.

~

Since the mid-nineteenth century, efforts to harness the power of Niagara Falls had always hit a dead end, whether it was design issues, costs, or lack of imagination. In 1886, Thomas Evershed, an engineer employed by the Erie Canal, proposed the most serious plan. But the millions of dollars needed to launch the project never materialized. Then in 1889, a group of moneyed and influential New Yorkers decided to try their hand at it. They formed the Cataract Construction Company resulting in a capital stock valued at $174,000, with a total of $308,500 paid for land and rights of way. The grand total invested at that point reached $482,500 (about $14.4 million today). Of that amount, just over $100,000 was paid in cash and the remainder was in bonds of the Niagara Falls Power Company.[38]

But the problems were just beginning to reveal themselves. In fact, no one knew just where to begin. They needed a direction and a bona fide plan. It seemed to some that the project was akin to summiting Mt. Everest without a guide. It was suggested that a day-to-day administrator was needed, as was a feasibility study; after all, this was not some short-term project or a paint-by-numbers venture. The choice to lead the effort was the respected banker and powerbroker Edward Dean Adams, who had an engineering background from his time as a student at Norwich University and MIT.

Adams realized early on that the real opportunities lay in sending power to factories in Buffalo some twenty miles away and then beyond to cities with even larger populations that required greater daily horsepower to operate their factories, stores, and living areas. Oddly though, Adams still had not decided on the best method for transmitting power over great distances. He even consulted Edison, who predictably recommended his direct current. Westinghouse himself had suggested compressed air could be used to power existing steam engines.[39] Flummoxed by it all, Adams called for the formation of a committee of experts. The International Niagara Commission was established in 1890, with its headquarters located in London.[40]

The Commission asked some twenty-eight firms in the United States and Europe to submit their proposals. To speed along the proposal process, they offered $20,000 in prize money. Westinghouse refused to play, convinced that his advice was worth many times the amount of the prize money. He commented, "These people are trying to secure $100,000 worth of information by offering prizes, the largest of which is $3,000. When the Niagara people are ready to do business, we shall make them a proposal."[41] The cagey Westinghouse knew how the process worked, and a man of his stature was not interested in contests or games of any kind. Even though contest prizes were awarded the next year, in the end, none of the fourteen proposals were thought to be comprehensive or viable enough to be considered.

While progress was being made regarding construction of the tunnel, the inlet canal, and the wheel pit for the powerhouse enclosure, the summer of 1891 presented the Cataract Construction Company officials with their most central issue: What was going to be the design of the water turbines and the type of power transmission? By the summer of 1892, basic designs had been accepted for 5,000-horsepower turbines. But what was the method of power transmission? After all possibilities were studied to the point of exhaustion, the obvious became more obvious: Tesla's AC polyphase power system was the obvious and only logical choice. Mind you, there were naysayers, such as England's Lord Kelvin. He was quoted as saying on May 1, 1893, "Trust you avoid [the] gigantic mistake of alternating current." His admonition fell on deaf ears, for later that same month the Cataract Construction Company had seen the trend toward alternating current building to the point that nothing else made sense anymore. It was going to be Tesla's alternating current polyphase system of power generation and distribution. As might be expected, by December, the Westinghouse Electric Corporation was prepared to submit its comprehensive proposal.[42]

Adams's leadership was proving to be just what the Cataract Construction Company needed, but he still had to seek counsel from Tesla himself regarding the latest developments in the field of electricity. As an example, he expressed the concern that European nations were switching from supporting polyphase AC to supporting single-phase AC. Tesla's answer was simple: "I have not the slightest doubt that all companies except Helios, who have acquired rights from my Company, will have to stop the manufacture of [poly] phase motors. . . . It is for this reason that our enemies are driven to the single phase system and rapid changes of opinion."[43]

Now technical specifics had to be negotiated, with Westinghouse and General Electric both jockeying for favored position and each wanting a piece of the huge project. It was finally decided that because of Westinghouse's greater knowledge and experience (its engineers improving AC products every day). with Tesla's alternating current polyphase system and having proven his expertise and wherewithal to complete large projects, which included the lighting and powering of the Chicago World's Fair, he was awarded the contract. After all, the Edison General Electric division of General Electric could never make those claims.[44] The aftermath suggests that General Electric's loss of the contract might also have been predicted by Adams selling his financial interest in the Edison Illuminating Company in late 1889. Moreover, the financiers behind General Electric were afraid to buck the judgment of the engineering advisors they counted on to give them the right answers, considering the massive investment they were facing should they win the contract.[45] Recognizing that Tesla's observable expertise was second to none, Adams often asked him to review and give him his thoughts about numerous articles on alternating current in technical journals. Tesla frequently dismissed, and rightly so, the plans proposed by others as lacking the necessary design and function, while his AC system resolved the issues.[46] Frankly, he was already working on his concept of wireless power transmission and found the entire matter an unnecessary distraction.

Admittedly, the two men's relationship was symbiotic. Tesla had a direct line to Adams, who would ultimately be the one to decide what system to use, while Adams needed to bounce his ideas off of Tesla and to ask technical questions when needed. In the end, Westinghouse had to share some of the general construction business with General Electric while keeping and controlling the most critical aspects of the project, beginning with the turbines, because Adams felt that he needed to keep a reasonable relationship with General Electric considering the size of the project.

Many Westinghouse executives were acutely aware that were it not for Nikola Tesla, they would not have been awarded the primary contract. One Westinghouse executive put it simply and accurately, "It must certainly be gratifying to you to think the largest water power project in the world is to be utilized by a system which your ingenuity originated. Your successes are gradually pushing to the front. . . . Let the good work go on."[47] The *New York Times* carried an article about the Niagara Falls project dated July 16, 1895, where it reported as follows: "Attracting little attention except in the scientific papers, there has been going on there for several years a work which stands completed to-day in its details, and which is the unrivaled engineering triumph of the nineteenth century." The article continued with: "Even now the world is more apt to think of him as a producer of weird experimental effects than as a practical and useful inventor. Not so the scientific public or the business men. By the latter classes Tesla is properly appreciated, honored, perhaps even envied. For he has given to the world a complete solution of the problem which has taxed the brains and occupied the time of the greatest electro-scientists of the last two decades—namely, the successful adaption of electrical power transmitted over long distances to the varied uses demanded in commercial work." The article concluded with: "To Tesla belongs the undisputed honor of being the man whose work made this Niagara enterprise possible, for without the possibility of long-distance transmission with practical motor service no way could have been found for utilizing profitably the almost boundless power of the great waterfall. Within a few years Tesla motors will be driving tools in every manufactory [factory] in the civilized world."[48]

Between 1893 and 1896, the Niagara Falls project was supervised by Adams and his confidante, lawyer William Rankine. The two gave noted architect Stanford White the responsibility of designing the building for the power-house and also relied upon him to design living accommodations for workers. Realizing that the new powerhouse would produce four times the amount of electrical power that any other such building did to date, once again, the logical choice was to employ Tesla's AC polyphase system. In fact, the two men were so convinced in "the daring promise of Nikola Tesla," that they even helped him set up a company to promote his wireless-power inventions in 1895.[49]

By November of 1896, the Niagara Falls powerhouse began to transmit electrical power to Buffalo, New York, and over the next ten years the power generated at the Niagara station was powering machines across the entire state of New York. Rankine, inspired by the success of the Niagara Project, formed a second company to construct a similar power plant on the Canadian side of the falls. As we now know, American and European utilities recognized the success of the Niagara Falls powerhouse and shifted immediately to the AC polyphase system invented by Tesla. Today that system is the gold standard of electrical power generation and distribution across most of the globe, and it marked a sea-change in human history from mechanical power to electrical power.[50] Together with the Cataract Construction Company Tesla made possible the world's first major hydroelectric power plant: Adams Station Power House #1. So began the electrification of the planet. And ponder this: Tesla had not yet reached the age of forty.

～

Even the curmudgeon himself, the man who started the War of the Currents, who even admitted to taking credit for the accomplishments of others, and who was essentially excommunicated from his own company, Thomas Alva Edison, had to tip his hat to Tesla.[51] In a New York City meeting of the National Electrical Exposition during the month of May 1895, Edison was quoted as saying: "To my mind it [AC] solves one of the most important questions associated with electrical development." And Alexander Graham Bell seconded the comment by adding: "This long distance transmission of electric power was the most important discovery of electric science that had been made for many years."[52]

～

As we now know, it was Tesla's desire to unburden the common people of their drudgery that inspired his work. Unfortunately, his constant struggle with his mental state was always an obstacle to that goal, and he knew it. Moreover, despite the Niagara Falls success, it too had its negative consequences. He did not enjoy the time away from his laboratory—his inner sanctum—and his experiments, the exploration of his many visions, and the comforting confines of his hotel room. Like the year he spent in Westinghouse's factory in Pittsburgh, deemed a "lost" year to him, the corporate-structured life and even being around the giant machines he encountered at the Niagara Falls power stations could be wearying. At a celebration of the Niagara Falls power station, with Tesla, Westinghouse, and other principals, an eager reporter asked Tesla if it was actually true what he had heard: that it was the first time he had ever been to the falls during its four years of construction, even though he was pleased with the many applications of his inventions. He answered the question thus: "Yes. I came purposely to see it. I am somewhat interested in the working of some machinery. But, it is a curious thing about me, I cannot stay about big machinery a great while. It affects me very much. The jar of the machinery curiously affects my spine and I cannot stand the strain."[53]

It does strikes one as strange for him not to have ever visited his creation at Niagara Falls during its construction, to see his childhood dream grow into reality in real time. But perhaps it can be explained in this way: If you know that the outcome of a project is a foregone conclusion, seeing it happen is not interesting but predictable and boring. Genius wants to be on to the next project. Moreover, he may have been having a hard time battling depression off and on since 1892 and his mother's death.

Today the Niagara Falls power stations on both sides of the falls generate some five million kilowatts. That is enough electrical power to light nearly four million homes, as 757,500 gallons of water per second drive the colossal Tesla turbines. Having accomplished his goals in this project, Tesla wanted to hide away again in his laboratory and continue his electrical research work, producing extremely high voltages and frequencies and utilizing the earth to transmit both information and power wirelessly. He still had dreams bigger and more costly than anyone else in the field, and money would become a problem.[54] But as his work and dreams continued, the fall of 1893 had something unforeseen waiting for him.

· *10* ·

The Protectors

Naiveté

NEW YORK CITY, NEW YORK, 1894

*B*y now Tesla had vanquished Edison, first by his revolutionary patents; second by, his lighting of the Chicago World's Fair; and third, by realizing his childhood dream—harnessing the endless power of Niagara Falls. It was the world's first large-scale water-to-wire power system. In the vernacular of today, he would be called the GOAT (Greatest of All Time). His reputation was now solidified for the ages. He was at his creative peak, in his glory, with everyone wanting his attention. Even money was no object—he dined regularly at Delmonico's where New York City's elite did the same. And he was at a point where he thought he had arrived at the enviable state of equipoise—a balance in his life that he thought was never possible. But did he?

As we recall, Thomas C. Martin, a respected writer for several technical journals and editor of *Electrical Engineer*, had become a true champion of Nikola Tesla and in many ways his personal manager. He had recognized the young inventor's talents immediately. It was after Tesla had been elevated to cult status that T. C. Martin decided it was time to introduce him to an even wider audience—kingmakers of culture and finance. Thankfully, in the 1880s T. C. Martin had developed a useful relationship with Robert Underwood Johnson, associate editor and eventually editor in chief of *The Century Magazine*. Johnson, well groomed from top to bottom—something Tesla always considered essential—had already met with publishing success, having convinced Ulysses S. Grant, with the help of Mark Twain, to write his bestselling memoirs.[1]

And now it was time for T. C. Martin to make his move.

$\sim$

T. C. Martin had already collaborated with Tesla on his book *The Inventions, Researches and Writings of Nikola Tesla*, an impressive assemblage of writings, nearly five-hundred pages, that explained AC motors, rotating magnetic field, polyphase motors, and other modern wonders. As one journalist wrote:

> Within the past three months a book has been published which tells the story and which is indeed a rather exceptional contribution to the electrical literature of the day, presenting, as it does, an elaborate description of all that Mr. Tesla has thus far given to the public in his patents and lectures. It includes not only descriptions of his numerous inventions, but his personal writings upon the lines of thought and experiment he has advanced, and so affords the opportunity for a brief review of what he has done.
>
> From both a literary and typographical standpoint the book is an excellent addition to the electrical library. It is tastily gotten up, and evinces care and literary skill in the editing; the illustrations, which are profuse, are good, and this is equally true of the frontispiece portrait of Mr. Tesla, which to many will largely enhance the value of the book.[2]

Certainly an unqualified success, another reviewer wrote glowingly that the book was: "a veritable bible for all engineers in the field."[3] It was in some ways a paean to Tesla, even though he was at this point still such a young man.

Now T. C. Martin decided to introduce Tesla to Robert Underwood Johnson and his wife Katharine, two of New York's most recognizable socialites. In 1876, Johnson took the hand of the comely Katharine McMahon in marriage, she a fiery personality of Irish descent. T. C. Martin entreated Johnson to let him write a feature article for his magazine on the electronic guru. Usually the popular magazine featured articles that served the Evangelical Christian community, but over time it changed its focus to appeal to a more educated audience, as it grew into the largest periodical in the country before it ceased publication in 1931. Johnson was familiar with inventors because he had visited Edison's laboratory in the 1880s as a reporter for *Scribner's Monthly Magazine*, the forerunner of *The Century Magazine*, and Martin was certain that Johnson was the right man to help him carry the torch for Tesla.

So the enthusiastic, handsome T. C. Martin marched off to Robert Underwood Johnson's editorial offices in Union Square, where the two exchanged cordialities among teetering stacks of mostly unread manuscripts atop Johnson's desk. On a breakfront behind him a mix of bric-a-brac, which included a photo of Johnson with President Harrison, consumed much of its top. Johnson had been selected to provide an original sonnet to celebrate the unveiling of the Washington Square monument arch, which was attended by President Harrison and other notables. The arch was designed by architect Stanford White, who would in time become a friend of Tesla's. Although

Johnson was no stranger to inventors, he had never met Tesla. After T. C. Martin's eloquent portrait of Tesla caught Johnson's attention, he invited him to dine at his elegant home at 327 Lexington Avenue. It was at the Johnson residence that the two would further discuss the specifics of T. C. Martin's article for the magazine. In the Christmas season of 1893, a date was set, and Johnson suggested that T. C. Martin bring the "wizard" with him to meet him at Johnson's home, thinking there might be an additional article to write.[4]

The cultured Johnsons had managed to craft a lifestyle that was the envy of high society.[5] The dinner arrangements were usually scripted, with place cards and six-piece silver place settings adorning the long dining room table, yet there was still a sort of informality to it all. And serious conversation was a central part of the gatherings. On any given evening, dinner might be with Theodore Roosevelt, Mark Twain, Rudyard Kipling, sculptor Augustus Saint-Gaudens, editor and author Mary Mapes Dodge, actress Eleonora Duse, or famed pianist Ignace Paderewski, among other impressive composers and musicians. Now Tesla was about to be invited into their ranks.[6]

Although the Johnsons were well known for their dinner parties, from time to time Katharine would exhibit a dark side to her personality and would secret herself away in her room, even refusing to take meals with her family or friends. Her daughter Agnes opined years later that perhaps her mother was simply bored with her husband. Compared to the Tesla she would come to know—a man of great intellect, worldview, and accomplishments—her husband seemed diminished. And as age so often brings things to a finer point, perhaps she started to see life differently. She even appeared to be jealous when Robert was occupying Tesla's time at her expense.[7]

The night Tesla and T. C. Martin appeared at the Johnsons' home, the two were greeted at the front door by Katharine and Robert with their two children, teenage daughter Agnes and young son Owen, standing obediently by their side. It was apparent that Tesla was not in good health. His fatigued face carried a feeling of total exhaustion and exasperation. Yet Tesla struck Katharine's fancy. Her eyes projected a sense of joie de vivre and playfulness compelling to all who met her. Even to Tesla, her physical beauty was obvious, despite his desire to always be alone with his work. He reluctantly accepted the hard fact that he needed to socialize with such important people as the Johnsons because funding for his more ambitious projects was becoming problematic. Also, it was a way to make his work known to the broader public. This evening was spent with Tesla recalling his successful European progress and fielding all manner of questions from the other invitees.

$\sim$

Tesla's growing fame had already attracted many of the city's glitterati, including J. P. Morgan, Edward H. Harriman, Thomas F. Ryan, John D. Rockefeller, Jay Gould, war-hero Richmond Person Hobson, and a gaggle of Astors and Vanderbilts, with the perplexing and powerful Henry Ford occasionally making an appearance. Tesla understood the need to promote his work, and since his bachelor lifestyle did not make room for entertaining at home, the laboratory was the obvious place where he would entertain and be in control. There were also celebrated actors, actresses, and musicians who would fill the magician's chamber, and Tesla would overwhelm them with all manner of electrified magic—but it was all real! Possibly Tesla's most famous and recurring guest was Mark Twain (Samuel Langhorne Clemens), whom, the reader will recall, Tesla credited with saving his life. "One day I was handed a few volumes of new literature unlike anything I had ever read before and so captivating as to make me utterly forget my hopeless state. They were the earlier works of Mark Twain and to them might have been due the miraculous recovery which followed. Twenty-five years later, when I met Mr. Clements [*sic*] and we formed a friendship between us, I told him of the experience and was amazed to see that great man of laughter burst into tears."[8]

One evening Twain, who had originally met Tesla at the fashionable Players' Club owned by actor Edwin Booth, came by the magician's chamber, as he did with regularity. But on this night Tesla described Twain's situation: "He came to the laboratory in the worst shape suffering from a variety of distressing and dangerous ailments."[9] While there, Twain scanned the bizarre-looking laboratory searching for something new, and he noticed Tesla was working with a device that produced intense vibrations—the study of mechanical vibrations had yet to be sufficiently explored, and Tesla knew he was on the scent of something new and useful. The actions of the oscillator drew Twain's attention. Suddenly a thought occurred to him, and he asked Tesla if he could invent a high-frequency electrotherapy device that he could sell to rich widows on his next trip to Europe.[10] Imagine that . . . Twain as a salesman for Tesla, but that was just what he was; after all, he was always an investor in the newest technologies—Twain was convinced that Tesla's AC induction motor was the greatest invention since the telephone.[11]

Knowing that Twain enjoyed a bit of comic relief now and again, Tesla told him that he had already invented such a device and that it would easily solve their digestive and other health problems, as it vibrated in sympathy with the peristaltic waves that moved foodstuffs through the alimentary channels. Twain immediately asked Tesla for a demonstration, so Twain stood on a platform as the oscillator was turned on. Initially, he was overcome with joy, stating, "This gives you vigor and vitality." Tesla knew what was about to happen if Twain stayed on the vibrating platform too long, so he insisted that Twain come down, but he refused; he was having a great time and told Tesla that his invention would be a godsend to all of humanity. After a few more minutes Twain's body tensed up, and he looked toward Tesla, who was smiling knowingly. Twain told him to turn it off, and Tesla purposely hesitated for a moment before shutting the machine down. Twain very gingerly stepped down from the platform, looking distressed, and Tesla pointed toward a small door in the corner. Several of Tesla's staff chuckled as Twain beat a path to the lavatory: the electrotherapeutic device had worked as advertised.[12] Despite the embarrassing event, Twain continued to use the vibrating platform more moderately, and within a few months he claimed its therapeutic benefits were worthwhile.[13] Even Tesla's staff availed themselves of the new therapy with Tesla noting: "Some of us, who had stayed longer on the (vibrating) platform, felt an unspeakable and pressing necessity which had to be promptly satisfied."[14] Tesla christened his electric generator an "electro-mechanical oscillator," but those who were familiar with its effects often called it "Tesla's earthquake machine."[15] (We will revisit Tesla's unique electro-mechanical oscillator and it potential to generate earthquakes later.)

Since it was the holiday season, and as a means of reciprocating the Johnsons' hospitality, Tesla invited them to the premiere of Antonín Dvořák's *New World Symphony*. He had secured the best available seats for a Saturday night performance—the fifteenth row. He even suggested that the seating would be better for Mrs. Johnson's active imagination. He also invited them to dine at Delmonico's after the performance. In return for the wonderful evening of entertainment, Katharine sent Tesla flowers on January 6 (January 7), the day Orthodox Christians celebrate their Christmas.[16] Tesla in turn wrote Robert, thanking Katharine for the spectacular flowers. He went on to admit that he had never received flowers and that he liked the way it made him feel. Tesla then sent Katharine a Crookes radiometer, or light mill, which measures electromagnetic intensity. Tesla simply thought it was the most beautiful looking of inventions.[17] These back-and-forth communiqués between Tesla and the Johnsons continued for many, many years.[18] More specifically, there exists myriad letters between the troika demonstrating the deep interests they shared with each other. Tesla was also fond of giving friends and acquaintances his calling card. One read: "Nikola Tesla, Consulting Engineer and Electrician, Waldorf-Astoria, New York."[19] Just like his impeccable dress, he believed that his personal presentation on all levels needed to reflect his unwavering belief in order, cleanliness, and rectitude.

As one could surmise, the Christmas season proved to be particularly socially active for Tesla, should he be convinced to attend. Formal gatherings were never to his liking, yet he felt his visits to the Johnsons' home were as close as he had ever gotten to experiencing family life in America. Such visits would often last many hours, even into the dead of night, while Tesla's hansom cab waited patiently outside, only to return him to his hotel just a few blocks away. When the Johnsons' children were present, Tesla would play the doting uncle and instruct his driver to give Agnes and Owen a ride through Central Park, while he carried on in conversation with their parents and any other luminaries who might be in attendance that day.[20]

Curiously, the Johnsons and railroad tycoon William "Willie" Vanderbilt became his surrogate family of sorts, and the only people to be on a first-name basis with him. This was most unusual, because Tesla generally made every effort to keep his emotional distance from others. The three men—Nick, Bob, and Willie—would while away the hours in deep conversation covering a series of serious topics.[21] Although Tesla enjoyed the Metropolitan Opera House, courtesy of Vanderbilt, and having a laugh or two attending theatrical comedies, he would in time stop going to either venue; instead, he opted for going to movies—where he could be alone. Once again, Tesla so valued his privacy that he eschewed most public appearances, unless they were for business.[22]

The Johnsons would arrange for dinners with potential investors, referring to them jokingly as "the millionaires." Efforts were even made to introduce Tesla to eligible woman, and even though he enjoyed their company, nothing ever came of such social liaisons.[23] Tesla always found a way to beg off when things became too personal, often responding with some version of: "I fear that if I depart very often from my simple habits [of daily laboratory work] I shall come to grief."[24] Tesla did take a liking to an English woman named Marguerite Merrington, a gifted pianist and author who excited his intellectual interests in the arts. But any efforts to engage Tesla with a woman in any serious way always came to ruin, for his interest was purely in science—his mistress. As one of his colleagues said, "I fear he will go on in the delusion that woman is generically a Delilah who would shear him of his locks."[25] Curiously, when Tesla encountered Swami Vivekananda—who lectured at the Columbian Exposition on the nexus between life force and the ether and modern science—he was convinced that Buddhist practices were the path to enlightenment and overall clarity and that chastity was required to achieve such a high level of personal illumination.[26] Moreover, he stated, when asked by a reporter, that marriage was a distraction to the work of an inventor or thinker or any other person of an artistic temperament. Yet he admitted that such a lifestyle was accompanied by great "loneliness."[27] He even slyly intimated to an interviewer that Edison's two marriages proved his point.[28]

Katharine Johnson's attraction and admiration for the man she saw as a super-hero continued to grow day-by-day. Yet, the Johnsons also felt compelled to protect him as if he were a child of theirs. In one instance, Robert recalled how ignorant people of all socioeconomic classes would treat Tesla at times as if they did not know who he was. He prefaced the example with: "The imaginative character of Tesla's work made him the prey of the sensational press, which, as in the case of Hobson, did everything it could to exploit him for its cruel and sordid purposes, with the result of making him ridiculous only to those who had neither knowledge nor the responsibility of sober judgment; but the general public remained ignorant of the principles of which he was a profound master and which technically were beyond their ken. I heard an English writer, a lady, say to him, 'And you, Mr. Tesla, what do you do?' Demonstrating his modesty and gentlemanly approach toward the lady, he answered politely, 'Oh, I dabble a little in electricity.' Clueless, the lady responded, 'Indeed! Keep at it, and don't be discouraged. You may end by doing something some day.'"[29] Robert concluded, "This to the man who had sold the inventions used at Niagara to the Westinghouse Company for a million dollars and lived to rue the bargain! Unsordid as he is, Tesla has used his fortune and the resources that he has won by his other patents in the furtherance of his scientific inventions and study and in the building of new laboratories to replace and extend the one that was destroyed by fire. If ever, in the interest of the public, a scientist deserved to be endowed, it is he."[30] We will remember that the central theme of Tesla's life was to "lift the burdens from the shoulders of mankind," and once again, it was recognized by Robert Johnson, who knew him better than most.

Robert, an accomplished poet as well as an editor, exchanged the joys of poetry with Tesla on a regular basis. On one occasion Tesla began reciting "Luka Filipov" by the famous Serbian poet Jovan Jovanović Zmaj. As he translated the poem in real time, he became consumed with the glory of it all. Zmaj's poem tells of the heroic exploits of Luka and his inevitable death in an incident of the Montenegrin War (1876–1878) against the offending Turks.

Luka Filipov

One more hero to be part
Of the Servians' glory!
Lute to lute and heart to heart
Tell the homely story:
Let the Moslem hide for shame,
Trembling like the falcon's game,

Thinking on the falcon's name—
Luka Filipov.

The battle rages on as Luka captures a pasha and marches him back to the prince. In the end, the prince is pleased, but the brave and gallant Luka Filipov loses his life.

We'd have fired, but Luka's hand
Rose in protestation,
While his pistol's mute command
needed no translation:
For the Turk retraced his track,
Knelt, and took upon his back
(As a peddler lifts his pack)
Luka Filipov!

How we cheered him as he passed
Through the line, a-swinging
Gun and pistol—bleeding fast—
Grim—but loudly singing:
"Lucky me to find a steed
Fit to give the Prince for speed!
Rein or saddle ne'er shall need
Luka Filipov!"

So he urged him to the tent
Where the Prince was resting—
Brought his captive, shamed and spent,
To make his true jesting.
And as couriers came to say
That our friends had won the day,
Who should up and faint away?
Luka Filipov.[31]

From that memorable moment onward Tesla always referred to Robert as "Dear Luka" and Katharine as "Mrs. Filipov."[32] Additionally, Robert was so taken with Tesla's florid recitation of the poem and his magnificent heritage that he included it in his anthology, *Songs of Liberty and Other Poems*. Robert also entreated Tesla to seek permission from Jovanovich, said to be the "Longfellow of that country," to have his poetry republished in *The Century Magazine*.[33] Permission was granted.

∽

As the troika's relationship matured, Katharine in particular was emphatic that she be connected to Tesla. The number of invitations, both formal and informal, were far more than his time and psyche could afford. One of her entreaties said: "Come and shed the radiance of your happy countenance upon us all, especially the Johnsons."[34] Tesla begged-off, saying that he cannot be distracted from his central purpose, and to that end, he must stay in his laboratory. He even turned a blind eye to invitations to go on holiday with the Johnsons even though he knew a few days away would do him good. After a while, his rejections of the endless invitations became a running joke. On one occasion Robert playfully scolded Tesla for not leaving his work to relax at his Hampton's retreat by saying that the scientist was "joined to your idols of copper and steel."[35] Tesla's riposte was, "I get all the nourishment I require from my laboratory. I know I am completely worn out, and yet I cannot stop my work. These experiments of mine are so important, so beautiful, so fascinating, that I can hardly tear myself away from them to eat, and when I try to sleep I think about them constantly. I expect I shall go on until I break down altogether."[36]

And still the invitations persisted, because, to Katharine, Tesla was of worldwide historical significance, and her tenacity would not let her stop; she wanted everyone to know of him. She was most likely a very driven, hence, selfish woman who knew she could excite men with ease. Her high social status and breathtaking physical beauty gave her license to do things that others simply could not, and she felt entitled. She had set her sights on Tesla, and she was determined to make the reclusive genius her own. Whenever she saw a possible opportunity, she would take it. As an example, in 1897, she made the simplest of requests to Tesla, "Come soon." Then again in 1898, she said in a missive to him, "Will you come to see me tomorrow evening and will you try to come a little early. . . . I want very much to see you and will be really disappointed if you do not think my request worthy [of] your consideration."[37] By, at the very least, November of 1898, Tesla's letters to her began with the intimate salutation: "My dear Kate."[38]

One wonders . . . did Katharine ever have her wishes granted by Tesla? In his personal letters that were kept archived for decades, we do have an idea of the closeness Katharine felt for Tesla. A billet-doux she sent to him from Italy illuminated her feelings for him:

What are you doing? I wish I could have news of you my ever dear and ever silent friend, be it good or bad. But if you will not send me a line, then send me a thought and it will be received by a finely attuned instrument. I don't know why I am so sad: I feel as if everything in life had slipped from me. Perhaps I am too much alone and only need companionship. I think I would be happier if I knew something about you. You, who are unconscious of everything but your work and who have no human needs. This is not what I want to say and so I am Faithfully yours, KJ.

In a postscript, she added: "P.S. Do you remember the gold dollar that passed between you and Robert? I am wearing it this summer as a talisman for all of us. Come, for a moment at least. I haven't seen you in centuries, although I'm always on your trajectory." While another letter demonstrated the obvious: "I stayed at home last Saturday and Sunday waiting for you to come. . . . I am tired of waiting for your reply. . . . How strange it is that we cannot do without you."[39]

Katharine was obviously in love with Tesla and proved over the years to be most sensitive to his depressive states. She was able to intuitively feel when he was progressing from mania to mania mixed with depression and exhaustion, and she knew he would suffer when depressed even if he did not suffer when manic. Tesla may have encouraged her companionship when he was depressed only to show no need for friends or love when reverting to a manic state.

It was not just Katharine and Robert who were under the electric spell of Tesla's charismatic power. Many, many others felt the same. Several chroniclers have suggested that the man who discovered him, the talented writer Thomas Commerford Martin, was also taken with Tesla, and by extension, he was taken with the Johnsons for protecting their mutual friend. All in all, they were most concerned about Tesla in a truly caring way. So much so that they believed that his loss would be humanity's loss. Although T. C. Martin understood the new article he was penning for *The Century Magazine* would certainly advance his career and bring Tesla's persona and accomplishments to the masses, he too was constantly concerned with the inventor's fragile health. He wrote to Katharine with a sense of alarm about Tesla: "I do not believe . . . that he will give up work at any very early date," T. C. Martin wrote. "Talking of California with him in a casual way elicited the fact that he had a couple of invitations to lecture there so that I don't want to jam his head into that lion's mouth. I believe he is going to take more care of himself and you may have done us all a great deal of service by your timely words. Yet in spite of that," he added, "If you can manage it, I believe it would be a good scheme to have that Doctor get hold of him. . . . My prescription is a weekly lecture from Mrs. RUJ [Katharine]."[40]

Of course, Tesla simply brushed off any talk of rest or slowing down or relationships with women. In fact, he went into overdrive, convinced that his life's purpose required it. He willingly sacrificed the joys of marriage and children for his work. Ever excited by what was next, he professed to Katharine that among the latest inventions he was developing was the wireless transmission of data and voice. As chronicled by Katharine's husband Robert, he told her, "The time will come when crossing the ocean by steamer you will be able to have a daily newspaper on board with the important news of the world, and when by means of a pocket instrument [cellphone] and a wire in the ground, you can communicate from any distance with friends at home through an instrument similarly attuned."[41] Johnson added that Tesla was convinced it was possible to direct the movements of an aëroplane or torpedo boat by wireless.[42] And now it was time, and Tesla knew wireless was to be his next frontier. Yes, he had set his own path in life long ago, and his own frontiers. Stopping was not an option for him. He would persist until there was no more breath.

With the success of T. C. Martin's knowledgeable prose portrait of Tesla receiving rave reviews, Robert was excited to have him write a follow-up article focusing on Tesla's laboratory work. As Robert described in his book *Remembered Yesterdays*, published in 1923, Tesla had produced a new form of light he called "cold light"—in which a glass bulb with gasses in it glows with a virtual white light when electricity is present. "When we first met him, his laboratory, in South Fifth Avenue was a place of absorbing interest. We were frequently invited to witness his experiments, which included the demonstration of the rotating magnetic field, and the production of electrical vibrations of an intensity not before achieved. Lightning-like flashes of electrical fire of the length of fifteen feet were an every-day occurrence, and his tubes of electric light were used to make photographs of many of his friends as a souvenir of their visits. He was the first person to make use of phosphorescent light for photographic purposes—not a small item of invention itself. I was one of a group consisting of Mark Twain, Joseph Jefferson, Marion Crawford, and others who had the unique experience of being thus photographed. At another time the company consisted of the Kneisel Quartet, Gericke, conductor of the Boston Symphony Orchestra, Madame Milka Ternina, the great prima donna, and ourselves. We took many of our friends to the laboratory, including John Muir, Captain Hobson and Maurice Boutet de Monvel the French painter. I was myself at that time the medium of exchange of an electric current of a million volts of the Tesla system of high frequency, whereas I believe twenty-five hundred volts of the ordinary current is sufficient to kill. Lamps were thus lit up brilliantly through my body."[43]

By now the connection was complete. Katharine was utterly absorbed with the essence of Tesla, and Tesla truly enjoyed her company and conversation. As we know, over the years Tesla's salutations, when writing Katharine, became more intimate—using her given name at times. He even wrote her in

French with the words "Cher Mademoiselle Johnson." And by the same to-ken, Robert's connection had deepened and strengthened. He even published a poem as a paean to his dear friend:

> "In Tesla's Laboratory"
> Here in the dark what ghostly figures press!—
> No phantom of the Past, or grim or sad;
> No wailing spirit of woe; no specter, clad
> In white and wandering cloud, whose dumb distress
> Is that its crime it never may confess;
> No shape from the strewn sea; nor they that add
> The link of Life and Death,—the tearless mad,
> That live nor die in dreary nothingness:
> But blessed spirits waiting to be born—
> Thoughts to unlick the fettering chains of Things;
> The Better Time; the Universal Good.
> Their smile is like the joyous break of morn;
> How fair, how near, how wistfully they brood!
> Listen! That murmur is of angels' wings.[44]

As we have learned, Tesla felt that he was not able to entertain the John-sons and other luminaries in his hotel suite at the Waldorf-Astoria, hence, he did so at his laboratory, where his work was fantastic entertainment for the uninitiated. Reluctantly, he often made it a point to speak to the Johnsons as well as reporters and potential financial backers, stating that millions in profit were imminent. It must have been very difficult for him to do so, because as a Serb, he was raised to always put "principle over profit," even to one's detriment. His courting the press when it came to financial matters was not his style, but the hard fact was that his work demanded great sums of money, and he had to do what was necessary to advance his scientific experiments, which were critical to humanity. He was so disturbed by his predicament that he wrote a newspaper reporter: "It is an embarrassment to me that my work has attracted as much attention, [in part] because I believe that an earnest man who loves science more than all should let his work speak for itself."[45]

As the end of 1894 rapidly approached, Tesla's image was still burning white-hot—he was featured in many of the nation's most widely read news-papers such as the *New York Times*, the *New York World*, the *New York Herald*, and the *Savannah Morning Times*. The many awards came fast and furious, including the Franklin Institute's Elliott Cresson Gold Medal for inventing a new and very significant form of light, which opened up a virgin field of artificial light. There were honorary degrees that came from England and France as well as Columbia and Yale.[46] Because his larger-than-life image was still in its ascendency and the demands on his time were without end, most

reporters never had a chance to interview him. So what did they do? They resorted to simply inventing stories, some quite creative, yet nonetheless bogus.
They even used his portrait in an attempt to give their fabulist stories an air
of authenticity. They pursued the elusive inventor as professional athletes and
rockstars are today. When he was onstage performing one of his lectures or
demonstrations, he was in deed in command of the audience. He was accused
of playing to the press at times, but he did so when he needed to . . . it was
part of the game. As the leading thinker in his field, it was difficult at times
for him to walk the tightrope between bragging and modesty.[47] But consider
a visit to the magician's chamber:

> Mr. Tesla spends his days on the fourth floor of a machine shop at No.
> 33 South Fifth avenue. His name does not appear anywhere on the build-
> ing [unlike Edison], and there is nothing about the place to indicate that
> it is one of the world's centers of electrical interest. The whole floor is
> occupied by Mr. Tesla's laboratory, except that one corner is partitioned
> off into the plainest of little offices containing principally a modest desk
> for his bookkeeper; a book case largely devoted to the "Official Gazette
> of the patent Office," and a small black board which hangs on the wall
> and bears the evidence of hard usage. The black is worn from this board
> in several spots, and the rest of it is covered with figures and cabalistic
> signs. No doubt the science of electricity would have been notably poorer
> but for some of the problems worked out on that shabby black board, for
> when the inventor is puzzled he goes to it and works on it nervously with
> a stubby piece of chalk.
>
> The laboratory itself looks commonplace to the uninitiated. It is filled
> with machinery and electrical appliances, and a stranger prowling about the
> building at will would surely mistake its fourth floor for part of the machine
> shops below. One who is not an electrician would find in the Tesla work-
> shop none of the marvels that make Edison's laboratory better than a circus
> for the sight-seer. An electrician, however, would find secrets there with
> which he could make and break colossal fortunes on the stock market, for
> the reason that will appear further on.
>
> One rarely meets a man more free from affectations and self-conscious-
> ness than Nikola Tesla. He does not like to talk of himself, and when that
> subject comes up he is sure to steer away from it as quickly as possible.[48]

But Tesla was looking toward his continued work with wireless communication—he remained laser-focused in pursuit of the answers. At the same
time, he was sworn to secrecy by Robert Johnson and T. C. Martin, because
the two were fast at work orchestrating the release of the unique photographs
that would be included in the second T. C. Martin article for *The Century
Magazine*. They wanted maximum impact, since these photographs would be
the first ever taken under Tesla's newest creation—cold light . . . by phos-

phorescence. The two men decided that "secrecy" must be the word of the day, and Tesla needed to adhere to it. Photographs taken with a new light source would undoubtedly leak to the press, thus diminishing their impact. The subjects of the new photographs were Mark Twain, actor Joseph Jefferson, novelist Francis Marion Crawford, and Tesla himself (in what can only be described as the first "selfie" with Mark Twain). The date of release in the magazine was to be April 1895.[49]

Tesla's meeting the Johnsons was most fortuitous and proved to be a turning point in his still young life. The Johnsons would in time show themselves to be Tesla's greatest protectors from opportunists the world over. Nevertheless, though the Johnsons desired to be lifelong and continuous friends, and maybe Katharine even wanted to have a love affair, Tesla only needed them when he was depressed. Their odd reward for rallying to him when he needed them was to neglect them when he was manic and no longer needed them. It was his way of preventing intimacy. Having friends of convenience is not a true form of friendship, something maybe Tesla was not capable of, especially during a manic state when he was entirely self-centered.

The year 1895 was to be a very pivotal year for Tesla. It would bring great glory and even greater grief.

The Great Conflagration

Broken Spirit

NEW YORK CITY, NEW YORK, 1895

*T*he newspaper headline screamed "Fruits of Genius Were Swept Away—Inventions in the Ruins—Years of Labor Lost." So said the *New York Herald*, March 14, 1895. The full report of the tragedy began with the words:

> Utterly disheartened and broken in spirit, Nicola Tesla, one of the world's greatest electricians, returned to his rooms in the Gerlach yesterday morning and took to his bed. He has not risen since. He lies there half sleeping, half waking. He is completely prostrated.
>
> In a single night the fruits of ten years of toil and research were swept away. The web of a thousand wires which at his bidding thrilled with life had been twisted by fire into a tangled skein. Machines to the perfection of which he gave all that was best of a master mind are now shapeless things, and vessels which contained the results of patient experiment are heaps of pot shards.
>
> By the fire which swept through the building at Nos. 33 and 35 South Fifth avenue in the early hours of yesterday [2:30] morning one of the most valuable laboratories in the world was destroyed. It was that of Nicola Tesla, the Servian electrician. In those upper lofts were apparatus and appliances which meant a revolution in electrical science.[1]

The report finished with, "The experiments which he has been carrying on there in the last few months were conducted in secrecy. He would say but little about them beyond the fact that he hoped soon to solve a vexed electrical problem."[2] What were the experiments? Will we ever know? He kept no official records. His secrecy demanded it. No notes, of what few there may have been, survived. No sketches, other than a possible pencil drawing here and there, survived. Not even the heavily used blackboard survived. Only he knew the totality of his work and what was lost to history.

After having been incommunicado for several days, as friends desperately searched for him across the city, it was his "protector," Katharine Johnson, who was able to contact and inform him of her concern for him: "Today with the deepening realization of this disaster and consequently with increasing anxiety for you, my dear friend, I am even poorer in tears, and they cannot be sent in letters. Why will you not come to us now—perhaps we might help you, we have so much to give in sympathy."[3]

In the end, the fire destroyed his laboratory, leaving only "four cracked and blackened walls."[4] It was bad enough that New York City was coming out of a harsh winter, that the world economies were fighting to survive the jaws of the Panic of 1893, and the fact that America was on the precipice of bankruptcy.[5] But the financial cost to Tesla was a massive setback, while the psychological cost was quite possibly insurmountable. He admitted to a friend that he was spent; that he had no more to give.[6] But that may have been his way of protecting himself against himself, as he would in the end rally again and would be "at work again with clenched determination while the ashes of his hopes lay hot."[7]

But what caused the great conflagration? The *New York Times* reported the day after the tragic event that "Gillis & Geoghegan, manufacturers of steamfitters' supplies, occupied the lower floors of the building in which Mr. Tesla had his laboratory. Night Watchman [John] Mahoney was employed to look after the building. He said that he was making his rounds as usual just before 3 o'clock yesterday morning when he discovered the fire. It had started on the ground floor, but from precisely what cause is not known. He tried in vain to put it out with pails of water."[8] By the time Chief Reilly arrived with Engine 83, the fire had taken root in the first two floors and was moving fast. One must also consider that the oil-soaked machinery in the laboratory and throughout the other floors proved to be a great hindrance to any attempt to put out the fire, which in the end consumed all six stories of the building.[9]

But there was more to the fire than the total loss of the inventor's pursuit of answers to God's secrets; it was a cataclysmic event. Tesla's friend T. C. Martin wrote the following:

> Of Tycho Brahe's famous observatory at Uraniburg remain to-day but a mound of earth and a couple of holes. The same inventory would aptly apply to all that is left of the upper floor where Nikola Tesla has of late years been carrying on his suggestive and beautiful demonstrations in electricity. Two tottering brick walls and the yawning jaws of a sombre cavity aswim with black water and oil were all that could be seen on the morning of March 13 of a laboratory which to all who had visited it was one of the

most interesting spots on earth. Beset by the squalor and clamor of plebeian South Fifth Avenue, the building was unpretentious, and its lower stories were given up to the prosaic details of an iron-pipe cutting business. Commonplace also was the origin of the disaster. A flaring gas jet, oil-soaked floors, and a "watchman" have been heard of before as an effective combination for midnight fires.

The Tesla laboratory was, in a sense, a private museum. The owner kept in it many souvenirs of bygone toil and experiment, and an important part of his records. During the past ten years Mr. Tesla has done an enormous amount of original experimentation, and it was all represented in the contents of the laboratory.[10]

T. C. Martin's reporting on the disaster only served to confirm that the loss of Tesla's laboratory had international implications the likes of which are difficult to understand fully even today. It forestalled Tesla's efforts at continuing with his wireless communications research and gave Guglielmo Marconi a running head start to some degree and handed others the hope that Tesla would never be himself again. The managing editor of the *New York Times* wrote: "The destruction of Nikola Tesla's workshop, with its wonderful contents, is something more than a private calamity. It is a misfortune to the whole world. It is not in any degree an exaggeration to say that the men living at this time who are more important to the human race than this young gentlemen can be counted on the fingers of one hand; perhaps on the thumb of one hand."[11]

~

As we now know, despite Tesla's setbacks in the past, he had always managed to overcome them, but how long would that last, given his state of depression as of the morning of March 13, 1895, which caused both physical and mental incapacitation that he had never experienced before. However, not wanting to stay stagnant and in a mental prison for fear of never getting out, he decided to employ his own electrotherapy research methods to heal himself.

It all began with Tesla's extensive work with high-frequency currents, wherein he observed that the heating effect of these currents could have health benefits for the treatment of arthritis and other ailments. In 1893, Tesla referred to such treatment as medical diathermy—cold fire—that could excite the brain.[12] So he spent the next several months giving himself "mechanical therapy" treatments of his own invention and prescription. He explained why he was convinced such treatments would heal him: "While investigating high frequency currents, I observed that they produced certain physiological effects offering new and great possibilities in medical treatment. . . . The currents furnished by them [apparatuses] have proved an ideal tonic for the human nerve system. They promote heart action and digestion, induce healthful sleep, rid the skin of destructive exudations, and cure colds and fever by the warmth they create. They vivify atrophied or paralyzed parts of the body."[13] He also believed that "electricity" was the best of all doctors, and he was convinced that the high frequencies produced anti-germicidal reactions.[14] He said he was responsible for the practice's development and eventual popularity, when he stated, "Leaders in the profession have assured me that I have done more for humanity by this medical treatment than all my other discoveries and inventions."[15]

But just when he began to believe there was a future again, more bad news blindsided him. He learned that the man he thought was his friend and business partner had deserted him in his time of need. It was only the month prior that the engineer and investor Edward Dean Adams, had entered into a business partnership with him, as well as Alfred S. Brown, another prominent engineer, together with William Rankine, yet another Niagara Falls promoter, and others to form the Nikola Tesla Company for the purpose of producing and selling all manner of electrical machinery and apparatuses invented by Tesla.

Adams had met Tesla during the Niagara Falls project, where he had cobbled together the necessary banks to finance it. Everything seemed perfect to Tesla. Adams felt himself more than capable of making things happen, but then he found himself far too busy to hold Tesla's hand and guide him through the financial minefields of the time. Tesla needed someone to temper his flights of fancy, as he slowly became unmoored from reality. Adams also understood that because of Tesla's unstable psyche and his constant desire to dream, many of his ideas would be difficult to commercialize. Moreover, because of the Panic of 1893, years later the country was still experiencing a recession, and as we know, many of the largest utility companies were struggling to stay solvent; hence, demand for new products was in decline.[16] One observer suggested that Tesla's tendency was to look wide, thus having difficulty zeroing-in on a problem at times.

Because Tesla's massive loss was uninsured, the overall cost was yet unknown. Nonetheless, within days he was out on the streets of New York City searching for a new place to set up shop and begin anew. Ironically, none other than Tesla's greatest nemesis, Thomas Edison, made an offer—seemingly recognizing his defeat in the War of the Currents—to let him use his New Jersey laboratory in the interim, while he sought out the necessary equipment to fill a new laboratory. Tesla accepted the offer, but what is more ironic is that his former partner, George Westinghouse, the man whom he made world famous and wealthy, began sending him bills for equipment lost in the fire and for new machinery he had ordered. During the next few months Tesla's efforts to cobble together a new workspace continued unabated. But then April came, and things changed.[17]

The *New York World*'s newest young-Turk reporter Arthur Brisbane had announced earlier that Tesla was "greater than Edison," thus, reigniting the War of the Currents—at least in the press, because it sold papers.[18] Brisbane had very quickly carved out a reputation for interviewing people of note, from prime ministers and potentates to actors and athletes. People were constantly telling him that it was a must that he interview the young inventor making noise all over the world: Nikola Tesla.[19] But where to find him? His daily haunt was the famous eatery Delmonico's on Madison Square in Manhattan. Brisbane said he first set eyes upon the enigmatic character while he was reading a newspaper at a window table in the hot summer of 1894. Tesla, he wrote, "is serious, he is earnest, and in all ways he commands respect."[20] Brisbane then observed, "Nikola Tesla is almost the tallest, almost the thinnest and certainly the most serious man who goes to Delmonico's daily."[21] It was not lost on the young Brisbane that this was the most significant interview of his lifetime: "His face cannot be studied and judged like the faces of other men, for he is not a worker in practical fields. He lives his life up in the top of his head, where ideas are born, and up there he has plenty of room. His hair is jet black and curly. He stoops—most men do when they have no peacock blood in them. He lives inside himself. He takes a profound interest in his own work. He has that supply of self-love and self-confidence which usually goes with success. And he differs from most men who are written about and talked about in the fact that he has something to tell."[22] It should be noted that no one, including Tesla, knew that he had reached the zenith (See: Timeline of Psychiatric Events) of his prowess after his meteoric rise unparalleled in scientific history.[23] Another reporter suggested, "Now that the rapid march of events in electrical development has brought the art (of alternating current) abreast of the times, Tesla finds himself lifted into prominence, not so much because of what he is to-day, but because of the hard and well-nigh unparalleled work that he has done during the last decade."[24]

Brisbane's cutting-edge interview ended with the following: "One wise man whom I knew used to say that the scientists with their jumble of laws governing the universe were ignoramuses, and that there weren't more than half a dozen fundamental laws all told. I asked Mr. Tesla what he thought about that. Tesla's response was most prophetic: 'I think that they could all be reduced to one.'"[25] Albert Einstein, Richard, Feynman, Stephen Hawking, and others, all chased a "Theory of Everything" (TOE)![26] The hunt continues today, but Tesla predicted its existence more than 125 years earlier.

As Tesla was in the process of building anew his laboratory, he thought it wise to shake hands again with Westinghouse, after all, the financial barbarians were at the gates, seeking blood from Westinghouse and anyone else in his orbit who owed money to them—the bankers. Tesla needed friends, money, and space to continue his work. Because of the paucity of cash in 1893, companies such as Westinghouse and General Electric found business accounts difficult to collect, and the best accounts not much easier.[27] As a consequence, Westinghouse in particular had found itself in financial trouble and its future uncertain. Finally, however, enough equipment had arrived at a new laboratory for Tesla to begin work at 46 and 48 East Houston Street—situated just east of Greenwich Village near Chinatown and it occupied two floors. He was elated to be back in New York City and in his own permanent laboratory. Although his mind was set on furthering his wireless transmission research, he nonetheless searched in all directions, entertaining any new prospect that made sense to him.[28] Money now was a central theme, which caused him to be ever vigilant for potential investors. This was anathema to Tesla and what he believed, but it was required if he was to realize his dreams for mankind's future.

With his wireless lighting system and oscillator not attracting the attention he had anticipated, he did what he always did, he rejected received wisdom and went his own way. What did that mean? He looked not to Hertz, Lodge, and Marconi, who believed that transmitting electromagnetic waves through the air was the way forward. Instead, he would expand his research with high-frequency AC and look toward powering the earth. He looked to transmitting power, not through electromagnetic waves piercing the atmosphere or ether, but through the earth by connecting his transmitter to an antenna and to the ground (currents).[29] His first experiment with ground currents resulted in a flood of electrical streamers. He knew he was heading in the right direction. Going his own way once again, he never looked back at the work of others. Now he felt supremely confident that he could transmit power and messages around the world.[30]

Let us take a moment and consider what one of the contemporary authorities, T. C. Martin, thought of this cutting-edge research:

> Part of Mr. Tesla's more recent work has been in the direction here indicated; for in his oscillator he has not simply a new practical device, but a new implement of scientific research. With the oscillator, if he has not as yet actually determined the earth's electrical charge or "capacity," he has obtained striking effects which conclusively demonstrate that he has succeeded in disturbing it. He connects to the earth, by one of its ends, a coil
> . . . in which rapidly vibrating currents are produced, the other end being

free in space. With this coil he does actually what one would be doing with a pump forcing air into an elastic football. At each alternate stroke the ball would expand and contract. But it is evident that such a ball, if filled with air, would, when suddenly expanded or contracted, vibrate at its own rate. Now if the strokes of the pump be so timed that they are in harmony with the individual vibrations of the ball, an intense vibration or surging will be obtained. The purple streamers of electricity thus elicited from the earth and pouring out to the ambient air are marvelous. Such a display is seen . . . where the crown of the coil, tapering upward in a Peak of Teneriffe, flames with the outburst of a solar photosphere.

After searching with patient toil for two or three years after a result calculated in advance, he is compensated by being able to witness a most magnificent display of fiery streams and lightning discharges breaking out from the tip of the wire with the roar of a gas-well. Aside from their deep scientific import and their wondrous fascination as a spectacle, such effects point to many new realizations making for the higher welfare of the human race. The transmission of power and intelligence is but one thing; the modification of climatic conditions may be another. Perchance we shall "call up" Mars in this way some day, the electrical charge of both planets being utilized in signals. Here are great results, lofty aims, and noble ideas; and yet they are but a beggarly few of all those with which Mr. Tesla, by his simple, modest work, has associated his name during recent years, the planets being utilized in signals.[31]

Mars! Imagine that!

~

Yes, there was a great conflagration that destroyed Tesla's laboratory. But there was an even greater psychological upheaval that may have destroyed Tesla's psyche. So much so, that he might never recover.

• *12* •

From Earth to Mars

Bewildered, Betrayed, yet Triumphant

NEW YORK CITY, NEW YORK, 1895

$\mathcal{O}$n February 18, 2021, the Mars rover *Perseverance* landed on the planet Mars in search of organic chemical signatures of microbial life. Beginning with its blastoff to its eventual landing some eight months later, it was in constant communication with NASA by "wireless communication"—Tesla's wireless communication to be specific.

~

In April of 1896, a little more than a year after Tesla suffered his momentous loss at his Fifth Avenue laboratory, he was still in a state of confusion, scrambling in all directions to right himself and set to work in his new laboratory located on East Houston Street. Although he cycled between mixed states, he nonetheless was making progress with new experiments, finding futuristic horizons, yet seeking solitude, for he felt most comfortable alone under such circumstances. His upbringing as a proud Serb would not let him show weakness or doubt to others. He would internalize his challenging thoughts, but he had to continue, and he would. Wireless transmission of electrical waves was his central focus, a subject that was on the minds of many people, as science-fiction authors and more than a few scientists pondered the possibilities of communication with other planets. Even Lord Kelvin was mesmerized by the prospect of such a phantasmagorical idea. Upon his arrival in New York City, he sought out Tesla to discuss the matter. The two discussed Tesla's work with wireless and Lord Kelvin wished to send a signal from New York City to Mars in an effort to let the "Martians" know that earthlings were real. Edison as well was caught up in the excitement, but his idea had to do with contacting the dearly departed.[1] But Tesla was very serious and intent upon reaching out into

the universe. After all, his development of wireless transmission was progressing quite nicely, and it surely gave him a reason to be energized by such an
endeavor. Even several New York daily newspapers sought his words on the
subject and just about anything else. Yes, the "lion hunters" were out. They
wanted a story. So Tesla had given them what they wanted, again. Interviews
carried headlines such as, "Is Tesla to Signal the Stars?" and "We May Signal
Mars." What follows is an excerpt from an interview with Tesla on this provocative subject of signaling Mars:

> I have had this scheme under consideration for five or six years, and I am
> becoming more convinced every day that it is based upon scientific prin-
> ciples, and is thoroughly practicable. We know that electric disturbances
> on the sun are productive of similar disturbances on the earth in the form
> of thunder showers. Now, why is it not equally conceivable that a distur-
> bance on the earth's surface should produce some tangible effect on other
> planets? The transmission of disturbances on the sun shows beyond doubt
> that waves of electricity are propagated through all space.
>
> But if there are intelligent inhabitants of Mars or any other planet, it
> seems to me that we can do something to attract their attention. This, of
> course, is the extreme application of this principle of the propagation of
> electric waves. The same principle may be employed with good effect for
> the transmission of news to all parts of the earth. It was formerly thought
> that in order to get any power out of electricity a conducting circuit must
> exist, but it has been shown that the same results may be accomplished
> with a single wire. It is perfectly possible to operate a motor or an electric
> light with just a single line of wire leading to the apparatus. It naturally
> occurred to me that, as the earth is a conductor, an electrical disturbance
> at one point, causing a change in the equilibrium of the earth's electricity,
> should be felt at all points on the earth's surface, and might be recorded by
> properly constructed instruments. The possibilities of such a transmission
> of intelligence cannot be exaggerated. Every city on the globe could be on
> an immense ticker circuit, and a message sent from New York would be in
> England, Africa and Australia in an instant. What a grand thing it would be
> in times of war, epidemic, or panic in the money market![2]

The newspaper also reported that "Mr. Tesla is at present engaged in finishing
his calculations and perfecting the apparatus necessary for experiment in his
new field of investigation, and he hopes soon to be able to show the practicability of his plan."[3]

As we can appreciate today, Mars fever was all the rage. Two years earlier
Tesla had offered his vision, convinced of the possibilities, when speaking to
his assistant Mr. Moore, that once he had sent a signal from any point on earth
to another point, the next logical step would be to do the same from earth to
other planets.[4] Even the renowned astronomer Percival Lowell was interested

in the "canals" on Mars. He subsequently made a very comprehensive study of these Martian canals that was published in *Nature*. Even American business magnate, millionaire John Jacob Astor IV, who authored a sci-fi thriller titled *A Journey to Other Worlds*, caught the Mars fever. He even gave a copy of his book to his friend Tesla.[5]

∼

With Mars fever still red-hot with excitement in the late 1890s, some of Tesla's investors were concerned about his latest theories and oscillation research taking away from the work that made them money, namely, his arc-lighting system.[6] On one occasion his oscillation research yielded a rather peculiar and unsettling result . . . an "artificial earthquake." Seeking to further understand resonance, he built a smaller, pocket version of his electromechanical oscillator. Some dubbed it "Tesla's earthquake machine."[7] Tesla recalled years later the particular event that took place in his Houston Street laboratory:

> I was experimenting with vibrations. I had one of my machines going [attached to a pillar] and I wanted to see if I could get it in tune with the vibration of the building. I put it up notch after notch. There was a peculiar cracking sound.
>
> I asked my assistants where did the sound come from. They did not know. I put the machine up a few more notches. There was a louder cracking sound. I knew I was approaching the vibration of the steel building. I pushed the machine a little higher.
>
> Suddenly all the heavy machinery in the place was flying around. I grabbed a hammer and broke the machine. The building would have been down about our ears in another few minutes. Outside in the street there was pandemonium. The police and ambulances arrived. I told my assistants to say nothing. We told the police it must have been an earthquake. That's all they ever knew about it.[8]

The bizarre event caused chaos in the streets as buildings began to shake, windows exploded, and people rushed onto the open streets in neighboring Chinese and Italian communities seeking safety.[9]

Some clever reporter asked Dr. Tesla what it would take to destroy the Empire State Building and the doctor replied: "Five pounds of air pressure. If I attached the proper oscillating machine on a girder that is all the force I would need, five pounds. Vibration will do anything. It would only be necessary to step up the vibrations of the machine to fit the natural vibration of the building and the building would come crashing down. That's why soldiers always break step crossing a bridge." His early experiments in vibration, he explained, led to his invention of his "Earth vibrating machine."[10]

∼

Once again, investors and many other opportunists did not possess the understanding to appreciate the way Tesla worked: from thought experiments to tangible experiments. Even his short excursion into the possibility of communicating with Mars, far-fetched to most, was all part of a larger picture that appeared quite clearly to him. And today, it is quite clear to most of us. He was on the right path and never wavered from his rejection of conventional wisdom because he knew it was often wrong. Ever since Tesla's work on reaching out into space, present-day scientists have picked up the gauntlet that he threw down more than a century ago and have continued his dream.

~

With the race for wireless communication in full swing, Tesla remained focused—until Wilhelm Roentgen made public his supposed discovery of the X-ray on December 28, 1895. Michael Pupin, physicist and fellow Serb, who grew to be a very minor nemesis of Tesla's, described the situation when he said every physicist who was curious "dropped his own research problems and rushed headlong," danger be damned.[11]

Let us for a moment look at the fact that in time Michael Pupin found Tesla's fame more than he could handle, despite his own doctorate in physics from the University of Berlin while Tesla had none. He was initially an admirer of Tesla's work, but soon he fell in league with Elihu Thomson, who, the reader will recall, made his own claim to being the inventor of alternating current, as did other pretenders. Sadly, in the end, Pupin made the wrong friends and fell afoul of Tesla, who ended up triumphing—Thomson finally admitted that Tesla held the clear advantage.

Pupin spent much of the rest of his life being an antagonist, and never achieving Tesla's level of success.[12] He recognized that Tesla would be a part of such an astonishing event in some way, so he took the offensive and regarded any connection between Tesla and X-rays as nonexistent.[13] Pupin had himself laid claim to having taken the first X-ray in America on January 2, 1896, wanting to partake in the glory. However, his claim, and the claims of others, proved specious, inasmuch as Tesla had preceded all of them.[14] To add a bit of absurdity to Pupin's unfounded claim, an article appeared in *The Electrical Review* at the time that stated in part: "That there should be a struggle for the honour of inventing so wonderful a Röntgen tube was to be expected, and a very pretty quarrel on this question between Dr. Pupin and Dr. Morton may be read in the pages of our contemporary *Electricity*, N.Y. Dr. Morton accuses Dr. Pupin of having seen the tube in his (Dr. Morton's) laboratory, and afterwards describing it as his own."[15]

Admittedly, Roentgen's announcement was earth-shattering to most. But interestingly, just prior to the great revelation, Tesla had been hard at work in his new laboratory on East Houston Street. There he always kept his antennae up and scanning, in search of the "new," when he noticed the effects of electrical discharge tubes on film. But he gave this new form of energy little thought. In hindsight, he knew he had missed it. "Then, too late, I realized that my guiding spirit had again prompted me and that I failed to comprehend his mysterious signs."[16] He added: "But while I have failed to see what others in my place might have perceived, it was always since my conviction, which is now firmer than ever, that I have not been forsaken by my kind spirit who then communed with me, but that, on the contrary, he has further guided me and guided me right in comprehension of the nature of these manifestations."[17] Here again we see Tesla's adherence and trust in his intuition. He forever made it a point to be open to new ideas, knowing that his internal compass knew the ultimate direction of his efforts.

So this begs the question: Was Tesla actually the first to discover and utilize the X-ray? There are those who suggest that his X-ray image of Mark Twain's hand was indeed the first such image by X-ray. In a magazine article that appeared in 1946, it was suggested that "Neither Tesla nor Hewitt [an American researcher] realized until a few weeks later, when Roentgen announced the discovery of X-rays, that the picture of Twain was in fact an example of X-ray photography, the first ever made in the U.S."[18] Admittedly, this does not settle the issue of who was first, but it certainly does demonstrate how advanced Tesla's research was over others. That said, Tesla never sought the credit others deserved, and he never claimed he predated Roentgen, but he did immediately send "shadowgraph" pictures to Roentgen upon hearing of the announcement. Roentgen's reply was somewhat curious: "The pictures are very interesting. If you would only be so kind as to disclose the manner in which you obtained them."[19] To the amazement of Roentgen, Tesla was producing "strong shadows at distances of 40 feet."[20] The results of one of his experiments with "Shadowgraphs" he described as follows: "An outline of the skull is easily obtained with an exposure of 20 to 40 minutes. In one instance an exposure of 40 minutes gave clearly not only the outline, but the cavity of the eye . . . the lower jaw and connections to the upper one, the vertebral column and connections to the skull, the flesh and even the hair."[21] He even tossed caution to the wind and exposed himself and an assistant to the radiation. He took notice of the bizarre effects of the mysterious force: "a tendency to sleep and the time seems to pass away quickly. There is a general soothing effect and I have felt a sensation of warmth in the upper part of the head. An assistant independently confirmed the tendency to sleep and a quick lapse of time."[22] At first blush, Tesla did not believe that exposure to radiation was deleterious, but the results of prolonged exposure produced noticeable headaches, severe blistering of the skin and eventual inflammation of the outer layers of the skin.[23]

Once Tesla became aware of the unwanted effects of exposure to radiation, he began musing about other possible uses of such a dynamic force. He postulated quite astutely that these powerful energy rays contained both wave-like and particle-like properties. He then concluded that X-rays were by extension "a wonderful gun, indeed, projecting missiles of a thousandfold greater penetrative power than that of a cannon ball, and carrying them probably to distances of many miles, with velocities not producible in any other way we know of."[24]

As a sidelight and not surprisingly, Edison had attempted to make use of the new discovery of X-rays, believing they could produce "sight" in the "blind." Tesla immediately scoffed at such nonsense and said so. "Is it not cruel to raise such hopes when there is so little ground for it? What possible good can result?"[25] Reporters had hoped the edgy retort from Tesla would reignite the war between the two titans, but it was not to be. After all, that war had already been decided: Tesla won!

~

BUFFALO, NEW YORK, 1897

With the rage of X-ray research gathering momentum and the amount of investors and well-funded companies joining in, Tesla decided that the room was far too crowded for him and decided to continue his pursuit of wireless transmission . . . that was where his focus should be. After all, his research had already yielded results. In fact, it was in the later months of 1896 and into early 1897 that the proud George Westinghouse had turned on the power at the Niagara Falls generating station—Niagara Power Station No.1—for which Tesla rightly received high praise at Buffalo's Ellicott Club.[26] It was at the very same club where Tesla, the man of the hour, was celebrated and gave a most curious and confusing speech called the "Power Banquet" speech.

He stood rod-straight, attired in an impeccably fitted black tuxedo, as he spoke before some four hundred of the nation's most prominent individuals from engineers and inventors to manufacturers and investors. Its content was at times confusing, at other times prophetic, and at other times it was apologetic. Think of it; just two months prior to the January 1897 speech, Tesla was successful in using his alternating-current polyphase system to transmit power from Niagara Falls—a momentous feat of ingenuity. The *New York Tribune* hailed it as "one of the triumphs of the century."[27]

Yet, Tesla's speech was an unexpected cataract of emotions pouring forth. As an example:

I have scarcely had courage enough to address an audience on a few un-avoidable occasions, and the experience of this evening, even as discon-nected from the cause of our meeting, is quite novel to me. Although in those few instances, of which I have retained agreeable memory, my words have met with a generous reception, I never deceived myself, and knew quite well that my success was not due to any excellency in the rhetorical or demonstrative art. Nevertheless, my sense of duty to respond to the request with which I was honored a few days ago was strong enough to overcome my very grave apprehensions in regard to my ability of doing justice to the topic assigned to me. It is true, at times—even now, as I speak—my mind feels full of the subject, but I know that, as soon as I shall attempt expres-sion, the fugitive conceptions will vanish, and I shall experience certain well known sensations of abandonment, chill and silence. I can see already your disappointed countenances and can read in them the painful regret of the mistake in your choice.[28]

What caused this inexplicable sense of inferiority from a man who, in his own time, was changing the world for the better, giving humanity the comforts it could not do without, then or now. No man before had done so, so dramati-cally. No da Vinci, no Edison, no Einstein. So why? Was this moment his op-portunity to tell the world that "I have made it; I have atoned for my brother's death; I have repaid my family; and finally, I am more than an inventor, I am a creator"? Could Tesla have fallen into the dark hole of bipolar disorder on that night? It had happened to him before.

Recall that Tesla's main reason for his great inventions and discoveries was to "lift the burdens from the shoulders of mankind," even at the cost of his sanity. To that end, he had succeeded. The closing remarks in his "Power Banquet" speech resonate with this purpose:

Gentlemen, some of the ideas I have expressed may appear to many of you hardly realizable; nevertheless, they are the result of long-continued thought and work. You would judge them more justly if you would have devoted your life to them, as I have done. With ideas it is like with dizzy heights you climb: At first they cause you discomfort and you are anxious to get down, distrustful of your own powers; but soon the remoteness of the turmoil of life and the inspiring influence of the altitude calm your blood; your step gets firm and sure and you begin to look—for dizzier heights. I have attempted to speak to you on "Electricity," its development and influence, but I fear that I have done it much like a boy who tries to draw a likeness with a few straight lines. But I have endeavored to bring out one feature, to speak to you in one strain which I felt sure would find response in the hearts of all of you, the only one worthy of this occasion— the humanitarian.[29]

Although his speech was troubling to many because it seemed so unlike Tesla, a man who relished the showmanship of his wares, he nonetheless spoke of his latest motivation: "I am glad to say that also in this latter direction [power] my efforts have not been unsuccessful, for I have devised means which will allow us the use in power transmission of electromotive forces much higher than those practicable with ordinary apparatus. In fact, progress in this field has given me fresh hope that I shall see the fulfillment of one of my fondest dreams; namely, the transmission of power from station to station without the employment of any connecting wire."[30] Once again, his dream of "wireless" was front-and-center.

~

Tesla's speech was a "tell" of sorts. It indicated that although he was certainly not manic, he did, however, exhibit some self-doubt, a sign of a state of depression—not experiencing his lowest lows, however. Most likely he was in a mixed state, since he was certainly adversely affected and "down" due to criticism, betrayal, and competition, yet able to continue innovating and filing patents. And by the end of this period, he was likely hypomanic again, wherein he created and demonstrated the radio-controlled boat.

BAR HARBOR, MAINE, 1896

August 6, 1896, found Tesla's protectors, the Johnsons, luxuriating in Bar Harbor, Maine. In a fit of pique, Katharine wrote Tesla wanting him more than ever to come to see her:

> Dear Mr. Tesla,
>   I am so troubled about you. I hear you are ill. . . . Leave work for a while. I am haunted by the fear that you may succumb to the heat. . . . Find a cool climate. Do not stay in New York. That would mean the laboratory every day. . . .
>   You are making a mistake my dear friend almost a fatal one. You think you do not need change and rest. You are so tired you do not know what you need. If somebody would only pick you up and carry you bodily. I hardly know what to expect to gain by writing you. My words have no effect, forgotten as soon as read perhaps.
>   But I must speak and I will. You do not send me a line? How delighted I should be if it bore an unfamiliar postmark.
>
> Sincerely yours,
> Katharine Johnson[31]

It was no secret that Katharine's "love" for Tesla was real to her. It was also no secret that Tesla was now on another flight path. His fame was unquestioned. The demand on his time had become too much—he even neglected writing his sisters. His desire to work had intensified. After all, he had competitors. Marconi was knocking at his door, inasmuch as he was close to demonstrating his own wireless device. And there were other pretenders to the throne. He had reminded himself of his fellow countryman, the treacherous Michael Pupin, so he knew they were out there.

NEW YORK CITY, NEW YORK, 1897

Robotics was where he was to fine-tune his efforts at present, although he called it "telautomatics." He was convinced that electromechanical devices could be operated over various distances completely "wireless," no physical connection at all. He also believed that such devices could think for themselves.[32]

Tesla's eagerness to turn his thought experiments with his wireless system into physical experiments of reality, resulted in his receiving eight patents in 1896 for oscillators, radio communication, and remote-control devices. All in all, during the next five years these important patents bloomed into thirty-three fundamental patents that covered virtually the complete essentials of transmitting electrical energy without wires.[33]

Now it was time to show the world what he was working on.

His first project was to demonstrate wireless using his unparalleled originality and complex thinking. The result was a radio-controlled model boat some four feet long and about three feet in height. It was a major invention that made its first appearance at the Electrical Exposition held at Madison Square Garden at the time of the Spanish-American War in May of 1898. However, he had already showed early examples of this technology before the Institute of Electrical Engineers in 1892.[34]

This bold invention established the fundamentals of what would come to be known in a few short years as the "radio." This one invention also formed the foundation upon which would become the wireless telephone, garage-door opener, drones, the car radio, the television, the facsimile machine, the cable-TV scrambler, satellite radio/television, and yes, the remote-controlled robots that are fast becoming a mainstay in modern-day industry.[35] But there was also another reason, perhaps far more important to Tesla, that he pursued this revolutionary technology. In an article he wrote for *The Electrical Review*, November 16, 1898, he best expressed his thoughts, first foreshadowed by the title of the article: "Tesla's Latest Invention: Details of An Invention Which May Assure the Peace of the World." Then he followed with: "Vessels or vehicles of any suitable kind may be used as life, dispatch or pilot boats or the like or for carrying letters, packages, provisions, instruments objects or materials of any description, for establishing communication with inaccessible regions and exploring the conditions existing in the same, for killing or capturing whales or other animals of the sea. and for many other scientific, engineering or commercial purposes. But the greatest value of my invention will result from its effect upon warfare and armaments, for, by reason of its certain and unlimited destructiveness, it will tend to bring about and maintain permanent peace among nations."[36]

So where did Tesla come up with such a revolutionary idea? After all, it has had far-reaching implications in everyday, modern life and continues to increase in importance every day. Reaching back into his early life, we find answers. As the reader is aware, Tesla struggled with nightmares, visions that frightened him as a young boy. He ultimately devised a method to combat his fear. His perceptiveness from an early age gave him the advantage in fighting off such fear. He described his belief in man's relationship to automata: "While I have failed to obtain any evidence in support of the contentions of psychologists and spiritualists, I have proved to my complete satisfaction the automatism of life, not only through continuous observations of individual actions, but even more conclusively through certain generalizations. These amount to a discovery which I consider of the greatest moment to human society, and on which I shall briefly dwell. I got the first inkling of this astounding truth when I was still a very young man, but for many years I interpreted what I noted simply as coincidences. Namely, whenever either myself or a person to whom I was attached, or a cause to which I was devoted, was hurt by others in a particular way, which might be best popularly characterized as the most unfair imaginable, I experienced a singular and undefinable pain which, for want of a better term, I have qualified as 'cosmic,' and shortly thereafter, and invariably, those who had inflicted it came to grief. After many such cases I confided this to a number of friends, who had the opportunity to convince themselves of the truth of the theory which I have gradually formulated and which may be stated in the following few words." He went on to say: "Our bodies are of similar construction and exposed to the same external influences. This results in likeness of response and concordance of the general activities on which all our social and other rules and laws are based. We are automata entirely controlled by the forces of the medium being tossed about like corks on the surface of the water, but mistaking the resultant of the impulses from the outside for free will. The movements and other actions we perform are always life preservative and tho seemingly quite independent from one another, we are connected by invisible links. So long as the organism is in perfect order it responds accurately to the agents that prompt it, but the moment that there is some derangement in any individual, his self-preservative power is impaired."[37]

Unlike most other physicists who languished in the arena of thought experiments, never able to venture beyond that boundary, Tesla's engineering ability also permitted him to actually build what he had conceived. "The idea of constructing an automaton, to bear out my theory, presented itself to me early but I did not begin active work until 1893, when I started my wireless investigations. During the succeeding two or three years a number of automatic mechanisms, to be actuated from a distance, were constructed by me and exhibited to visitors in my laboratory. In 1896, however, I designed a complete machine capable of a multitude of operations, but the consummation of my labors was delayed until late in 1897."[38]

By 1898, Tesla's work on radio-controlled boats was well underway when he penned an article for *The Sun* newspaper in New York City: "Referring to my latest invention, I wish to bring out a point which has been overlooked. I arrived, as has been stated, at the idea through entirely abstract speculations on the human organism, which I conceived to be a self-propelling machine, the motions of which are governed by impressions received through the eye. Endeavoring to construct a mechanical model resembling in its essential, material features the human body, I was led to combine a controlling device, or organ sensitive to certain waves, with a body provided with propelling and directing mechanism, and the rest naturally followed."[39]

Although his control mechanisms could be adapted to virtually any device, vehicle, or flying machine, he decided to construct a boat for the United States in direct response to the arms race that was happening in Europe. His autonomous torpedo boat could carry an explosive charge and be controlled by electromagnetic signals.[40]

~

So the scene was set at the immense Madison Square Garden. Tesla had erected, with the help of the architect Stanford White, a rainbow room flooded with neon lights at the entrance beckoning all to enter the "magician's chamber." Tesla's famous showmanship was about to begin—he was ready to summon charisma on demand. Numerous celebrities were invited to the electric extravaganza. Among the glitterati were Vice President Garret Hobart and Thomas Edison's son, Thomas Edison Jr., he a representative of the Marconi and

Edison companies—the liaison having been organized by T. C. Martin.[41] Yes, there were other scientists and the regular gaggle of investors and journalists who always appeared at a Tesla function of any type. As the demonstrations increased in complexity, onlookers witnessed the battery-powered, iron-hulled boat dance about, antennae protruding upward, in the special indoor, manmade pond erected in the private auditorium. Utilizing a skein of transmitters and frequencies, the boat was directed to start, stop, move, steer, and operate other functions such as turning lights on and off. Everyone was nonplussed. They thought their eyes had deceived them. Then Tesla went a step further in demonstrating the boat's abilities. He let a member of the audience ask a question: What is the cube root of sixty-four? Tesla immediately manipulated levers on a small box off to one side. The answer: The boat's lights flashed four times.[42] At a time when just a handful of people had any concept of radio waves, some thought Tesla was controlling the boat just with his mind.[43] Actually, the demonstration signaled the dawn of remote-control applications, robotics, and a radio transmitter that could operate on multichannels.[44]

Picture yourself in the year 1898, and you were witness to this astounding event. You too would be breathless, as everyone in attendance was . . . even Tesla to a degree. He joyously said, " . . . it created a sensation such as no other invention of mine has ever produced. In November 1898, a basic patent on the novel art was granted to me, but only after the Examiner-in-Chief had come to New York and witnesst the performance, for what I claimed seemed unbelievable. I remember that when later I called on an official in Washington, with a view of offering the invention to the Government, he burst out in laughter upon my telling him what I had accomplished. Nobody thought then that there was the faintest prospect of perfecting such a device."[45]

The U.S. Patent 0,613,809 ("Method of and Apparatus for Controlling Mechanism of Moving Vehicle or Vehicles") was granted to him on November 8, 1898—it is worth appreciating that there are patents, then there are consequential patents. In the modern technical sense, Tesla's first boat was not "remote controlled" but rather "radio controlled," inasmuch as the former name describes situations in which different signals are sent to execute different functions. Nonetheless, it was an astonishing feat of technical and engineering prowess.[46]

As a sidebar, few computer engineers are aware today, recalls engineer Leland Anderson, that when computer manufacturers attempted to patent digital logic gates after World War II, the U.S. Patent Office informed them that Tesla's priority in the electrical implementation of them for secure communications, control systems, and robotics had already been established by his own patents. As a consequence, no new patents could be issued and technology for digital logic gates remained in the public domain.[47]

∼

Tesla's presence at the Electrical Exposition proved to be a great success for him. Such success buoyed his spirits, and he sought to make ever greater use of his miraculous invention, the *telautomaton*, in an effort to end all wars. Of this, the *New York Herald* of November 8, 1898, carried multiple headlines that read in several font sizes: "Tesla Declares He Will Abolish War"; "Magician of Electricity Announces That He Has Perfected an Application of the Current Which Will Make Possible the Destruction of Battle Ships at Any Distance without the Intervention of Wires"; "Projects Controlling Impulses Through Atmosphere"; "Exhibits a Working Model, the Propelling, Steering and Exploding Mechanism of Which Are Controlled at Will from a Distance without the Interventions of Wires"; "Says Will Render Useless The Navies of the World"; and "At the Paris Exposition the Inventor Intends to Exhibit a Model of a Vessel, All the Manoeuvres of Which Will Be Directed by Him from His Office in New York."[48]

Tesla could not have been more clear-eyed about his wanting to protect his adopted country that he so cherished, yet he did say, with great emphasis: "But I have no desire that my fame should rest on the invention of a merely destructive device, no matter how terrible. I prefer to be remembered as the inventor who succeeded in abolishing war. That would be my highest pride."[49] His work on what he would at times refer to as "devil automata" continued in earnest.[50] He was further buoyed by the failure Marconi experienced with his own wireless detonation system. At the same exposition, in a separate pool, he had placed a model frigate that was to explode the moment it came near a Spanish ship. Regrettably for Marconi, he had not understood how to send separate messages on different frequencies. This problem was made worse by the fact that his assistant, Thomas Edison Jr., suddenly pressed a button on his control device, setting off an explosion that rocked a storage room where Marconi had kept his miniature bombs.[51] Needless to say, Marconi's embarrassment put a smile on Tesla's face.

What appeared to some as Tesla's braggadocio, was not well received by a certain ilk. They flinched at his belief that he could indeed end all wars; no, that he could indeed prevent future wars . . . unfortunately such a belief set off a firestorm of verbal assaults aimed at him. The *New York Journal* published an article penned by Tesla, a portion of which further upset many scientists and other detractors:

> My submarine boat, loaded with its torpedoes, can start out from a protected bay or be dropped over a ship side, make its devious way below the surface, through dangerous channels of mine beds, into protected harbors and attack a fleet at anchor, or go out to sea and circle about, watching

for its prey, then dart upon it at a favorable moment, rush up to within a hundred feet if need be, discharge its deadly weapon and return to the hand that sent it. Yet . . . it will be under the absolute and instant control of a distant human hand on a far-off headland, or on a warship whose hull is below the horizon and invisible to the enemy.

I am aware that this sounds almost incredible, and I have refrained from making this invention public till I had worked out every practical detail of it. In my laboratory I now have such a model, and my plans and description at the Patent Office at Washington show the full specifications of it.[52]

T. C. Martin, a friend and supporter of Tesla's, joined in the dispute when he allowed an editorial condemning his "friend" to be published in his journal *The Electrical Engineer*. The editorial scoffed at Tesla as a dreamer who had not accomplished much the past ten years, while it praised Marconi for telegraphing from balloon to balloon sans wires, and it discredited his wireless torpedo. The fusillade of vitriolic verbal abuse continued when sadly his friend Mark Twain questioned his desire to sell his invention to other countries.[53] In Tesla's mind some type of "mutually assured destruction" could be established, thus lessening the prospect of war. But before the crazy kerfuffle could run its course, T. C. Martin took aim at his friend once again, claiming he had not come up with anything new or had completed "in-progress" inventions. To pour salt on Tesla's wounds, he hurriedly published a paper Tesla delivered at the American Electro-Therapeutic Association without Tesla's permission. It was not so much that permission was not given, but that his friend had questioned his honesty. As a Serb, who always put principle over profit, this attack was a call to arms, verbal arms. So in response, Tesla fired back with:

Your editorial comment would not concern me in the least, were it not my duty to take note of it. On more than one occasion you have offended me, but in my qualities both as a Christian and philosopher I have always forgiven you and only pitied you for your errors. This time, though, your offence is graver than previous ones, for you have dared to cast a shadow on my honor.

No doubt you must have in your possession, from the illustrious men [i.e., Professors Brackett and Dolbear] whom you quote, tangible proofs in support of your statement reflecting on my honesty. Being a bearer of great honors from a number of American universities, it is my duty, in view of the slur cast upon them, to exact from you that in your next issue you produce these, together with this letter which in justice to myself, I am forwarding to other electrical journals. In the absence of such proofs, which would put me in the position to seek redress elsewhere, I require that, together with the preceding, you publish instead a complete and humble apology for your insulting remark which reflects on me as well as on those who honor me.

On this condition I will again forgive you, but I would advise you to
limit yourself in your future attacks to statements for which you are not
liable to be punished by law.[54]

Tesla had given T. C. Martin an opportunity to do the right thing, but to no
avail. It seems there is no accounting for jealousy, not only from his friend and
de facto manager, but from many others who were possessed of a malignant
envy of Tesla. Seemingly many were misguided and motivated by their inability to understand and appreciate such a groundbreaking creation. The attacks
continued, assailing his character because he was a big dreamer, who made
his dreams reality in his own time while others watched in astonishment. As
we now appreciate, he did not seek sales of his inventions to other countries
for profit; no, he sought such arrangements in the hopes of preventing war.
Perhaps his scientific contemporaries had nothing to sell, and this was the only
way they could join in a chorus of incorrigible, petty persons of little accomplishment to make themselves feel worthy.

T. C. Martin permitted several additional critiques of Tesla to appear in
his journal that fueled the hatred and envy. The challenge to his originality
was what angered Tesla the most. The most unforgiving of attacks occurred
when the following appeared in *The Electrical Engineer*: "One of the foremost
electrical inventors of this country [possibly Elihu Thomson], whose name is
known around the world, has been kind enough to say that The Electrical
Engineer made Mr. Tesla. This is an attribution we naturally put aside, for it
is a man's own work that makes or unmakes him, but we do plead guilty to
the fact that for these ten years past we have done whatever mortals could do
to bring Mr. Tesla forward and secure for him the recognition that was duly
his. Not only in the columns of this and other journals, but in magazines and
books we have striven with all the ability we possessed to explain Mr. Tesla's
ideas. The record is before all men. If there is a line or a word in it that seeks
to do Mr. Tesla 'serious injury,' who says we have ever in word or deed or
thought tried to do Mr. Tesla any sort of injury, lies.

"Within the last year or two Mr. Tesla has, it seems to us, gone far
beyond the possible in the ideas he has put forth, and he has to-day behind
him a long trail of beautiful but unfinished inventions. By mild criticism and
milder banter, not being able to lend Mr. Tesla the cordial support of earlier
years of real achievement, we have only very lately endeavored to express our
doubts and to urge him to the completion of some one of the many desirable
or novel things promised. We believe this to be true friendship."[55] It is interesting to note that *The Electrical Review*, a direct competitor to T. C. Martin
at *The Electrical Engineer*, to whom Tesla had been sending articles, hailed his
radio-controlled boat as one of the most significant contributions to the advancement of civilization.[56]

Ever the creator, the imagineer, he would not be cowed. As he was still engaged in a war of words with T. C. Martin in the press, he managed to begin work on an even larger radio-controlled boat (six feet long), feeling that the boat should more accurately respond as a person would to input from outside. In the end, the constructive criticism coming from his friends was not what he wanted to hear, and perhaps it was not meant to be constructive after all; just maybe it was jealousy in disguise. Finally, the imbroglio ended with Tesla and T. C. Martin parting ways.[57] It should be remembered that T. C. Martin's greatest achievement, the book *The Inventions, Researches, and Writings of Nikola Tesla with Special Reference to His Work in Polyphase Currents and High Potential Lighting*, was accomplished because of Tesla and it proved to be a great success. The exclusive granted to T. C. Martin with regard to the book and the numerous articles by Tesla published in his journal *The Electrical Engineer* also served to engender an unhealthy competitiveness among journalists as well as resentment toward both men. Without Tesla there would have been no T. C. Martin, as he came to be known in the world of science.

It cannot be emphasized enough that Tesla's view of his remote-controlled boat was unlike anyone else's. He considered it humanlike, if not the first nonbiological life-form on Earth, and its ability to "think" was what separated it from all other inventions.[58] He wrote eloquently in defense of his new creation:

> The automatons so far constructed had "borrowed minds," so to speak, as each merely formed part of the distant operator who conveyed to it his intelligent orders; but this art is only in the beginning. I purpose to show that, however impossible it may now seem, an automaton may be contrived which will have its "own mind," and by this I mean that it will be able, independent of any operator, left entirely to itself, to perform, in response to external influences affecting its sensitive organs, a great variety of acts and operations as if it had intelligence. It will be able to follow a course laid out or to obey orders given far in advance; it will be capable of distinguishing between what it ought and what it ought not to do, and of making experiences or, otherwise stated, of recording impressions which will definitely affect its subsequent actions. In fact, I have already conceived such a plan.
>
> Although I evolved this invention many years ago and explained it to my visitors very frequently in my laboratory demonstrations, it was not until much later, long after I had perfected it, that it became known, when, naturally enough, it gave rise to much discussion and to sensational reports. But the true significance of this new art was not grasped by the majority, nor was the great force of the underlying principle recognized. As nearly as I could judge from the numerous comments which appeared, the results

I had obtained were considered as entirely impossible. Even the few who were disposed to admit the practicability of the invention saw in it merely an automobile torpedo, which was to be used for the purpose of blowing up battleships, with doubtful success. The general impression was that I contemplated simply the steering of such a vessel by means of Hertz'ian or other rays. There are torpedoes steered electrically by wires, and there are means of communicating without wires, and the above was, of course an obvious inference. Had I accomplished nothing more than this, I should have made a small advance indeed. But the art I have evolved does not contemplate merely the change of direction of a moving vessel; it affords means of absolutely controlling, in every respect, all the innumerable trans-latory movements, as well as the operations of all the internal organs, no matter how many, of an individualized automaton. Criticisms to the effect that the control of the automaton could be interfered with were made by people who do not even dream of the wonderful results which can be ac-complished by use of electrical vibrations. The world moves slowly, and new truths are difficult to see.[59]

As a coda to Tesla's original work with his invention of remote-controlled devices, the United States government did not take seriously such a series of inventions until 1918, after Tesla's patents had already expired.[60]

By the closing weeks of 1898, a long-chain of events served to place Tesla once again in a dark place. He had been mercilessly, verbally assaulted by friends, causing great disappointment. His bank account had dwindled to near cents after years of costly research on radio and robotics, and legal fees that were constant, both to defend standing patents and to apply for new patent protec-tions. But it all had a single purpose to him when he said, "Money does not represent such a value that men have placed upon it. All my money has been invested into experiments with which I have made new discoveries enabling mankind to have a little easier life."[61] And to that end he would never stop.

Yes, the central purpose of his life had just been reaffirmed yet again: to lift the burdens from the shoulders of mankind!

He was completely mentally and physically exhausted after a very complicated few years, and he found himself still struggling with bouts of melancholy. In fact, it had been stalking him for years. Often exhaustion brought it on, but then again, sometimes it just appeared. No warning. Perhaps he was signaling to himself that he needed time away. In the late summer of 1896, he took his first visit to Niagara Falls to view the construction project at work, and it was there with the roar of tumbling water as a backdrop that he was interviewed by an eager reporter, hoping to get an exclusive. "I came to Niagara Falls," said Tesla, "to inspect the great power plant and because I thought the change would bring me needed rest. I have been for some time in poor health, almost worn out, and I am now trying to get away from my work for a brief spell and at the same time see the great results of electrical development within the last half dozen years. Those results have been wonderful, have far surpassed the expectations of the people generally, but they are what those who have made electricity their study for years and their life work have expected and have labored so hard to bring about."[62] Perhaps he was hoping to engender sympathy from the reporter or to put his feelings out into the ether in the hopes of lighting his mental load.

It was not long after that a newspaper correspondent caught him in a café in the darkest hour of night. There he once again appeared physically and psychologically spent. Then out of nowhere, the intrepid correspondent asked the impertinent question about marriage and why he was not with a woman. Tesla answered by saying that he had successfully engineered, much like his Superman, romance and love from his important life. And in a 1932 interview Tesla stated emphatically, "I destroyed my sexuality at the age of thirty-three because a certain French actress kept coming to me making it impossible for me to concentrate."[63] Like a multitude of specific characteristics about Tesla, including his sexuality, nothing is certain . . . not the tenor of his voice nor the color of his eyes nor the shade of his hair. Like so much about the man, even the fundamentals of his human makeup were viewed differently by those who knew him.

However, several biographers have formed the opinion that handsome men caught Tesla's attention. First it was his attraction to his early assistant Szigeti, then it was a young college graduate named Emile Smith, who died before any relationship could have come about. What is not known, other than someone's conjecture, was whether Tesla's attraction to men was platonic or if it were platonic and physical. On the hundredth anniversary of Tesla's birth, the American Institute of Electrical Engineers honored Tesla. It was said that several of the older members were pleased that such an honor just might sweep under the carpet the fact that Tesla never married; that he avoided women in any serious way; and that he had a propensity for voyeurism, all of which, if discovered by the public, could bring shame upon the institute.[64]

It must be understood, given the temper of the time, that what would pass today for a homosexual bent was not thought of in the same way over a century ago. There is nothing in the literature about Tesla that would link him to his old friend Szigeti, but a Mr. Richmond P. Hobson, a very hand-some Spanish-American war hero, just might be an exception. It was Robert Johnson who introduced Tesla to the married Mr. Hobson. Over the next several months the two would encounter each other at the one of the many Johnson soirees. It was obvious that the genius inventor took a fancy to the dashing naval officer, thinking him to be "a fine fellow."[65] Then in late 1898, Hobson was called to duty in Hong Kong and the meeting of the two men ended for a period, although they exchanged letters. Once back in the states, the two maintained a social relationship scheduled by the Johnsons, and in time Hobson suggested that the Navy should look at Tesla's radio-controlled boat. Because of personal differences between officers who would be involved in any acquisition of Tesla's boat, Hobson's efforts on Tesla's behalf were for naught.[66]

So what is the truth about Tesla's sexuality? Like the color of Tesla's eyes or the tenor of his voice, it all depends on one's personal interpretation. However, one of Tesla's earliest biographers had this to say about his sexuality: "The superman that Tesla designed was a scientific saint. Tesla's superman was a marvelously successful invention—for Tesla—which seemed, as far as the world could observe, to function satisfactorily. He eliminated love from his life; eliminated women even from his thoughts. He went beyond Plato, who conceived of a spiritual companionship between man and woman free from sexual desires; he eliminated even the spiritual companionship. He designed the isolated life into which no woman and no man could enter; the self-sufficient individuality from which all sex considerations were completely eliminated; the genius who would live entirely as a thinking and a working machine."[67]

∽

As we leave Tesla in the complicated milieu of 1896 New York City life, he readied himself for a drastic move that no one had anticipated with any certainty. It would prove to be a seminal change in his highly productive life that was being battered from all sides by friends, foes, and himself. A change of scenery was required.

Lightning and Thunder Redux

Explosions in the Mind

COLORADO SPRINGS, COLORADO, 1899

$\mathcal{A}$s we know, Tesla's last few years of the 1890s were both productive and profoundly poignant. He filed the most patents at the fastest rate in his career between 1896 and 1901. Although he needed a change, he also needed to stay focused on his wireless transmissions, and to that end he was convinced that the best method for sending power and messages was by electrostatic forces, optimally through the earth, believing it was the "ideal method of transmission."[1] He compared, without hesitation, his method of transmission to the Hertzian-wave system: "It will be of interest to compare my system as first described in a Belgian patent of 1897 with the Hertz-wave system of that period. The significant differences between them will be observed at a glance. The first enables us to transmit economically energy to any distance and is inestimable value; the latter is capable of a radius of only a few miles and is worthless."[2]

The ultimate success of Tesla's system of wireless transmission was predicated on higher and higher voltages and frequencies from his magnificent coils. At one point in his Houston Street laboratory he had generated an output of some four-million volts that produced breathtaking, cinematic displays of electrified sparks exceeding sixteen feet in all directions. But that was not enough, he needed to exit the confines of the laboratory and reach out into the wide-open spaces of America; he had to scale up to build a bigger system.[3] That would cost more money than he had. But he was not discouraged, for he had convinced himself that funding would come his way. After all, he had friends, very wealthy friends, and they would come to his aide—they had to for humanity's sake. But why would they invest in what many thought were his latest schemes? Men with money to invest did not acquire wealth by be-

Tesla at twenty-three

Tesla at forty

Tesla in his laboratory

Tesla lamp, lit wirelessly

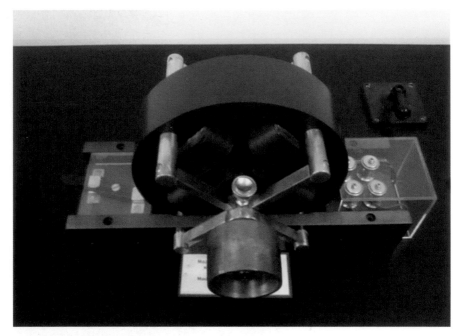

Tesla's AC induction motor

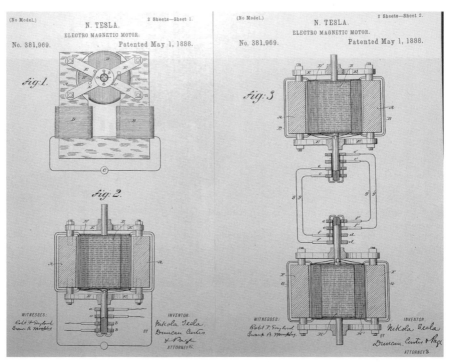

Tesla's induction motor patent

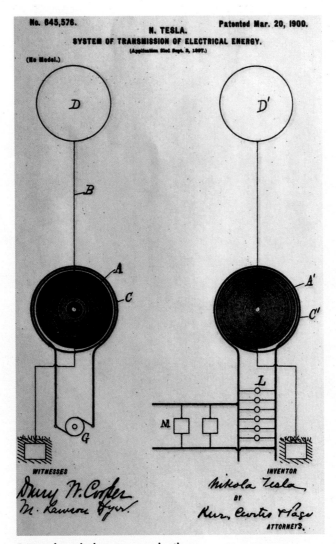

Patent for wireless communications

Tesla's radio-controlled boat

Tesla's bladeless turbine engine

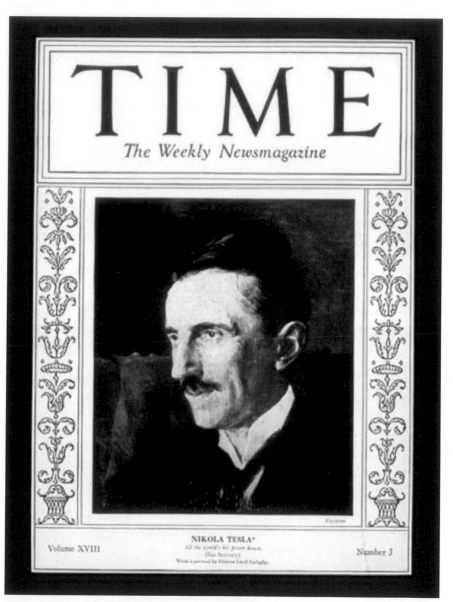

TIME
The Weekly Newsmagazine

NIKOLA TESLA*
All the world's his power house.
(See Science)
*From a portrait by Princess Lwoff-Parlaghy.

Volume XVIII   Number 3

Tesla on his seventy-fifth birthday, *Time* magazine cover

Nikola Tesla Museum, Belgrade, Serbia

The golden urn containing Nikola Tesla's ashes

ing stupid or spendthrifts. But they do invest in visionaries such as Tesla; they were lion-hunters who could act on their impulses. Tesla had long held a "cult status," and his compelling charisma was always at his command. These men were realistic and understood that no amount of wealth could purchase genius, good looks, or charisma. Either you were born with one or more of these "X factors" or you were not. So how does one get within the orbit of a person of such great consequence as Tesla? Invest in his dreams and have a front seat to watch them come true. Although Tesla had had a difficult time raising funding in the past few years for his various projects through his Nikola Tesla Company, he did have other options, most specifically, those wealthy people who wanted in, who wanted to be associated with the man who had achieved more to advance the human condition than anyone else to date.

Since he had a lifelong distaste for amassing wealth just for the sake of amassing wealth, having to sell himself, so to speak, was anathema to him. Nonetheless, he held his breath and set his sights on Colonel John Jacob Astor IV (1864–1912), heir to a $100 million fortune (approximately $3.12 billion today), a member of Teddy Roosevelt's Rough Riders, and the author of a science fiction novel, a copy of which he signed to Tesla.[4] As a member of New York City high society, Astor became a successful real estate developer and investor like many in his family. In time, he would team up with his cousin William Waldorf Astor, who had built the Astor, a luxury hotel. He followed suit and erected his own luxury hotel in 1897, naming it the Astoria. Shortly thereafter the two hotels merged to form the Waldorf-Astoria, the largest hotel in the world. He also fancied himself an inventor of sorts as well as an author and aficionado of science and technology. Hence, Astor joining with Tesla seemed like the logical fit.[5]

After a series of complicated stock buybacks and other accounting maneuvers and machinations, Tesla was able to gain back control of his company. Aware that there was still a "powerful clique" who opposed him, he revealed to Astor that his friendship and financial support would go a long way in producing great profits for his company. He stated that in the past several of his earlier patents had paid back some $1,500 for every $100 invested. He now believed that the return on funds invested would far exceed that number.[6] As an example, he produced a better light than the incandescent lamp, requiring less energy and essentially zero maintenance, thus lasting forever.[7]

Tesla thought he would further entice Astor—after having failed to bring him in as an investor three years earlier—more quickly if he played upon his military background, telling Astor that his oscillators, automatons, and other inventions could neutralize large cannons and make the building of battleships unnecessary, thus, bringing peace to the world. This idea fell upon deaf ears, for it threatened many industries and Astor would have none of it. Astor did not need to bring upon himself the ire of his contemporaries like Tesla had with his fellow scientists. He was satisfied with the production of oscillators and cold light and wanted to leave the saving of the world to others. Tesla then tried to bring Astor into the fold by informing him that he already signed contracts with the Creusot Works in France, Ganz and Company in Austria, the Helios Company in Germany, and several other firms.[8]

Tesla continued to pander to Astor in every way he could bring himself to do so, calling him a prince among wealthy men and a patriot who would spend his life for his country and whose word was his bond. In the end, he succeeded. He promised Astor to lower the stock price, to stay focused on what was already producing profits, and to bring in Westinghouse and Adams as smaller investors. And for that, on January 10, 1899, Astor finally committed to an investment of $100,000 (equivalent to approximately $3,000,000 today) and contracts were signed, giving Astor five hundred shares of Tesla Electric Company and a seat on the board of directors.[9] However, when Astor learned that Tesla had redirected funds to his wireless power project, he stopped additional payments after the first $30,000.[10] Tesla was not quite forty-three years old.

The chess pieces had now been moved to their appropriate squares, permitting Tesla to make his final move and checkmate Marconi, the press, and other scientists who rejected his heterodoxic scientific views. Now Colorado Springs, Colorado, would become his home base where he could work out his plans on a massive scale. He had the necessary financing in the bank; his basic patents on his wireless communication and radio-control inventions had been granted; he knew what type of energy was required to disrupt the electrical conditions of the earth; and he was in negotiations with the U.S. Navy. And yes, he even moved into the grand Waldorf-Astoria Hotel, where the social elite would meet.[11]

～

On the train to Colorado, Tesla made a stop in Chicago to speak before the Chicago Commercial Club, a conservative organization composed of sixty of the city's most influential businessmen. The Saturday evening that Tesla spoke represented the club's 150th event in its thirty-two years—a red-letter day in the society's history.[12] The members present were excited to hear Tesla speak on electricity, and the club's president, Cyrus McCormick, gave a rousing introduction of the great inventor: "We shall be much interested to hear from Mr. Tesla of the realm [electricity] which he dwells, which seems to us so incomprehensible." Mr. McCormick then proceeded to predict that Tesla had much more to accomplish, achieving greater results yet to be.

Tesla boldly took to the podium and began with the unexpected: "There is no one who does not speculate about the questions of his existence, asking whence he comes, whither he is going and what is reality. Soul and matter and their relations heave eternal interest for human beings. On the other side, there is always the desire to comprehend the marvelous manifestations of nature in all its phases. Let us talk about these."[13] The audience of businessmen looked on, amazed by the demonstrations and the powerful words uttered by a man who was still a mystery to them and others. It was reported that Tesla wandered off the topic of electricity several times into areas of the divine, nature, and what the ancient Greeks referred to as episteme. His lecture was at times confusing, at other times illuminating, and still at other times possibly self-serving and even incoherent to the audience. Ironically, it was only a few years before that he had made history when the very same city hosted the World's Fair of 1893, which was lit by his alternating current.

He spoke of utilizing the forces of air, which he called his greatest invention. He said, "The air, which is an insulator, becomes a conductor by means of the millions of volts I am able to produce. It means the using of the air for all purposes and uses of men. Not only for the transmission of messages without wires, but for the transmission of power in any amount to any distance across the earth's surface."[14] He continued with talk about the distinction between the living and the dead, stating there was no such difference, and he went on to criticize the ordinary definition of life, while he believed "he was nothing more than an automaton in every act and sense."[15] He even broached the sensitive subject of communicating with Mars. One can easily understand the reactions from the very conservative businessmen who lacked any level of imagination or scientific knowledge that remotely approached Tesla's mindset. They were nonplussed by nearly every word he spoke, understanding little of the lecture.

Tesla's closing words reflected his desire to make it known that although the invention of the system of "wireless telegraphy" was his, people mistakenly associated it with Marconi (that would change). The Italian Guglielmo Marconi (1874–1937) was making great progress with his (Tesla's) wireless system and Tesla knew he had to move fast to intercept Marconi before it was too late. The situation became more existential when Marconi began demonstrating his wireless before the public, thus, giving him what is called in the modern parlance of business today, "first mover advantage." It was said in the press in both England and the United States, that Tesla was concerned about Marconi and one gossip columnist even suggested that while Tesla talked about wireless, Marconi was demonstrating wireless. It seemed that Tesla had once again engendered some harsh words from the press and other scientists.[16] But it was not over, and Tesla knew it.

Tesla ended his peculiar remarks with the following: "I am going to perfect the ideas of transmission of energy which I have found in the new and measureless ocean of the atmosphere about us. . . . Distance, in the new method of transmission, amounts to little or nothing."[17]

~

Tesla's thoroughly confusing lecture before the Chicago's Commercial Club caused him to leave the Windy City empty handed. He had squandered an opportunity to acquire additional funding, and he had to know it. Next stop, Room 207 [222] at the Alta Vista Hotel, just a short walk from his newly constructed Colorado Springs laboratory. His hotel room window looked out onto an empyrean vista of the Rocky Mountains—with an unobstructed view of Pike's Peak—where he felt free and able to begin unpacking his emotional suitcase and get to work. It also afforded him the opportunity in the wide-open spaces to take his eight- to ten-mile walks to ruminate when time permitted.

On May 18, 1899, a local newspaper announced his high-noon arrival in the city sited at an altitude of 6,000 feet above sea level, where he came to "harness the power of the upper heavens."[18] It was further reported that he came to Colorado Springs to conduct "extensive investigations in the realm of his chosen science, and with a view of solving the problem of sending a telegraphic message around the world."[19] Asked why he chose Colorado Springs, Tesla replied, "I might as well tell you the truth. I have come here to carry on a series of exhaustive experiments in regard to wireless telegraphy. I will be here perhaps several months. I come here to work."[20] Tesla would be joined by two assistants (Kolman Czito and Fritz Lowenstein) and a shipment of his instruments and other equipment while George Scherff, who served as the company's assistant and accountant, was left behind in his New York office to supervise the craftsmen. The two men worked in concert with each other, communicating via letters or telegrams on a daily basis—the communiqués often contained detailed instructions, technical schematics limned to perfection, and complex calculations from Tesla. But there were those times when communication between the two men concerned finances—which were often tight—and sometimes personnel matters. But the majority of the correspondence comprised business matters.[21]

Tesla would instruct Scherff in terms of what he expected on the New York City end of the operation and everything had to be done first class. No exceptions.[22] All was arranged for by Tesla's patent attorney Leonard E. Curtis, a longtime adviser and friend as well, who stood by him during the war years with Edison.[23] In fact, he had also arranged for the actual laboratory (a barn that measured sixty-by-seventy feet, had ladders on the roof for access, and two small offices next to the main experiment room) located at Knob Hill within a cow pasture. There was also a school for the deaf and blind just east of the immense structure of wood and nails. Local carpenters, led by Joseph Dozier, built a telescoping mast that could rise 142 feet and topped it with a thirty-inch diameter copper-covered ball. To give the tall, spindly structure functional support, a twenty-five-foot tower was added that had a retractable roof, should Tesla's work create fiery streamers that could threaten the laboratory below. There was also a smaller, ancillary tower with a ball attached to calculate how capacitance changed with distance from the earth. The entire wizard's laboratory—especially the Tesla coil—and everything that went on with his experiments would be photographed by Dickenson Alley, an employee of the photographic studio of Tonnele & Company. Moreover, it would all be orchestrated by Tesla and Charles W. Price, editor of *The Electrical Review*, the direct competitor to T. C. Martin's publication, *The Electrical Engineer.*[24]

A periodical of the west some decades later reported on what the laboratory also contained:

> The major part of the interior was taken up with a variety of Tesla innovations. The electrical wizard was pioneering virgin fields and his apparatus, yet untried and exhibiting all the characteristics peculiar to an H. G. Wells fantasy, had to be constructed by highly trained technicians and shipped from the east. High-voltage transformers, dynamos, resonant-tuning devices, capacitor-discharge apparatus, oil-insulated capacitors (a Tesla invention), and a large metered control panel were among the items neatly spaced about the hall.
>
> At one end of the laboratory was the secondary coil of a giant Tesla transformer, which the inventor termed a "magnifying transmitter." Its primary coil (buried underneath the floor) was fifty-one feet in diameter and wound with heavy copper bars. In the center of the secondary was another coil with a diameter of ten feet. It carried 100 turns of wire and served to function as an extension of the secondary. The 200-foot mast [142 feet] extended up through the center and supported a large copper cable, which connected to the one-meter copper sphere. Using these devices, Tesla intended to determine if the earth possessed an electrical charge (it does) and to institute experiments that would alter its magnitude (he did). Who but Tesla would be so bold as to undertake a scientific investigation of such proportions?[25]

The finishing touch to the facility was a large sign that read: "Keep Out. Great Danger."[26] Curtis, a man about town, also managed to convince the El Paso Power Company to provide free power to the famous inventor during nighttime hours.[27] After all, Tesla informed Curtis in no uncertain terms that the tests would be secret and occur under the cover of night.[28]

Tesla's move to such a remote place to work also gave him a greater level of secrecy from snooping journalists and competitors. It was in view of the breathtaking Rocky Mountain range where he could be free from distractions and detractors. However, local residents were curious as to why such a famous man would come to their humble little area of Colorado. He spoke of sending a message from Pike's Peak to Paris.[29] This immediately endeared him to the locals, thinking that their town would become renowned. However, the statement, yet again, drew attention from his detractors. Thankfully, he paid them no mind and set to work.

∿

This was an exciting time for the inventor. The rotating magnetic field aside, wireless communication of data and power was the central focus of his life for many years. With the thought of Marconi lurking around every corner, Tesla's was pushed to succeed. Moreover, this was a pivotal moment in the history of electrical sciences, for it would ultimately usher in much of the modern era . . . and Tesla knew it. From reducing the burden placed on the common worker to increasing the distribution of data (knowledge), it was the future!

Tesla's purpose was to determine if the earth held an innate electrical charge. A few years after his Colorado Springs experiments, he recalled to a reporter: "In the middle of June, while preparations for other work were going on, I arranged one of my receiving transformers with the view of determining in a novel manner, experimentally, the electric potential of the globe and studying its periodic and casual fluctuations. This formed part of a plan carefully mapped out in advance. A highly sensitive, self-restorative device, controlling a recording instrument, was included in the secondary circuit, while the primary was connected to the ground and an elevated terminal of adjustable capacity. The variations of potential gave rise to electric surgings in the primary; these generated secondary currents, which in turn affected the sensitive device and recorder in proportion to their intensity. The earth was found to be, literally, alive with electrical vibrations, and soon I was deeply absorbed in the interesting investigation."[30]

Tesla began each day, often with a walk about the surrounding area with the glorious Rockies in full view. He would ponder the area's unique characteristics that suited his experiments: "Lightning discharges are accordingly, very frequent and sometimes of inconceivable violence."[31] He even took a closer notice when he calculated about twelve-thousand discharges in two hours and within a fifty-kilometer radius from the laboratory.[32] Although he never saw fire balls, "I succeeded later in determining the mode of their formation and producing them artificially."[33]

The evening of July 3, 1899, proved to be an unforgettable time in Colorado Springs for the inventor. During a ferocious storm in the mountains off to the west that ultimately moved over Colorado Springs and then suddenly roared off to the east, he observed myriad discharges in rapid succession. As Tesla recalled the formative event a few years later: "It was on the third of July—the date I shall never forget—when I obtained the first decisive experimental evidence of a truth of overwhelming importance for the advancement of humanity. A dense mass of strongly charged clouds gathered in the west and towards the evening a violent storm broke loose which, after spending much of its fury in the mountains, was driven away with great velocity over the plains. Heavy and long persisting arcs formed almost in regular time intervals. My observations were now greatly facilitated and rendered more accurate by the experiences already gained."[34] After the watershed moment had passed, he stated: "Subsequently, similar observations were also made by my assistant, Mr. Fritz Lowenstein, and shortly afterward several admirable opportunities presented themselves which brought out, still more forcibly, and unmistakably, the true nature of the wonderful phenomenon. No doubt whatever remained: I was observing stationary waves."[35] Stationary waves are the result of two waves of the same frequency but moving in opposite directions and interfering with each other. With the earth's innate electrical vibrations, he came to realize that his earthbound discharges in the form of artificial lightning resulted in stationary waves.[36]

What was the import of these stationary waves? He wrote to Scherff: "Impossible as it seems, this planet, despite its vast extent, behaves like a conductor of limited dimensions." He went on to say with immense joy: "When the great truth accidentally revealed and experimentally confirmed is fully recognized, that this planet, with all its appalling immensity, is to electric currents virtually no more than a small metal ball and that by this fact many possibilities, each baffling imagination and of incalculable consequence, are rendered absolutely sure of accomplishment; when the first plant is inaugurated and it is shown that a telegraphic message, almost as secret and non-interferable as a thought, can be transmitted to any terrestrial distance, the sound of the human voice, with all its intonations and inflections, faithfully and instantly reproduced at any other point of the globe, the energy of a waterfall made available for supplying light, heat or motive power, anywhere—on sea, or land, or high in the air—humanity will be like an ant heap stirred up with a stick: See the excitement coming!"[37]

Tesla became convinced that with the help of resonance, anyone could plug into the earth and use some of the pulsing waves triggered by his oscillators to run machines or light lamps. On one occasion in his early experiments, as he turned on his oscillator—while one terminal of the coil was grounded, the other was in space to dangle freely—he witnessed a gushing forth of what can only be called electrical streamers summoned from the earth, as if a mythical dragon had breathed fire into the air. But what caused this marvelous result? Was he tapping into what the earth had to offer? It was resonance. He described it by using an odd metaphor: By injecting electrical oscillations into the ground at the earth's precise resonant frequency, Tesla came to the astonishing conclusion that he might be able to broadcast power and messages around the entire planet quite easily. According to T. C. Martin, Tesla described this ambitious plan by using a rather odd analogy for Tesla: "With this coil [Tesla] does actually what one would be doing with a pump forcing air into an elastic football. At each alternate stroke the ball would expand and contract. But it is evident that such a ball, if filled with air, would, when suddenly expanded or contracted, vibrate at its own rate. Now if the strokes of the pump be so timed that they are in harmony with the individual vibrations of the ball, an intense vibration or surging will be obtained." Here Tesla realized, in the simplest of ways, the idea of a "hidden trigger mechanism," that had come to him during his time in Lika, as he played with snowballs on the side of a mountain—once thrown, watching them roll, gathering speed as they continued to roll downhill. It demonstrated that a small force, properly utilized with precision, could be used to harness the limitless forces of the earth. Tesla believed that he would not need to pump large amounts of electrical energy into the earth. He was convinced that only a minimal amount was needed, at the precise frequency, to serve as the triggering mechanism, and the principle of resonance would do the rest. With the entire earth pulsing like his metaphorical football, Tesla knew he could eliminate the distance between point A and point B, thus sending power and messages around the world."[38]

~

We must not forget that although the event was critical to Tesla's work, once again, the reader should take note: his reference to "humanity." One would have to peruse myriad books on other scientists to find even a single quote from a scientist working for humanity rather than himself. This was Tesla's overriding purpose in life. It was one of the many characteristics that separated him from other scientists. His quest to help humanity was what in the end would define him.

~

He spent time each morning in his office sending instructions to Scherff in New York City while Czito and Lowentstein worked building the massive magnifying transmitter (a photo of which is the most famous photo ever taken of Tesla and his famous coil) that would draw its AC power from the streetcar line at the boundary of the Knob Hill prairie. As the laboratory was filling up with all manner of equipment (some from Westinghouse), Tesla would sometimes dine with Lowenstein, a twenty-five-year-old ex-patriot from Czechoslovakia. The two would discuss ideas. But one day in September of 1899, they had a major disagreement. Some biographers have said it was due to the fact that Lowenstein missed home; others speculated that Tesla saw Lowenstein as more than an assistant; and still others have said that it was Lowenstein's wanting to get married that upset Tesla. Whatever the reason, by the fall of 1899, he was gone from the Colorado Springs project. Strangely, he did return for a few years. But Tesla never called him as a witness in any legal (patent) proceedings.[39] As a sidebar, *Nikola Tesla: Colorado Springs Notes, 1899–1900*, Tesla's comprehensive notes of his time in Colorado Springs, indicates the following regarding the Lowenstein departure: "Lowenstein remained until the end of September, when family matters obliged him to return to Germany. Tesla was satisfied with him as an assistant and asked him to return later, which he did, again becoming Tesla's assistant in February 1902."[40]

Much of the experimentation began in the late afternoon. When electrical power surged through the laboratory, sparks shot out, horrendous noises penetrated ears—despite being filled with cotton—and crashing sounds bounced about . . . a general cacophony reigned over the facility. As his detailed notes showed, "Tesla devoted the greatest proportion of his time [in the Colorado Springs project] (about 56%) to the transmitter, i.e. the high-power HF generator, about 21% to developing receivers for small signals, about 16% to measuring the capacity of the vertical antenna, and about 6% to miscellaneous other research. He developed a large HF oscillator with three oscillatory circuits with which he generated voltages of the order of 10 million volts. He tried out various modifications of the reciever [*sic*] with one or two coherers and special preexcitation circuits. He made measurements of the electromagnetic radiations generated by natural electrical discharges, developed radio measurement methods, and worked on the design of modulators, shunt-fed antennas, etc."[41]

The intense summer heat on many days wreaked havoc with his instrumentation, but he always thought the "purity of the air" was a fair trade-off and the surroundings provided the ideal environment for scientific research. Then on other days accidents served as unnecessary distractions, causing the laboratory to catch fire at times, the destruction of special equipment and financial loss as well. But he was willing to handle all situations because he had a specific purpose, and he would not waver from it. Too much was at stake—his integrity and humankind's future. As indicated in a technical journal report authored by Tesla, he came to Colorado Springs to accomplish three objectives: "Towards the close of 1898 a systematic research, carried on for a number of years with the object of perfecting a method of transmission of electrical energy through the natural medium, led me to recognize three important necessities: First, to develop a transmitter of great power; second, to perfect means for individualizing and isolating the energy transmitted; and, third, to ascertain the laws of propagation of currents through the earth and the atmosphere."[42] His experiments were all leading to "wireless telegraphy," later to be called "wireless communications." It was no Panglossian view, it was no chimera, it was real.

~

What was needed now was a stronger, more precise Tesla coil. His New York City laboratory had already been generating volts that were in the four-million-volt range. However, that was simply not enough for his Colorado experiments. He willingly jumped into the mélange of electrical equipment, wires, and the like and set to doing a great deal of tinkering: adjusting equipment; switching out various capacitors and coils; increasing the height of the spherical terminal on the roof of the laboratory; and increasing power from the outdoor transformer.[43] Once the adjustments were completed, it was time to test his theories with the highest voltage experiment to date.

Tesla directed his assistant Kolman Czito to man the special switchboard through which current was delivered into the laboratory from the El Paso Power Company of Colorado Springs using an elevated transmission line that stretched some two miles. For such an auspicious event, Tesla was properly attired in an exquisitely tailored Prince Albert coat draped over his tall, thin frame; his de rigueur linen gloves were in plain view, and a black derby hat served to complete the ensemble—the proud Serb was ready to make history, again.

Tesla looked to Czito, who stood at the ready. "When I give you the word, you close the switch for one second—not longer."[44]

Tesla stood near the large door of the laboratory with his giant coil in plain sight, positioned in the center of the laboratory within the wooden barn. He made it a point to keep his distance from the actual coil, lest a stray bolt of his man-made lightning find its mark on his body. To avoid such an unwanted shot of electricity, he wore shoes with four-inch-thick soles. Then Tesla looked up to the top of the 142-foot-tall mast where the three-foot-diameter copper ball was attached. Satisfied with what he saw, he gave the order, "Now." Czito flipped the large switch, sending out the powerful current—the secondary coil sparked and cracked, and a variegated-blue halo formed around it. Tesla looked up again to see his experiment produce artificial bolts of electrical fire, shooting out some 135 feet from the top of the tall mast. Massive electrical arcs burst forth as crackling sounds, some resembling the snapping of tree branches, echoing throughout the laboratory. Noxious fumes wafted across the laboratory. It was all a feast for the eyes. It had worked. He had created his very own lightning.[45] All the technical changes did indeed result in longer electrical ribbons that seemed to glide across the widening spaces between coils and reach out into the ether in every direction.

He was overjoyed and commanded Czito to try it again, but this time Tesla would go outside for a better view of what was about to happen atop the mast where the copper ball was attached. "When I give you the signal I want you to close the switch and leave it closed until I give you the signal to open it." Czito was at the ready. He repeated the process until Tesla said, "Czito, close the switch—Now!" Czito answered the command and did so, not knowing what would happen next.[46] At first there was no response as the device built up its electrical load. Czito had anticipated that a short circuit might be the result of the heavy current surging. But nothing. The snarling, vicious sounds and fiery blazes had ceased. An irate Tesla yelled to a panicky Czito, "Czito! Czito! Czito! Why did you do that? I did not tell you to open the switch. Close it again quickly!" Czito motioned to his boss to look at the still closed switch.[47]

Both men knew that the power company had shut off the power to the laboratory. It is not clear whether it was Tesla who called the power company or whether it was Czito, but the result was: "'Cut off your power, nothing,' came the gruff reply from the other end of the line. 'You've thrown a short circuit on our line with your blankety-blank-blank experiments and wrecked our station. You've knocked our generator off the line and she's now on fire. You won't get any more power!'" And he would not get more power unless he had the burned-out generator repaired. Easy enough for him to do. Less than a week later the repaired dynamo (generator) was working again.[48]

As the spectacular displays orchestrated by Tesla in his laboratory produced results he expected, there were still other results happening outside that he had not expected—visual phenomena of all types. Thankfully, he had worn his thick cork-soled shoes inside, as did his assistants, but the wild outdoors had reacted differently. He noticed that the ground beneath him was so heavily charged that it made cracking sounds when he walked. As recorded in his Colorado Springs notes, he said, "A curious observation is that all horses are shy. It is due to sound or possibly to current action through the ground to which horses are highly sensitive either owing to greater susceptibility of the nerves or perhaps only because of the iron shoe establishing good ground connection."[49]

It was observed that the horses trotting or simply grazing about the field would suddenly go berserk, darting about with bolts of lightning shooting out from their iron shoes. Butterflies were unwillingly drawn into the whirlpool of the transmitter coil. Even spectators at quite a distance claimed to have seen small sparks fly between grains of sand and their shoes and experienced the same from the highly electrified ground as they sauntered about.[50] Even lightning rods on buildings flashed with each coil discharge, and water faucets spit out small sparks—mini lightning. But what seemed most impressive was that three bulbs positioned some sixty feet from the laboratory glowed![51]

Once he understood the complexity of his discovery of "stationary waves," he wrote about his thoughts on such an important observation: "Stationary waves in the earth mean something more than mere telegraphy without wires to any distance. They will enable us to attain many important specific results impossible otherwise. For instance, by their use we may produce at will, from a sending-station, an electrical effect in any particular region of the globe; we may determine the relative position or course of a moving object, such as a vessel at sea, the distance traversed by the same, or its speed; or we may send over the earth a wave of electricity traveling at any rate we desire, from the pace of a turtle up to lightning speed."[52]

As a sidelight, one summer night while in Colorado Springs, Tesla noticed strange "repetitive static signals" coming through on his low-frequency radio receiver. He subsequently noted the event: "My first observations positively terrified me, as there was present in them something mysterious, not to say supernatural, and I was alone in my laboratory at night." Not believing they were coincidental, he added, "The feeling is constantly growing on me that I had been the first to hear the greeting of one planet to another. A purpose was behind these electrical signals."[53] Newspapers immediately seized the opportunity to speculate in what became known as the "Martian Messages." Tesla's observations of signals from outer space seemed jokingly absurd to most in 1899, but modern scientists think otherwise. In fact, the "search for extraterrestrial intelligence" (SETI) project lists Nikola Tesla as a pioneer.[54]

With the majority of his experimental work complete at his Colorado Springs laboratory, he was now certain that his wireless system would reach beyond anything Marconi could offer, despite Marconi having transmitted a message (using a Tesla oscillator) across the English Channel from Wimereux, France, to South Foreland Lighthouse, England, in March of 1899. Tesla fired back with this response: "The people of New York can have their private wireless communication with friends and acquaintances in various parts of the world. It will be no great wonder to have a cable tower [with a balloon tethered to it] than it is now to have a telephone in your house. You will be able to send a 2,000 word dispatch from New York to London, Paris, Vienna, Constantinople, Bombay, Singapore, Tokio [sic] or Manila in less time than it takes now to ring up 'central.'"[55]

While Tesla's latest nemesis was nipping at his pant leg, he would proudly pronounce himself "Father of the Wireless," because his work on wireless telegraphy began nearly a decade before Marconi had done anything of consequence—Tesla had accomplished a great number of revolutionary improvements to the art. He declared: "I gave to the world a wireless system of potentialities far beyond anything before conceived. I made explicit and repeated statements that I contemplated transmission, absolutely unlimited as to terrestrial distance and amount of energy. But, altho I have overcome all obstacles which seemed in the beginning unsurmountable and found elegant solutions of all the problems which confronted me, yet, even at this very day, the majority of experts are still blind to the possibilities which are within easy attainment."[56]

Tesla was so convinced that his position was on solid ground, he challenged anyone to deny the obvious: "First, my method of oscillatory conversion by means of condensers; second, the so-called 'Tesla transformer'; third, my apparatus for the transmission of energy without wire, comprising grounded, resonant circuits; fourth, my methods and apparatus for individualizing signals; and, fifth, my discovery of the stationary waves."[57] There were no takers.

∿

To Tesla it was the War of the Currents revisited, but this time it was against the younger Marconi. Tesla had already applied for his basic radio patents in the United States in 1897 and was granted patent protection three years later, just when Marconi's first patent application in the United States was rejected. Marconi tried to submit several revised applications over the next three years, but all were rejected because of the primacy of Tesla and other inventors. In fact, in 1903 the U.S. Patent Office had tired of Marconi's efforts so much that it made a very telling comment about his want to obtain a patent: "Many of the claims are not patentable over Tesla patent numbers 645,576 and 649,621 of record, the amendment to overcome said references as well as Marconi's pretended ignorance of the nature of a 'Tesla oscillator' being little short of absurd. . . . The term 'Tesla oscillator' has become a household word on both continents [Europe and North America]."[58]

The title of the actual U.S. Patent—one of the most important in history—for Tesla's "System of Transmission of Electrical Energy" reads as follows:

UNITED STATES PATENT OFFICE.
NIKOLA TESLA, OF NEW YORK, N.Y.
SYSTEM OF TRANSMISSION OF ELECTRICAL ENERGY.
SPECIFICATION forming part of Letters Patent No. 645,576, dated March 20, 1900.
Application filed September 2, 1897. Serial No. 650,343. (No model.)[59]

∿

Curiously, an assistant of Marconi's admitted sometime later that the Italian "also" did not comprehend the earth's significant role in conveying electrical energy. "We knew nothing," he stated, "about the effect of the length of a wave transmitted governing the distance over which communication could be affected."[60] However, as is all too often the case, that did not matter. Marconi managed, because of his dramatic presentations; his family's social connections to rich and influential Englishmen; and the inexplicable move by stock markets to drive up the Marconi Wireless Telegraph Company's stock value, to grab the spotlight. The irrational reaction to Marconi's showmanship and what others thought he had accomplished seemed to mimic the "tulipmania" craze (1634–1637).[61] Marconi also attracted the investments of American industrialists and even some advice from Thomas Edison himself. And for unknown reasons, in 1904, the U.S. Patent Office reversed its original decision to reject the Marconi patent applications and in the same year, it granted him a radio patent.[62] It was four years after Tesla's radio patents.

Tesla had vanquished Edison in a previous technological war and was still fighting himself, trying to remain focused and not go to his dark place. Then Marconi arrives on the scene, and it seems to the embattled warrior Tesla that the last decade of his life would now be a rerun. Add to his present situation the strange fact that even Edison himself played a role, a minor role, but a role, nonetheless. However, Tesla knew the war was not over; he knew he was indeed the "Father of the Wireless"; and as a Serb, he would fight to the very end.

As Marconi's fame was on the ascent, Tesla refused to say anything negative about the Italian, who used his Tesla oscillator to make his very first wireless transmission across the English Channel. Tesla even took the time to meet with the burgeoning scientist to explain how his namesake transformer facilitated the sending of power over great distances. But in time, Tesla made it very clear that he held the high ground when it came to wireless transmission of power. No matter that Marconi's currency rose exponentially in short order, Tesla nevertheless came to view him as "that donkey."[63]

By 1901, Marconi's system of wireless transmission was showing that it lacked the security required by both industry and the U.S. military. His radio broadcasts could be easily intercepted, and that was unacceptable, while on the other hand, Tesla promoted his "individualization" advances that he achieved while he was engaged in his robotic boat research and experiments. By virtue of his transmitter's complexity—it was not a one-note wonder—only a corresponding receiver could respond to its communication.[64]

The Marconi "experience" was now in full swing, while Tesla was still secreted away in Colorado Springs, where networking was a virtual impossibility. His remoteness resulted in very little communication with technical journals that praised his genius, newspapers that once hailed him as the "second coming," and friends, particularly the Johnsons—except through a few letters. There were also those who remembered Tesla's last few years in New York City, when he was being attacked as more of a dreamer than a doer, never mind the Mt. Everest of accomplishments he had already achieved by the time of his Fifth Avenue laboratory fire in 1895. As the reader knows, Tesla was also scoffed at and mocked for having claimed that he had received some sort of communication from beyond Earth, perhaps Mars. In an article penned by Tesla in 1901 for a weekly magazine, an excerpt will serve to explain his thoughts on a matter that so many eschewed at first blush: "The idea of communicating with the inhabitants of other worlds is an old one. But for ages it has been regarded merely as a poet's dream, forever unrealizable. And yet, with the invention and perfection of the telescope and the ever-widening knowledge of the heavens, its hold upon our imagination has been increased, and the scientific achievements during the latter part of the nineteenth century, together with the development of the tendency toward the nature ideal of Goethe, have intensified it to such a degree that it seems as if it were destined to become the dominating idea of the century that has just begun. The desire to know something of our neighbors in the immense depths of space does not spring from idle curiosity nor from thirst for knowledge, but from a deeper cause, and it is a feeling firmly rooted in the heart of every human being capable of thinking at all [even if they do not admit it]."[65] It must be noted that Marconi made the same claims some twenty-five years later. And Edison, claimed that he had devised a "spirit phone" to communicate with the dead.[66]

So how does one accurately access Nikola Tesla's time spent in Colorado Springs? You begin with his original purpose: (1) to develop a transmitter of great power; (2) to perfect the means for individualizing and isolating the energy transmitted; and (3) to ascertain the laws of propagation of currents through the earth and the atmosphere. To develop a transmitter of greater power, he did just that by building a Tesla coil that produced fifty million volts. He created "artificial lightning" that was heard and witnessed miles away. He also showed the power of high-voltage transmissions, but there was more to do. Some sixty-eight images, most spectacular in what they represented, especially the most amazing photo, taken in double-exposure, of Tesla sitting with a book in hand as fantastical streamers of wild electricity explode from the top of the Tesla coil. This image (see book cover) is easily the most reproduced image of Tesla ever taken! Inasmuch as the photograph was the result of an experiment, let us let Tesla explain the photographic legerdemain that went into the making of such a famous photograph:

> Some sparks passed also to the roof causing, as usual, considerable concern, the fireproof paint notwithstanding. Some streamers in the same positions as others relative to the camera are very weak, this probably due to their red color. Many streamers show clearly the phenomenon of splitting up or ramification and the sparks, where they strike the floor, develop increased luminosity. The feature of "splashing" on the floor upon striking the same is also well illustrated. To give an idea of the magnitude of the discharge the experimenter is sitting slightly behind the "extra coil." I did not like this idea but some people find such photographs interesting. Of course, the discharge was not playing when the experimenter was photographed, as might be imagined! The streamers were first impressed upon the plate in dark or feeble light, then the experimenter placed himself on the chair and an exposure to arc light was made and, finally. to bring out the features and other detail, a small flash powder was set off. It was found necessary to sit in the chair during the exposure to arc light as, otherwise, the structure of the chair would show through the body of the person, if the same were exposed merely to the light or the flash powder. As the weather during these experiments, which were carried on late at night, was most generally far below zero, I tried to overcome the above necessity but neither I nor Mr. Alley could, a practical remedy.[67]

Tesla had also claimed to have placed two hundred 50-watt incandescent lamps in the ground charged with a high-frequency current some twenty-five miles away from his laboratory in the mountains. Tesla had done this experiment in a limited fashion at his laboratory, but there is no documented evidence that it happened elsewhere. Some passed it off as apocryphal, while others would do or say anything to diminish his legendary achievements.

Such an event begs the question: Why no witnesses? Some have argued it was because Tesla had not yet achieved his dream of absolute wireless transmission of power. However, Tesla believed that his notes and photographs were the best witnesses and not his assistants, because at least the photographs were constant, unlike a witness, who, over time can change what he saw: memories fade with time. Still others postulate that Tesla was tiring. He had traveled a very long distance in a very short time, for he was not yet forty-five-years of age and had achieved more than a thousand other scientists in advancing the human condition. And an even more basic explanation could be that he was so focused on his experiments, possibly in a Zen state-of-mind, that he did not see the need or did not think it that important. And then again, perhaps Tesla's inventing process, which began as a child, did not rely on anyone to give him confirmation that he was on the right track of a problem to be solved as he invented new devices and discovered laws of nature. Once he gained the answers he needed, he did not care that he should be concerned about investors who wanted hard evidence; he wanted the joy of invention and discovery to be his remuneration.[68] The truth will never be known.

He had also invented his own lightning, calling it from the heavens by way of his devices, and making it dance on his command and produce kaleidoscopic lightshows. He had injected the earth with power and used it as a piece of experimental equipment. He had created his Tesla coil, an instrument of science unequaled by others. Aleksandar Marinčić's commentary, remarking on Tesla's final entry (January 7, 1900) in *Colorado Springs Notes*, notes that the inventor was acutely aware that all had not been completed, for there was much more to accomplish: "This is the last entry in the diary. Apart from the usual description of photographs, Tesla writes about experiments he intends to carry out on his return (where?). He qualifies the experiments to date as satisfactory, considering that his aim was 'to perfect the apparatus and make general observations.' The apparatus which he was then envisaging for future experiments was to be an improved oscillator which would enable better results than any he had so far obtained."[69]

~

As the year 1899 was fast coming to a close, Tesla had readied himself to return home. He had been away for nearly eight months. He was mentally and physically exhausted, as he often was following a burst of productivity while hypomanic, but thankfully, a December 22, 1899, letter certainly put a smile on his forlorn face:

Dear Mr. Tesla,

We will keep your memory green Christmas day. . . . How lovely it would be if you should suddenly appear in our midst . . . to spend it with us. . . .

I sometimes wonder if you could make me glad again, just to see you, it is so long since gladness has been in my days. Everything that once was has disappeared. It is as if one had gone to sleep in soft moonlight and had anchored out of place and out of time to find himself in the stone age, himself a stone.

What does it all mean? . . .

Sometimes I have a little sign of you through Robert by way of the office. I am hoping the New Year may bring you what you most desire and that it may bring to us my dear friend.

Faithfully yours,
Katharine Johnson[70]

The Johnsons were about to reenter his life, and no one was more pleased than Katharine. It was not that the troika had not been in contact with each other during Tesla's absence, for they would occasionally exchange letters, with Tesla writing from the new Alta Vista Hotel, his residence while in Colorado Springs. But there would be no dinners nor parties nor soirees into the darkest hours of the night until he was once again in New York City. In an effort to feel close to his friends, his protectors, he would address them with his regular salutations: My Dear Luka and My Dear Mrs. Filipov. The letters were his only way of being in touch with New York City, because he rarely did interviews with scientific journals or newspapers while in Colorado Springs.

The time had come to return home. Although he held hope that one day he would find his way back to Colorado Springs, it was not to be. The secret laboratory was shut down forever!

The Wardenclyffe Wonder

Narcissism, Hypomania, Vanity

NEW YORK CITY, NEW YORK, 1900

$\mathcal{D}$eep inside a very exhausted Nikola Tesla, he was being tested—it was the dark thought that his return to Colorado Springs was not to be. Yes, he had hope, but no, it was not a realistic hope, it was in truth nothing more than notional—his $100,000 investment would turn fallow. Moreover, in the ensuing years he would be stalked by debt collectors seeking sums ranging from an unpaid $180 electric bill for power used at his Colorado Springs laboratory to a lawsuit filed by his laboratory's chief watchman C. J. Duffner and a second watchman who were told by Tesla that he would return and to keep things as they were: in 1905 C. J. Duffner was granted a judgment for $928.57, paid to him by the sale of Tesla's laboratory property.[1] Tesla denied any indebtedness whatsoever.

What is important to understand at this moment in Tesla's life was not that his indebtedness of some $1,108.57 (nearly $35,000 today) was of great concern, but rather it was an unfortunate signpost along his journey to eventual bankruptcy. Such pecuniary matters became a daily issue in his life for the rest of his life. Some will pass it off as just what mad, absentminded scientists do, but he was no such scientist, he was a brilliant one, and no one could rationally say otherwise. However, as his experiments became more and more grand and complicated, so did the requirement for funding. He went into a state of protective denial over such matters, however, thus allowing himself to take hold of the city where he had already experienced so many successes—he was at the zenith of the science of electricity at that very moment. As a major electrical journal reported at the time: "Nikola Tesla had gone to the top of a high mountain in Colorado to experiment not only in wireless telegraphy, but also in wireless transmission of energy over a long distance."[2] Tesla stated in the same article: "My experiments have been most successful, and I am now convinced that I shall be able to communicate by means of wireless telegraphy not only with Paris during the Exhibition, but in a very short time with every city in the world."[3] It was said that he had produced in Colorado Springs "the greatest point-to-point electrical discharge ever achieved—sparks 100 feet in length."[4]

Another factor in his mounting indebtedness and inability to obtain funding for projects with as much success as he had in the past was that it was becoming apparent that Tesla was nearing the end of his productive years (namely, 1886 through 1894; and 1896 through 1901) when he not only created and received almost all of his patents, but also the years in which he was spending most of his time in a hypomanic or manic state. After 1901, as his debts soared, his charisma faded, his ability to charm funds from patrons collapsed, and he was entering a terminal era of over forty years when he had no convincing signs of infectious, attractive, confidence inspiring mania—only depression and negativity, isolation and increasing oddness. While the ideas flowed, the money did not and neither did the patents. The year 1900 was pivotal, presaging the end of his productive scientific life, and the end of Tesla as we most want to remember him.

Since he always lived in hotels, as we now know for his own peculiar reasons, it seemed fitting that the Waldorf-Astoria was where he should lie his head down after a difficult but exciting several months in Colorado Springs. After all, one of his investors was an owner of the Waldorf-Astoria Hotel, located on Fifth Avenue and Thirty-Fourth Street. It was the world's tallest hotel—with the first radio tower—whose luxury exuded class and was the first to offer en suite accommodations. Moreover, it was completely electrified and boasted a staff of nearly one thousand. Nothing was left to chance, as the mission statement was represented by the highest quality in everything the hotel had to offer, be it service, food, drink, dinnerware, décor, and employees—especially management. And as an elitist himself, Tesla felt at home in a place that attracted the likes of the world's most accomplished and celebrated people, after all, he was one of them. So, when Tesla signed the investment deal with John Jacob Astor IV to help fund the Colorado Springs project, part of the deal was that Tesla would live in Astor's hotel rent-free for the next twenty years.[5]

Since his return on January 7, 1900, he had the Waldorf-Astoria, Delmonico's, and the Johnsons' support, to look forward to—Tesla now had all he needed to be comfortable in the city. There is no question that regardless of his financial situation, be it feast or famine, he was a cognoscente of the high life. He renewed friendships at such uptown watering-holes as the Players' Club; he was once again lionized by the "Four Hundred"; and his private table at both Delmonico's and the Waldorf-Astoria Palm Garden was yet again a place where individuals sought their chance to meet the very secret inventor.[6] He usually dined alone, and he never did so with a woman privately. Even access to his table at these eateries was such that he could enter without being the cynosure of all eyes.[7] With the basics behind him, he set to work building a public relations plan that would accomplish several things, not the least of which was to reaffirm to himself and the public that his theories and experiments were on the right track; to attract new investors; and to counter any additional attacks, principally from Marconi, challenging his primacy in wireless telegraphy. In fact, when an article written for *The Century Magazine* appeared regarding Marconi's latest tuned transmitter, Tesla wrote comments on his copy of the magazine article pointing out Marconi's flagrant infringement of Tesla's patents of 1896–1897.[8] Immediately, Robert Johnson came to his aide by suggesting that he write a treatise on his recent accomplishments and what he had planned for the future. As this was happening, Marconi was also in New York City pandering to investors and possibly to speak on his inroads into wireless. Tesla caught word of this and responded: "When I sent electrical waves from my laboratory in Colorado, around the world, Mr. Marconi was experimenting with my apparatus unsuccessfully at sea."[9]

Tesla's treatise for *The Century Magazine* caused great consternation among those involved: Tesla as author; Robert Johnson as assistant-editor; and Robert Gilder as editor in chief. Tesla insisted on writing from a metaphysical point of view, which Johnson was not pleased with. Despite his great fondness for Tesla, he knew what the public wanted. Tesla resisted any attempts to change his mind. He insisted on writing about subjects that ranged from the metaphysical to the scientific: race, artificial intelligence, personal hygiene, robotized humans, telautomatics, green energy, wireless communications, and the idea of parallel worlds. Tesla would not give in. Ultimately, he presented a twelve-thousand-word masterpiece of prose, which demonstrated Tesla's erudition on all levels of human thinking. Moreover, it has been postulated that the treatise, as eloquently written and colored with masterful prose was an indicator of something more than simply the product of a talented writer. The same could be said for some of his letters to the Johnsons. It was all an expression of a subconscious, noticeable change in his state of mind when contrasted with his usually basic style of writing. Graphologists state that "the paper [is] frequently treated as a substitute object. . . . [Thus] the graphically expansive [writers] . . . usually are the same who not only dominate the paper, but also their environment."[10]

Johnson and Gilder decided to take the magnificent, original-concept treatise and have Tesla explain in more detail his inventions; add the compelling photographs from the Colorado Springs experiments (especially the one showing Tesla sitting calmly next to a gargantuan Tesla coil, reading quietly as long, deadly streamers of his lightning pierced the air around him); and utilize subheadings to make the treatise more digestible to the reader.[11] The end result of the effort was "The Problem of Increasing Human Energy," which appeared in the June 1900 issue of *The Century Magazine*.

Inspired by John William Draper's *History of the Intellectual Development of Europe*, the treatise became an immediate, boffo success, and Tesla's intellectual credentials were once again solidified, although, he did have his regular detractors, including his former friend T. C. Martin.[12] Scientists, newspapers, and scientific journals attacked Tesla, as they had done before, claiming he was too much of a dreamer, he was off the mark. Ironically, today what these detractors thought were absurd claims have become reality—robotics, focus on Mars, artificial intelligence, the internet, etc.

The success of *The Century Magazine* treatise followed Tesla's filing for three patents involving wireless communication (685,012; 787,412; 725,6050) just months earlier. In addition, later in the year he would receive even greater news: his primacy as the sole inventor of alternating current was upheld, after years of patent infringements both in the United States and Europe. Westinghouse had spent countless thousands of dollars defending the patents that he had purchased from Tesla years before, and now it had all come to fruition in a major decision by the U.S. Circuit Court for the District of Connecticut. The Court sustained three broad Tesla patents: 381,968; 382,279; and 382,280 of May 1, 1888. It proved to be a triumph for both Tesla and Westinghouse. Judge Townsend spoke with the utmost clarity:

> It remained to the genius of Tesla to capture the unruly, unrestrained and hitherto opposing elements in the field of nature and art and to harness them to draw the machines of man. It was he who first showed how to transform the toy of Arago into an engine of power; the "laboratory experiment" of Bailey into a practically successful motor; the indicator into a driver; he first conceived the idea that the very impediments of reversal in direction, the contraindications of alternations might be transformed into power-producing rotations, a whirling field of force. What others looked upon as only invincible barriers, impassable currents and contradictory forces he seized, and by harmonizing their directions utilized in practical motors in distant cities the power of Niagara.[13]

The treatise continued to create such a stir that nearly a century later a celebrated historian was quoted as saying: "Here is a man who is moving out of the conventional realm of invention into one that is more dramatic, heroic, and in a way tragic; where he's playing with the fire of the gods . . . almost a Prometheus trying to steal the fire of the gods. . . . Because of this he lost his way as an inventor. He became a heroic experimenter in search of the unknown."[14]

As Marconi continued to have a lot of ink spread across newspapers touting his exploits, Tesla paid it no mind and soldiered on. By November of 1900, *The Century Magazine* was still paying dividends, for Tesla's treatise caught the discerning eye of John Pierpont Morgan (1837–1913), the undisputed financial heavyweight of Wall Street. Morgan, son of a prominent financier, was in the throes of creating a billion-dollar steel industry, having already purchased Carnegie Steel as well as shipping lines and iron fields from John D. Rockefeller. As a savvy investor, Morgan had appreciated the success of the Niagara Falls power project, so he surmised that if Marconi failed in the fight to control wireless power and communications, he honestly believed Tesla would take the lead.

Soon Tesla found himself accepting an invitation to attend the Thanksgiving festivities at Morgan's mansion. Thankfully for Tesla, he was then invited to the post-holiday dinner the next day. It was after dinner that the two men met privately. Although there is no full record of what transpired, we do know that Tesla proposed a scheme of a "world system" of wireless that included all aspects of communications. Tesla wrote: "When wireless is perfectly applied the whole earth will be converted into a huge brain, which in fact it is, as all things being particles of a real and rhythmic whole. We shall be able to communicate with one another instantly, irrespective of distance. Not only this, but through television and telephone we shall see and hear one another as perfectly as though we were face to face." Although this was beyond Morgan's understanding, it nevertheless piqued his interest, believing faster access to stock market reports, communicating with ships at sea, and speaking with world banks would be to his advantage. So this edge over competitors caused him to strike a deal with the inventor.[15] Moreover, Morgan did understand that the success of the Niagara Falls power project was due to Tesla's superior AC patents.[16] And if Tesla was promising world telegraphy, he would take a chance on the dreamer.

One day during the frosty Christmas season, Tesla hailed a cab to Morgan's office at 23 Wall Street, and it was there that the deal was put to paper. Morgan informed Tesla that he was not enamored of his disagreeable ability to attract negative publicity and his notorious project cost overruns. Morgan was emphatic that the deal would not change, meaning that no additional funds would be available. In the end, the House of Morgan would write a check in the amount of $150,000 (equivalent to about $4.7 million today), and the stock-split was what Tesla said should be in Morgan's favor, hence, it resulted in 51 percent for Morgan and 49 percent for Tesla.[17] Unfortunately, as any unscrupulous wheeler-dealer would be expected to do, Morgan had buried in the fine print that with Tesla's signature, he, Morgan, would control the fundamental patents and the (cold) lighting concern as well—this absolute control was what many referred to as "Morganization."[18] Theodore Roosevelt called such a man one of the "Malefactors of Great Wealth."[19] This put Tesla in a very tenuous position with his original backer John Jacob Astor IV. However, he had assumed that Astor would be interested in his new project, but he had stepped away. With Morgan's money Tesla was able to begin the Wardenclyffe project and repay Westinghouse for a previous loan of $3,045; now he was debt-free—for the moment.[20]

Although Tesla was not pleased with the final deal, having been both intimidated by the powerful Morgan and grateful at the same time for what he saw as a sort of philanthropy, he kept quiet. After all, now his final dream would come alive! It would be his encore to a truly magnificent life lived.

WARDENCLYFFE, SHOREHAM, NEW YORK, 1901

The Wardenclyffe Tower facility was to be Tesla's wireless research center on Long Island. It derived its name from James S. Warden, a lawyer and banker, who thought that Tesla would be a draw to those who wanted to be near the great man, as well as the extension of the Long Island Railroad's Northern Branch. Warden purchased sixteen hundred acres believing that a real estate boom would ensue, after all, it was located only sixty-five miles from New York City on the Atlantic Coast. He named his land investment Wardenclyffe, although in 1906 the name was changed to Shoreham. He then approached Tesla, offering him a two-hundred-acre plot of wooded land for his laboratory to be situated just opposite the train station. The low-cost mortgage offer was accepted and construction began in August of 1901 in the wooded land on the north shore of Long Island.[21]

The immense research facility would comprise the mammoth tower that soared into the heavens, a plant, large laboratory, and additional outbuildings. Tesla's friend, famed architect, gadfly, and infidel Stanford White, offered his services free of charge. He would design the research center's laboratory, which would measure ninety-six-feet square. He estimated the cost to be some $14,000. It would be a one-story, redbrick-sided structure with a chimney placed in the center and finished off with a wellhead platform made of cast iron.[22] The wood-framed tower was designed in such a way that if a piece needed to be replaced, it could be without disturbing the overall structure. Reaching 187 feet upward, the tower was topped-off by an iron dome sixty-eight feet in diameter. Tesla had wanted the tower to rise some six-hundred feet, but major cost over runs ($450,000) put an end to that idea. This ungranted wish of a taller tower was yet another indicator that Tesla was becoming unmoored from reality and fighting paroxysms while searching for grander and grander expressions of his final dream. The cupola or globe, which weighed fifty-five tons could be problematic in high winds—could act like a sail—so White designed a clever eight-sided, lattice-like structure that slowly narrowed as it rose above the forest floor. All girders for the tower were fashioned from rough-cut pine timbers to ensure that it had a nonconductive quality.[23] The electromagnetic oscillator could be adjusted to emit a wide range of extra-low and high frequencies and the globe was purposed to store electrical energy for use on demand, much like a large capacitor. The tower and powerhouse were linked by two separate channels: one carried compressed air and water directly to the tower and the other channel dedicated for electric mains.[24]

Tesla wanted the system to transmit power through the ground, so it needed a special connection to the earth. As he said: "You see the underground work is one of the most expensive parts of the tower. In this system that I have invented it is necessary for the machine to get a grip of the earth, otherwise it cannot shake the earth. It has to have a grip on the earth so that the whole of this globe can quiver, and to do that it is necessary to carry out a very expensive construction. I had in fact invented special machines. But I want to say this underground work belongs to the tower."[25] There was also a very curious shaft, accessible by a spiral staircase, that reached 120 feet into the ground with sixteen iron pipes driven some 300 feet deeper to allow currents to pass through them and grab hold of the earth.[26] Beneath the earth there was also an extensive, honeycomb-like network of stone-lined tunnels that traveled in specific directions, with each ending some 100 feet away with an igloo-like exit that opened to the surface. It is not known what purpose they served.[27] In addition, there were also several other curiosities of the complex for which there are no answers. It was classic Tesla.

The laboratory was sectioned-off into four distinct rooms: a fully-equipped machine shop, an engine and dynamo room, a boiler room, and an electrical room. Power was provided by a 400-horsepower Westinghouse steam engine connected to a specially designed dynamo. Each room was immaculate at all times and finished in polished wood treatments.[28] Tesla also made it his duty to meticulously photograph every aspect of the complex.[29] The remainder of the research center comprised some five additional buildings, and a real estate development called Radio City, which was to house several hundred employed at Wardenclyffe.[30] The majority of the equipment Tesla required had to be custom made; hence, the price of the project escalated yet again. Even his special tubes used for transmitting and receiving his broadcast programs had to be handmade. Tesla's special tube predated Lee de Forest's invention of the radio tube that was not in use until at least a dozen years later. Tesla took the secret of his special tubes with him to the beyond.[31]

By September of 1902, the tower had reached its final height, but Tesla's funding was at its lowest low! Thousands of dollars spent without concern for tomorrow. Once again, Tesla's profligate expenditures were unfortunately all too consistent, and it would ultimately lead to the ending of the project—it was an ineluctable fact—at some point, unless he could obtain a fresh injection of capital. Tesla had undertaken a project of monumental proportions without the adequate funding or the charismatic mania to obtain the funding he needed.

To begin that process, he laid off most of his twenty highly skilled employees, and he also sold off a land holding for $35,000. Although not sufficient to keep it all going as he wished, he still managed to retain a small work crew. To exacerbate an already difficult situation, the war between Tesla and Marconi was still a major distraction for him. Tesla soon came to learn the painful fact that Marconi had also planned a competing tower higher than the Wardenclyffe Tower. A local newspaper reported the following: "By February 22, 1902 *The [Port Jefferson] Echo* printed the following announcement: 'The immense wireless telegraph plant now being built at Wardenclyffe marks the beginning of the real war between Marconi and Nikola Tesla. Marconi has so far found only one way to send messages by wireless telegraphy—through the air. Tesla will try two methods. By means of his great tower he will send messages through the air. By means of his great well he will send messages through the ground. It is the latter method that Tesla thinks will achieve the greatest success.'"[32]

Tesla's funding had devolved to a drip-drip stage, yet he never shorted himself when it came to his fashionable lifestyle while the Wardenclyffe project was still in development. During the late spring and summer, he made daily trips cocooned in a luxury-train car from the posh Waldorf-Astoria to Wardenclyffe. He was accompanied by a Serbian manservant who was burdened with a large basket that contained a magnificent lunch prepared by the hotel's many chefs. Each morning at approximately 11:00 a.m., the train would pull into Shoreham—he would return to the city each mid-afternoon. But as soon as the facility became somewhat operational, he quickly arranged to rent the

Bailey bungalow on the shoreline of Long Island. There he would spend the next year.[33]

On December 12, 1901, the Italian inventor Guglielmo Marconi and his coworkers were amazed after receiving a Morse code message for the letter "S." It had been sent during a lull in another hailstorm from Poldhu in Cornwall, England, to Signal Hill in St. John's, Newfoundland. In spite of there being "no" genuine eyewitnesses, the clever Marconi reported the event two days later to international acclaim—it was the first trans-Atlantic communication by means of wireless telegraphy. The *New York Times* reported a few days later with the headline: "WIRELESS SIGNALS ACROSS THE OCEAN; Marconi Says He Has Received Them from England. Prearranged Letter Repeated at Intervals in Marconi Code. The Italian Inventor Will Now Leave St. John's, N.F., and Will Go to Cornwall to Continue the Transatlantic Experiments from His Station There."[34]

In that instant, Marconi cemented his celebrity status, and in 1909 he was rewarded the Nobel Prize for Physics.[35] And this would ignite the "Radio Wars." In truth, there were many skeptics who doubted Marconi's claim. However, Tesla's former friend, T. C. Martin, arranged for a celebratory feast to be held in Marconi's honor at the AIEE. Tesla did not attend, but his name was mentioned, and he did send his regrets.[36] Martin subtly attacked Tesla yet again by saying: "I am only sorry, therefore, that Mr. Tesla, who has given the matter so much thought and experimentation, and to whose initiative so much of the work is due, should not also have been able to accomplish this wonderful feat."[37] Such unwarranted vitriol only caused others, detractors of Tesla, to find confirmation in T. C. Martin's remarks. Strangely enough, in 1903 he did dedicate a book to Tesla. Another of Tesla's former friends, the fellow Serb Michael Pupin, also fired a verbal slap-in-the-face at Tesla—jealousy knows no bounds. The entire affair carried the theme of Marconi's feat, imagery and all.[38]

～

What was Tesla's next move in the ongoing war with Marconi, as he was trying desperately to keep The Wardenclyffe Wonder alive? A bitter Tesla responded with remarks that ranged from factual to emotional. He still believed the Italian to be a donkey who would not admit that all he may have accomplished came at Tesla's expense, because it was based on Tesla's own oscillators, coils, and other designs. Marconi did not agree, believing his subsequent Nobel Prize was proof enough; but was it? The year 1943 just might prove otherwise.

Marconi was accumulating more fame and fortune, while Tesla was accumulating more debt. In fact, the debt clock was spinning out of control.

With great trepidation, Tesla sought a larger loan from Morgan. Not only did Morgan refuse a second loan, reminding Tesla that the original deal was the only deal, but he took an extra two months to make the final payment on the original contracted loan. Crestfallen at the turn of events, Tesla took to producing a very slick prospectus known as the "Tesla Manifesto." It was a reading of his "world telegraphy" enterprise and his expectations. It was presented to the wealthiest members of New York City high society in an elegant, colored vellum folder, and it enumerated his many successes, valuable patents, a broad-strokes explanation of his overall plan (much of it we moderns utilize today, e.g., the cellphone, email, GPS, World Wide Web, and texting), and his availability as a consultant for others.[39]

One can only wonder what it must have been like for Tesla to have to stoop to such depths to keep his dream alive. Years before he had torn up his contract with Westinghouse, giving away considerable control over his genius and significant royalties. Why? Then along comes Morgan, and he finds himself in an unenviable position wherein he gives up control again. Why? Frankly, the answer has always been the same: money was not his remuneration, but rather only his tool to lift the burdens from the shoulders of mankind by his ingenious inventions. He was always true to his personal mission statement, which served to define his life from beginning to end. That said, needing additional funding was indeed an existential threat to his Wardenclyffe project and his future.[40]

His mental state was now growing frayed. After all, this was a man whose nonlinear approach to science—up to this point—had served him well. And this approach almost always occurred in the context of mania/hypomania. His rejection of received wisdom and bravery in challenging existing norms had made him famous, but it also invited mockery from those whose level of understanding was limited by those very same norms. His attackers never ceased, and even Edison had miraculously resurfaced to lead the charge against the man's integrity after having been vanquished by him years before.[41] He became one of only two members of the Marconi Wireless Telegraph Company Board of Technical Directors, with a possible third member in the near future—Tesla enemy Michael Pupin. Edison also offered some of his patents to assist the Marconi effort.[42] Learning of such a betrayal, in a desperate moment, Tesla even considered building a second Wardenclyffe at Niagara Falls to transmit its excess power directly to New York City.[43] Such a thought at that time was nonsensical to most, but not to Tesla. He still held to his hard-earned confidence that he would see the Wardenclyffe project to its successful end, no matter the forces arrayed against him.

Tesla's emotional naiveté would not let him give up on Morgan. He penned several additional pleas for Morgan's financial support, despite the panic that had ensued on Wall Street because of Morgan's several attempts to manipulate the market. At one point, Tesla dunned Morgan for additional funds, thus revealing his hand . . . hoping to appeal to his altruistic side—not knowing he had none. "Mr. Morgan, what I contemplate and what I can certainly accomplish is not a simple transmission of messages [like Marconi] without wires to great distances; it is the transformation of the entire globe into a sentient being, as it were, which can feel in all its parts and through which thought may be flashed as through a brain. . . . Will you help me or let my great work—almost complete—go to pot?" Morgan replied flatly: "I have received your letter . . . and in a reply would say that I should not feel disposed at present to make any further advances."[44]

Over the next several months, which turned into years, Tesla continued his pleas to Morgan via personal letters, but to no avail. It was a fait accompli, to be sure, but Tesla would not accept the obvious. Finally, and sadly, Tesla resorted to an uncouth, horrific letter filled with veiled threats: "What a dreadful thing," he wrote, "it would be to have the papers come with your name in red letters—A MORGAN DEAL DEFAULTS. It would be telegraphed all over the globe." Tesla truthfully admitted to Morgan: "Financially, I am in a dreadful fix."[45] Could anything be more humiliating to such a man as Nikola Tesla? Yes! Tesla then resorted to his verbose prose with a six-thousand-word article that appeared in a major technical journal. The hope was to make Morgan see the error of his ways. It read in part:

> It is not a dream, it is a simple feat of scientific electrical engineering, only expensive—blind, faint-hearted, doubting world! . . . Humanity is not yet sufficiently advanced to be willingly led by the discover's [discoverer's] keen searching sense. But who knows? Perhaps it is better in this present world of ours that a revolutionary idea or invention instead of being helped and patted, be hampered and ill-treated in its adolescence—by want of means, by selfish interest, pedantry, stupidity and ignorance; that it be attacked and stifled; that it pass through bitter trials and tribulations, through the heartless strife of commercial existence. So do we get our light. So all that was great in the past was ridiculed, condemned, combated, suppressed—only to emerge all the more powerfully, all the more triumphantly from the struggle.[46]

Tesla's goal of the betterment of humanity was a constant that could not be abandoned. The letters continued until Morgan finally shot back for the last time, having had enough. By third party, he was notified that "Mr. J.P. Morgan wishes me to inform you that it will be impossible for him to do anything more in the matter."[47]

~

Bowed but not broken, Tesla continued his hunt for new investors, not so much in a depressed state, but lacking the charm and self-confidence of his past mania that made men who could not understand him nevertheless believe in him. Men such as Henry Clay Frick (a neighbor in the Waldorf-Astoria), Wall Street banker Jacob Schiff, Oliver Payne (an associate of Rockefeller), and others, joined together in a chorus of "NOs." They all asked the same question: "Where was Morgan?" There was also another sticking point for these well-heeled men: What was Tesla talking about? They just did not understand the concept; they possessed an alarming level of nescience. They were hard-facts-oriented men.[48] Moreover, with Morgan still controlling 51 percent of Tesla's patent rights regarding any wireless transmission system, new investors saw no real profit in such a deal.

To make matters worse, Tesla even sought financial aid from several family members to whom he had sent generous amounts of money during his most productive years. His nephew Nikola Trbojevic answered with several thousand dollars from a royalty deal (for an invention) with General Motors, but eventually that funding stopped. Tesla's acerbic response indicated that he felt it was an affront to his integrity and that his nephew challenged his *Serbian-ness*. His continued appeals for the funding of his greatest vision had reached an excruciatingly bizarre state of mind, and he was on the verge of a mental breakdown. The handwriting demonstrated in his letters to friends and others continued to devolve into some sort of chaotic, almost chicken-scratch style of cursive script. Hence, his letters became more and more unreadable, as his desperation was causing him to slide deeper into the abyss.[49] Then, as he looked at the dying of his dream, the inevitable happened. He described it in his autobiography several years later: "No subject to which I have ever devoted myself has called for such concentration of mind and strained to so dangerous a degree the finest fibers of my brain as the system of which the Magnifying Transmitter is the foundation. I put all the intensity and vigor of youth in the development of the rotating field discoveries, but those early labors were of a different character. Although strenuous in the extreme, they did not involve that keen and exhausting discernment which had to be exercised in attacking the many puzzling problems of the wireless. Despite my rare physical endurance at that period the abused nerves finally rebelled and I suffered a complete collapse, just as the consummation of the long and difficult task was almost in sight."[50] He was undoubtedly in the depression phase of his bipolar disorder.

Now it had become a sort of psychotic attempt to win. In a letter, three times longer than his manifesto, Tesla informed Morgan that other funding possibilities were not available to him because he, Morgan, had deserted him, leaving others to doubt that investing in Tesla was a good idea. Tesla closed the eloquent letter by saying that if Morgan would support him to the very end, he would see that Tesla's reasoning was faultless in every way.[51] Rejection was swift.

It is interesting to note that there have been suggestions over the years that Morgan's eventual lack of financial support for the inventor was purposeful. Morgan might have come to believe that financially supporting Tesla would not be in his best interest: Tesla's system, if workable, could very well threaten his interest in General Electric, and his copper mines were supplying the necessary wire essential to the transmission of electrical power. This would also have a tangential effect by challenging his primacy as a capitalist-monopolist—this would not stand. To put a fine point on it, a story had been circulating, maybe apocryphal, that Morgan was characterized as a gambler by another Wall Streeter, Bernard Baruch. Morgan flinched at the idea, emphatic that he never gambled, and quite possibly for that reason begged-off of any further investment with Tesla. The story has a denouement. The Yugoslav Andrija Puharich, who was somewhat involved with helping to get Tesla's papers to the Tesla Museum in Belgrade, Yugoslavia (now Serbia), and who knew Tesla's first recognized biographer John J. O'Neill, recalled a conversation with O'Neill: "Now, I [Puharich] always got this second hand; you won't find it anywhere in print, but Jack O'Neill gave me this information as the official biographer of Nikola Tesla. He said that Bernard Baruch told J. P. Morgan, 'Look, this guy is going crazy. What he is doing is he wants to give free electrical power to everybody and we can't put meters on that. We are just going to go broke supporting this guy.' And all of a sudden, over-night, Tesla's financial support was cut off, the work was never finished."[52] In truth, giving away free energy was always Tesla's ultimate objective, and if it cost the end of all monopolies, then so be it! The monopolies did not end, but Tesla did.[53]

To be fair, Morgan may have not thought Tesla was crazy, but he did know that in the nearly three years he had invested in Tesla's promise to cross the Atlantic Ocean with his world system of wireless, he had not succeeded, while Marconi had made good on his promise, albeit far less optimistically. Morgan's desire was not as comprehensive as Tesla's plan; he just wanted to communicate with his ships and the stock exchanges and send basic Morse code messages to the Continent. Moreover, the Banker's Panic of 1907 did not help, as it resulted in less investments all-around.[54]

~

In mid-December 1903, Tesla, yet again, sent another hopeless communiqué to Morgan, "the biggest Wall Street monster"—Tesla's words. Mind you, during this time his personal debts at the Waldorf-Astoria continued to accumulate at a rapid rate. Although braggadocio to some, what he did say in the letter had a ring of truth to it: "I have more creations named after me than any man that has gone before not excepting Archimedes and Galileo—the giants of invention. Six thousand million dollars are invested in enterprises based on my discoveries in the United States today."[55] Despite any plea to Morgan that could guarantee him a profit, he was over with Tesla.

Tesla was forty-seven years of age. Having seen the previous fifteen years of his efforts produce profound accomplishments that became a distant memory to others, he was now thinking and acting irrationally. He was thus also having to deal with continued attacks on his sanity and credibility by scientists of all stripes. Moreover, the press, who always thought him to be a genius and a great showman who could sell newspapers, now looked at him as nothing more than an aging crank. It was Tesla's belief that to counter the growing negative publicity, it would be most effective to produce "something commercial without delay."[56] This was Tesla without the mania, hypomania, frenetic energy, and charm necessary for raising funds. Tesla's magic was gone although his ideas still flowed. His credibility had faded, and his mania dissipated forever.

By the summer of 1903, Tesla had managed to form the Tesla Electric Manufacturing Company with a $5 million capitalization needed to produce small Tesla coils for use in scientific laboratories and to power X-ray tubes. However, investors were nowhere to be found, and the company failed. A Serbian bank was on the list of prospective investors, but that too resulted in no investment. His next mark was Astor, who also begged-off of any future funding of a Tesla company—he was still disappointed that Tesla had spent his money on his Colorado Springs project and not lighting. There were several other rejections, but Tesla never seemed to lose the faith that he would prevail.[57]

The spiraling downward continued. In May of 1905, his unique patents for his polyphase system expired, leaving him without the royalties and licensing fees that had produced a significant cashflow. As we recall, he had also lost a lawsuit for past debts while in Colorado Springs, causing his research facility to be dismantled for its lumber.[58] Without regard for his sorry financial state, however, the elitist inventor did not limit himself to bread and water and used clothing. He still held court at the Waldorf-Astoria, regularly dining at its luxurious Palm Room, where he was still the most famous of the famous that ate there. Every evening he would make the long walk down the corridor named Peacock Alley. He still wore his bespoke suits with the frame of a much younger man, and his gait was that of an athlete. All eyes would be on him when he took his seat at his private table—set only for one—which occupied a secluded area next to a wall. It was a ritual that played out for some eighteen years at the very same table.[59]

It goes without saying that his eating habits were as peculiar as his preference to always eat alone, save on formal occasions. There were single-use napkins and sterilized eating utensils and a special requirement that no one else ever dine at his table. The sight of one fly near him would cause the staff to provide an entirely new meal. Over several years he transitioned to veganism. These unusual habits were seen in every aspect of his daily life and exercised in every way possible—demonstrating his deep-seated obsessive-compulsive disorder (OCD).[60] After a sumptuous meal, he would always return to his laboratory for more work, sometimes with celebrities, who would be privy to a private show of miraculous feats of electricity.[61]

The Johnsons were still his regular friends who never left his side, whether socially or spiritually. The connection between the three was unbreakable, and now-and-again dinners and soirees with the city's glitterati were always something he looked forward to. It was his way of distracting himself from his never-ending financial woes. The failed Wardenclyffe project left behind a mountain of debts, including $30,000 owed to Westinghouse for equipment purchased for the project. Colleagues witnessed a brilliant man declining in every way. George Scherff, his secretary and assistant, stated: "I have scarcely ever seen you so out of sorts as last Sunday and I was frightened."[62] Tesla would frequently communicate with Scherff via letter, sometimes asking him for specific supplies he needed and at other times he would draw the schematics of a device he needed built.[63]

Some chroniclers have intimated that Tesla would occasionally see Richmond Hobson, a war hero and naval engineer. Their most recent meeting proved beneficial to Tesla's psyche, but in May of 1905, he learned of Hobson's betrothal to Grizelda Hull, a young lady from a well-respected Kentucky family. The two men did admire each other and Hobson was not above telling Tesla that he thought of him fondly. Tesla's response was to try to dissuade Hobson from marrying, but he ended up serving as best man at his wedding.[64] Tesla admitted to anguishing over the marriage for several years.

As Hobson's wife reported years later, her husband and Tesla would often meet to attend a movie, have a pleasant walk in the park, and engage in intellectual conversations until the darkest hours of the night. Tesla once bragged to Hobson that he could, if he wanted, shake the earth out of its orbit. It all ended when Hobson was laid to rest as a rear admiral and former member of the U.S. House of Representatives in Arlington National Cemetery in March of 1937. He was sixty-six years old. Tesla sent a personal note and flowers to the widow.[65]

By December of 1905, Tesla had demonstrated that his failed Wardenclyffe project and financial problems were of his own making and led to his ultimate undoing. He became more reclusive, secretive, and unpredictable. The breakdown ensued after his business partner William B. Rankine died at the age of forty-seven. He began having flashbacks about dead family members, bemoaning their leaving him alone. But the mental torture was not to stop there. In June of 1906, his close friend and the architect who designed much of the Wardenclyffe project, Stanford White, fell into trouble. He was known as a man about town who always had a mistress in tow; this time it was Evelyn Nesbit, whose wealthy husband, Henry Thaw, became enraged when he learned of the affair. In a moment of pique, he took aim and fired a kill shot between White's eyes. Already bedridden from exhaustion and depression, Tesla managed to attend the funeral when others refused to be associated with their former friend who lost his life in such an unseemly way.[66]

WARDENCLYFFE, SHOREHAM, NEW YORK, 1906

Postmortem: In 1906, the Wardenclyffe project was closed forever, just like the Colorado Springs project, but unlike the Colorado Springs project, it never became fully operational. The tower was demolished for scrap ($1,750) in 1917, and the property was ordered into foreclosure in 1922. For the next five decades Wardenclyffe served as a processing facility producing photography supplies. Several buildings were added to the site, and the land it occupies has been trimmed down to sixteen acres from its original two hundred acres. The original brick building designed by Stanford White is still standing . . . awaiting Tesla's return.[67]

Today the Tesla Science Center at Wardenclyffe (also known as TSCW) occupies the site. It is a nonprofit organization established to promote science and technology to the masses. Funding has come from a crowdfunding effort and a generous donation from Elon Musk. In 2018, the Wardenclyffe site was listed on the National Register of Historic Places.[68]

~

After his breakdown in 1905, which continued into 1906, Tesla never again attempted another major project. Despite living for nearly four more decades, the daring, go-it-alone inventor and creator would not accept that his phenomenal career had come to an ignominious end. So, what was next? He did not know. But what he did know was that he would never give up on his ideas and on what he knew was true.[69]

III

TO LIFT THE BURDENS

· *15* ·

Principle over Profit

Grandiosity, Humanity

NEW YORK CITY, NEW YORK, 1906

The Wardenclyffe project was still with him. It would never leave his thoughts. Yes, he was thoroughly disappointed. And yes, he was in a dark place, but his photographic and eidetic memories were intact, and his grand ideations were still flowing while the paroxysms appeared now-and-again, but no matter, he knew he was right.[1]

It was about a year before that Tesla gave an interview to a Pittsburgh newspaper reporter. During the conversation, the reporter observed the following: "Tesla was a peculiar worker. Failures do not trouble him. After he undertakes a thing and decides that it should come out a certain way he keeps on experimenting and experimenting, believing in his success. He says that if he doubted his ability it would make him crazy."[2] Believing that he was always practical, because he was an infallible instrument of response to ideas that presented themselves before him, he would always continue; he would soldier on. He may have not worn his faith on his sleeve, but typical of Serbs, his belief was confirmed by his Serbian Orthodox Christian faith and his confidence in the logic of science; thus, he would be led to the basic truths of nature. Moreover, discovering the rotating magnetic field when others never saw it or rejected the concept . . . gave him the strength and understanding that his theory of transmitting power through the earth must be right.[3]

With his manifesto and other stunning publications behind him, having utterly failed to raise additional funding, he was in quite a fix.[4] In 1907, the year of the stock market panic, he penned an article to the *New York World*, wherein he said the following: "Personally, basing myself on the knowledge of this art to which I have devoted my best energies, I do not hesitate to state here for future reference and as a test of accuracy of my scientific forecast that

233

flying machines and ships propelled by electricity transmitted without wire will have ceased to be a wonder in ten years from now. I would say five were it not that there is such a thing as 'inertia of human opinion' resisting revolutionary ideas."[5] Here again, Tesla reconfirms his belief in the accuracy of his extraordinary abilities, some dating back to childhood, and the tenor of his tone demonstrated that in some ways he still possessed the zeal of 1884, the year he came to America.

~

Once the realization hit him that his dream of transmitting wireless power and messages (data) across the world might very well remain a dream, his bitterness surfaced. It was further incited when some suggested that his Wardenclyffe project would have helped the enemy during World War I. As a loyal countrymen who cherished his citizenship, which he had obtained early on, he recoiled at the idea, stating emphatically in his autobiography:

> The tower was destroyed two years ago but my projects are being developed and another one, improved in some features, will be constructed. On this occasion I would contradict the widely circulated report that the structure was demolished by the Government which owing to war conditions, might have created prejudice in the minds of those who may not know that the papers, which thirty years ago conferred upon me the honor of American citizenship, are always kept in a safe, while my orders, diplomas, degrees, gold medals and other distinctions are packed away in old trunks. If this report had a foundation I would have been refunded a large sum of money which I expended in the construction of the tower. On the contrary it was in the interest of the Government to preserve it, particularly as it would have made possible—to mention just one valuable result—the location of a submarine in any part of the world. My plant, services, and all my improvements have always been at the disposal of the officials and ever since the outbreak of the European conflict I have been working at a sacrifice on several inventions of mine relating to aerial navigation, ship propulsion and wireless transmission which are of the greatest importance to the country. Those who are well informed know that my ideas have revolutionized the industries of the United States and I am not aware that there lives an inventor who has been, in this respect, as fortunate as myself especially as regards the use of his improvements in the war. I have refrained from publicly expressing myself on this subject before as it seemed improper to dwell on personal matters while all the world was in dire trouble. . . .
>
> I am unwilling to accord to some smallminded and jealous individuals the satisfaction of having thwarted my efforts. These men are to me nothing more than microbes of a nasty disease. My project was retarded by laws

of nature. The world was not prepared for it. It was too far ahead of time. But the same laws will prevail in the end and make it a triumphal success.[6]

~

The vernal equinox had come and gone. Tesla was feeling refreshed by the spring of a new year, 1906. His idea for a brand-new invention that took advantage of his mechanical engineering expertise was a momentary diversion for him. He was quoted as saying: "I felt certain there must be some means of obtaining power that was better than any now in use."[7] It would be as simple to him as his induction motor. It was the "bladeless turbine" or Tesla turbine engine. It was the result of analyzing the properties of water and steam as they passed through the propeller, and hence, the association of viscosity and adhesion to the blade's corresponding spin. The Tesla turbine utilizes the "boundary-layer effect," wherein the boundary layer is the layer of fluid in the immediate area of a bounding surface where the effects of viscosity are significant. The liquid or gas in the boundary layer has a tendency to stick to the surface.[8] In addition, Tesla's turbine was unique in that it utilized a patented series of disks rather than propeller blades. Each disk had a perforation for removing the incoming fluid, thus causing the shaft to rotate. He also understood the standard dynamics of friction impeding the movement of a ship being counterproductive and turned the negative into a positive. Simply put, the rotation of the turbine was enhanced by the adhesion and viscosity of the medium.[9] This ingenious invention also made sense to the inventor because it was practical and could make use of the patent-promote-sell business model, which had worked before. He tried to sell the idea to Astor, but it was a pass almost before Tesla had finished his first sentence of the pitch. He had hoped the money would help him complete the Wardenclyffe project.[10]

The world was once again excited in a Tesla invention. In March of 1909, the result was the Tesla Propulsion Company, and subsequent companies followed.[11] Tesla was invigorated, and the patent (U.S. Patent 1,061,206) for his Tesla turbine was issued in October of 1913. However, what had happened with Morgan now happened again: His new Tesla turbine engine was not welcome in industry because it was a direct challenge to another technology. Now it was the "war of the bladed turbines." Where industrialists invest money to make more money, Tesla invested money to introduce new technologies. All tests of Tesla turbine engines produced phenomenal results (16,000 RPM claimed, 16,000 RPM produced) and its use in other means of travel, such as automobiles, showed great promise. That said, a staff of evaluators at the Edison Waterside Power Station in New York City rejected the turbine as unsatisfactory. Tesla continued to make inroads into other companies, even designing a gasoline-powered version, but in the end, industry saw to it that it would never become reality, and the Tesla turbine died an ignominious death, like several of his other inventions—twenty years of work on his turbine engine was all for naught.[12]

But perhaps the greatest blow to Tesla's psyche and his integrity came in late 1909, when his latest nemesis, Guglielmo Marconi was awarded the Nobel Prize in Physics (along with a co-recipient Karl Ferdinand Braun) for their "invention" of the radio. Tesla had claimed that "the Italian tinkerer had 'abandoned the old devices of Hertz and Lodge and substituted mine instead. In this manner the transmission across the Atlantic was effected.'"[13] This was a direct slap-in-the-face to Tesla and hastened his slide into deeper depression mixed with irritability, disappointment, and resentment.

~

Bipolar depression with mixed features is characterized not only by depressive symptoms of depressed mood, sleep disturbance, and feelings of worthlessness, but also by a few symptoms of mania/hypomania including agitation, irritability, distractibility, and racing thoughts/flight of ideas. The record indicates that Tesla had signs of many of these symptoms at this time, while lacking the high energy, self-confident, grandiose, inflated self-esteem, and the elevated, expansive, overactive mood of pure mania. Tesla would spend the rest of his life in a state of fluctuating depression, often with mixed features of mania but never back to the heights of mania associated with his most productive years and greatest ability to attract investments. What were all the years of experiments about? What was his creation of the Tesla coil all about? What was Wardenclyffe about? The argument has raged on as to who is the rightful inventor of the radio—that is, who holds the patent—but frankly, as we shall see, the issue was adjudicated some decades ago.

~

With his emotions being thoroughly whipsawed about by one insult after another, Tesla lashed out again, this time, Edison and Marconi were the targets. One would think that at the very least Edison would have deemed it prudent to remain in the background after having been humiliated years ago in the War of the Currents. And Marconi's comeuppance was still years away, but such are most men of power. Tesla made his views clear, once again, about Edison, considering that he, Tesla, through his many profound inventions had, unlike Edison, truly bettered the human condition. He wrote a letter of complaint to a New York newspaper stating his case:

> Had the Edison companies not finally adopted my invention [AC electrical power] they would have been wiped out of existence, and yet not the slightest acknowledgment of my labors has ever been made by any of them,

a most remarkable instance of the proverbial unfairness and ingratitude of corporations. But the reason is not far to see. One of their prominent men told me that they are spending $10,000,000 every year to keep Edison's name before the public, and he added that it is worth more to them. Of course, in all that unceasing and deafening shouting from the housetops any voice raised to apprise people of the real state of things is like the chirp of a little sparrow in the roar of Niagara. So it comes that very few have a clear idea of the situation.[14]

Tesla's remarks were not a gasconade, they were flatly true. Virtually every individual in the United States today and across the globe has a better life by virtue of just one of Tesla's many inventions—alternating current power. As for Edison, were it not for what Tesla said regarding Edison's adoption of his invention of alternating current power, Edison's name would not be remembered by most, except possibly in history books. Even Westinghouse was not above stealing patents from the man who made him very rich. Sadly, though, he died on March 12, 1914, and his friend, Scottish naturalist John Muir followed by year's end. Tesla had already experienced the death of his dear friend Mark Twain in 1910, a man who in a literary way mirrored Tesla's life. Twain constantly lost money on failed business ventures, which forced him to keep writing. Much of his best work was fast becoming a distant memory.[15] The parade of Tesla's friends who broke their earthly bonds did not end there. Astor had lost his life aboard the *Titanic* in April of 1912. In a few years most of his closest friends were gone—he felt very alone. And yet he would live on for nearly another three decades.

His emotional baggage continued to grow in size, for in the spring of 1913 his confidants and protectors, the Johnsons' were having their own problems when a scandal broke out at *The Century Magazine* that made Robert's position very tenuous. Tesla offered his help, thinking his name would lend legitimacy and hopefully get the powers that be to reconsider Robert's fate. However, ultimately Robert was forced to resign as editor over the "little embarrassment." The cause of the scandal was never revealed, yet Robert managed to become the new permanent secretary of the American Academy of Arts and Letters. Although the couple had to pair back their lifestyle, and time together with Tesla became infrequent, the troika nonetheless remained close friends. Katharine continued to ask to see Tesla as much as possible. Her expression in a note to Tesla said it all: "I want to see if you have grown younger, more fashionable, more proud. But whatever you may be you will always find me the same."[16]

~

If there was a central theme, other than Tesla's desire to make people's lives easier, it was that Tesla never had enough money or time to complete many of his most important projects especially when his manic output and cockiness had deserted him. It is likely Tesla asked himself, as do many bipolar patients, why can't I sustain my mania and do so at a level that is just short of dangerous? It was not to be for Tesla, nor is it for any patient with bipolar disorder. Such a recurring situation would have pressed lesser men to simply give up. But as a proud Serb, giving up was not in him. Because he was always physically fit and blessed with a long, lean frame, which is a very Serbian physical trait, he still looked younger than his years. As one reporter remarked: "Nikola Tesla is fifty-nine years old. Where he keeps his years, no one knows. They are not in his face, for his face looks like forty. They are not in his hair, for his hair is black. If they are anywhere, they are in his eyes, for his eyes are sad. This lone man, who always dines across from a vacant place, has given to the world a long series of wonderful inventions. Tesla's system of electric power transmission is used by all the world, and he has a new turbine which, he says, could drive the *Lusitania* across the Atlantic at the rate of fifty miles an hour."[17]

Tesla added this thought during the interview: "If I cannot send a message to a ship at sea, send it without wires, and make the message understood to those aboard the ship, I am willing to lay my head on the block."[18]

∼

In 1913, "the biggest Wall Street monster," J. P. Morgan, died. Just a few months thereafter, Tesla presented the new head of the House of Morgan, his son Jack Morgan, with the opportunity to finance his Tesla turbine engine project, which showed great promise, inasmuch as Sigmund Bergmann in Germany expressed great interest. Morgan took the carrot, pledging $20,000 in four equal installments. This allowed Tesla to move into new offices in the very stylish Woolworth Building. But with World War I breaking out, the Bergmann connection was no more, and therefore Morgan redirected his energies toward helping the French and British win the war with his money.[19]

Another disappointment: His spirits had been dashed once more. "My enemies have been so successful in representing me as a poet and visionary," he said, "that it is absolutely imperative for me to put out something commercial without delay." In the years to follow, he would time-and-again dig himself out from under one heap of debt only to be faced with another, always believing that his next invention could be commercialized.[20]

His contemporaries, including those in scientific fields and other regular critics, often took great pleasure in attacking him with reckless abandon. It was a classic case of "schadenfreude" that these naysayers had succumbed to over the years. But Tesla was still thinking on the cutting-edge. He had foreshadowed so many of the twenty-first century's indispensable tools—the internet, cell phones, computers, flight, jet power, and so on. The criticism of his only prophesizing and not producing more inventions was at this point in time very difficult for him to shake off. Most people have very short memories (What did you do for me today?), so for Tesla to constantly surprise people with new, amazing inventions became more and more difficult. Not even his Tesla turbine engine could satisfy the hungry hordes.

As we recall from an earlier chapter, the Tesla turbine had been rejected at the Edison testing facility. Nevertheless, he had created several versions of the Tesla turbine, ranging from 100 to 5,000 horsepower—his turbines had the distinct advantage of producing more power per pound than any "bladed" turbine. In an interview in a New York newspaper in the fall of 1911, Tesla stated: "I have accomplished what mechanical engineers have been dreaming about ever since the invention of steam power. . . . That is the perfect rotary engine. It happens that I have also produced an engine which will give at least twenty-five times as much power to a pound of weight as the lightest weight engine of any kind that has yet been produced."[21] Because the necessary alloys had not been available to Tesla, his brilliant engine was decades ahead of its time. However, he did sell a license to the Alabama Consolidated Coal & Iron Company to use his turbine in reverse mode, thus creating a highly efficient pump. He was energized by this. He had again created an invention that became a commercial product. However, the denouement was quite different. As of today, the Tesla turbine has not seen widespread commercial use since its invention and subsequent patent issued in 1913.

Unfortunately, the debts never stopped piling up. He was left no choice but to move his office out of the Woolworth Building to more modest surroundings at 8 West Fortieth Street. To make matters worse, he had no retainers left to do his personal bidding and carry out his wishes. He was alone. Then in 1916, the city he loved and would spend some sixty years living and working in, sued him to force him to make good on his debts of $935 in back taxes. When questioned under oath before the Court how he lived from day to day, he said on credit and that he was penniless. Most of his patents had expired; nine-tenths of the Tesla Company's stock had been pledged to bankers, creditors, and even friends—the company's secretary, George Scherff, was one of the recipients. Tesla even admitted to the Court that he barely earned $350–$400 per month, and that he owned no tangible assets. His finances were a mess. Imagine, Tesla was once a very wealthy man, and now he was reduced to pauper status. Justice Finch of the Supreme Court appointed Robert McC. Marsh receiver to look after Mr. Tesla's earnings.[22] Not having the funds to pay his debts, he knew bankruptcy was imminent. It was a crushing feeling not to have sufficient finances to continue his work, but it was not the first time the inventor was in such a position.

In an effort to somewhat ameliorate his situation, at least psychologically, the august American Institute of Electrical Engineers decided to award him its highest award, the Edison Medal. Although he rarely participated in the organization's affairs, other than serving as its vice president in 1892–1893, the award was made public in December of 1916. His friend, colleague, and the AIEE's award chairman, the prominent Westinghouse engineer B. A. Behrend, visited him to give him the good news and to make arrangements for the award dinner and ceremony. However, the award itself was not presented until May of 1917. Why? Because Tesla thought it an affront to his integrity, and besides, probably three-quarters of the AIEE's membership owed their jobs to Tesla's inventions.[23] Behrend continued to cajole Tesla into receiving the medal that he so richly deserved, telling him that it would help his future work. But he resisted, for it was "principle over profit" to him. Behrend asked for an explanation as to why he refused. Finally Tesla responded, with vitriol in his voice: "Let us forget the whole matter Mr. Behrend. I appreciate your goodwill and friendship, but I desire you return to the committee and request it to make another selection for a recipient. It is nearly thirty years since I announced my rotating magnetic field and alternating current system before the Institute. I do not need its honors and someone else may find them useful."[24] Tesla's words made sense to Behrend, but he wanted more from his friend.

Finally, Tesla answered the request, not backing off from his fiery words: "You propose to honor me with a medal which could pin upon my coat and strut for a vain hour before the members and guests of your institute. You could decorate my body and continue to let starve for failure to supply recognition of my mind and its creative products which have supplied the foundation upon which the major portion of your Institute exists. And when you would go through the vacuous pantomime of honoring Tesla you would not be honoring Tesla but Edison, who has previously shared unearned glory from every previous recipient of this medal."[25]

Admittedly, this was a man who was profoundly disappointed by his colleagues. This was a man who, in the modern parlance, was said to have not only talked the talk but, unlike most in the Institute, had certainly walked the walk. Tesla's unabashed claim to greatness was no psychological tic—it was true—and Behrend understood Tesla to be within his rights to reject the award. However, after several more visits to Tesla's office, the tenacious Behrend finally convinced Tesla to accept the medal.

The day of honor had finally arrived—May 18, 1917—and Tesla appeared at the Engineers Club, attired in white tie and midnight-black swallowtails, looking as if he were a very young leading man. He sauntered about, chatting lightly with friends and courtiers during dinner. When the witching hour arrived, a time for the mystical Tesla to be honored, those in attendance were asked to walk across the street from the Club that faces Bryant Park to the stately United Engineering Societies building for the formal addresses. Suddenly, Behrend realized that Tesla was nowhere to be found. Several committee members searched for him, even backtracking to the Club, thinking he was still there. Behrend suddenly had the painful thought that Tesla had changed his iron mind and returned to his office, knowing that he truly did not want or seek the medal. Then it occurred to him, several times in the past he and Tesla would take long walks in Bryant Park, just across from the Club. There they would discuss all manner of scientific issues.

As dusk signaled the end of the day, Behrend entered the park. Bryant recalled Tesla's love for pigeons and how he would take great care in feeding them wherever and whenever he could. Then a sight that startled and mesmerized Behrend was before him, the elegantly dressed, statuesque Tesla stood on the New York Public Library's plaza . . . feeding pigeons! As observers looked on in amazement, Tesla had his long arms stretched outward, with pigeons alighting on them from all sides, as if he were a scarecrow. Even his head was a crown of pigeons and more were arriving to take part in the great man's generosity—for they were the messengers of God—of food to all feathered friends. Behrend went to approach Tesla when something stopped him. His instinct was correct, for Tesla reached out a hand to warn him not to move any farther. Then, without explanation, all the birds suddenly flew to the ground and encircled his black formal cap-toe pumps. Behrend begged Tesla to come into the building and accept the award. The irony of it all, Tesla would have rather communed with the pigeons, were it his choice at that very moment.[26]

All was set, and Behrend began his opening remarks, noting that it was exactly twenty-nine years ago, to the very day and hour, that Nikola Tesla presented and demonstrated his original polyphase alternating-current system before the AIEE. He then continued with his remarks:

> We asked Mr. Tesla to accept this medal. We did not do this for the mere sake of conferring a distinction, or of perpetuating a name; for so long as men occupy themselves with our industry, his work will be incorporated in the common thought of our art, and the name of Tesla runs no more risk of oblivion than does that of Faraday, or that of Edison.
>
> Nor indeed does this Institute give this medal as evidence that Mr. Tesla's work has its official sanction. His work stands in no need of such sanction.
>
> No, Mr. Tesla, we beg you to cherish this medal as a symbol of our gratitude for a new creative thought, the powerful impetus, akin to revolution,

which you have given to our art and to our science. You have lived to
see the work of your genius established. What shall a man desire more
than this? There rings out to us a paraphrase of Pope's lines on Newton:
"Nature and Nature's Laws lay hid in night / God said, Let Tesla be, and
all was Light."[27]

Tesla's acceptance speech included thoughts on his future work; a re-
luctant tip of the hat to others; a recounting of his difficult times when many
gainsayers attacked him; his mortality; and his still-alive dream of taking his
wireless communication system worldwide. Let us for a moment listen in on
the great inventor:

> I do not speak often in public, and wish to address just a few remarks di-
> rectly to the members of my profession, so that there will be no mistake in
> the future. In the first place, I come from a very wiry and long-lived race.
> Some of my ancestors have been centenarians, and one of them lived one
> hundred and twenty-nine years. I am determined to keep up the record and
> please myself with prospects of great promise. Then again, nature has given
> me a vivid imagination which, through incessant exercise and training,
> study of scientific subjects and verification of theories through experiment,
> has become very accurate and precise, so that I have been able to dispense,
> to a large extent, with the slow, laborious, wasteful and expensive process
> of practical development of the ideas I conceive. It has made it possible
> for me to explore extended fields with great rapidity and get results with
> the least expenditure of vital energy. By this means I have it in my power
> to picture the objects of my desires in forms real and tangible and so rid
> myself of that morbid craving for perishable possessions to which so many
> succumb. I may say, also, that I am deeply religious at heart, although not
> in the orthodox meaning, and that I give myself to the constant enjoyment
> of believing that the greatest mysteries of our being are still to be fathomed
> and that, all the evidence of the senses and the teachings of exact and dry
> sciences to the contrary notwithstanding, death itself may not be the ter-
> mination of the wonderful metamorphosis we witness. In this way I have
> managed to maintain an undisturbed peace of mind, to make myself proof
> against adversity, and to achieve contentment and happiness to a point of
> extracting some satisfaction even from the darker side of life, the trials and
> tribulations of existence. I have fame and untold wealth, more than this,
> and yet—how many articles have been written in which I was declared to
> be an impractical unsuccessful man, and how many poor, struggling writ-
> ers, have called me a visionary. Such is the folly and shortsightedness of
> the world![28]

Lamentably, most members in attendance had no real appreciation for
who stood before them. He was the man who defined the field of the electri-

cal arts, and whose major work—in alternating current generation and distribution—was announced some thirty years before. And sadly, their schools rarely if ever taught of Tesla's revolutionary work that is so prevalent today.[29] Before the night's festivities had come to an end, several other scientists stood in praise of Nikola Tesla.

Tesla kept the Edison Medal under lock-and-key for the rest of his days. After his passing, it inexplicably disappeared, and it has never been found.[30]

$\sim$

In the midst of World War I, and just months after the Edison Medal event, Tesla spoke about his fundamental concepts that would lead to "radar." "Now we are coming to the method of locating such hidden metal masses as submarines by an electric ray. . . . That is the thing which seems to hold great promises. If we can shoot out a concentrated ray comprising a stream of minute electric charges vibrating electrically at tremendous frequency, say millions of cycles per second, and then intercept this ray, after it has been reflected by a submarine hull for example, and cause this intercepted ray to illuminate a fluorescent screen (similar to the X-ray method) on the same or another ship, then our problem of locating the hidden submarine will have been solved."[31]

In 1934, seventeen years later, Dr Emile Girardeau and his French team constructed and installed radar systems on both land and aboard naval ships at sea. Dr. Girardeau's response to the positive outcome was to say that he used "precisely apparatuses conceived according to the principles stated by Tesla." He added, "On the subject of Tesla's recommendation, concerning the very great strength of impulses, one must also recognize how right he was."[32]

Tesla's *imagineering* never seemed to wane. He even foreshadowed what moderns know as "vertical-takeoff-and-landing" (VTOL) aircraft. He also managed to develop it to such an extent that a patent (1,655,113) was issued to him in 1928 as a new "method of aerial transportation," although he had first filed for protection in 1921.[33] It was his last patent, which he received at the astonishing age of seventy-two. Once again, he could not raise the funds to build his creation; however, today it has become a much utilized aircraft in the U.S. Navy and Marines arsenal of troop movers.[34]

Then the nightmare that became Wardenclyffe came back to haunt him on Independence Day 1917, when, as we recall, the mighty tower was brought to the ground for salvage dollars to pay debts. The intimation that the tower might have had a connection to the enemy was also something that deeply troubled Tesla, and he had made his case on more than one occasion. One would think that his suggestion of radar to stop the German enemy would have been enough to "stop" such nonsense. As the tower was falling to earth, one local junkman sadly witnessed Tesla's papers blowing helter-skelter down the deserted street.[35] The dream was officially over.

⁓

Tesla still made every effort to socialize with the Johnsons when possible. They had always thought Tesla would be awarded a Nobel Prize, if not several such awards. Needless to say, Marconi's Nobel Prize in 1909 made their dream seem just that, a dream, and to Tesla, as we know, it was a near fatal blow to him. The day that Tesla heard, he simply took a long walk, and along the way he fed Manhattan's many pigeons. His favorite place to feed them was at Bryant Park just behind the New York Public Library, where once he was asked by a reporter what he was doing. Tesla replied, "These are my sincere friends." The exact location in the city is now identified by a street sign that reads: "Tesla Corner."[36]

Celebrated reporter and Tesla biographer John J. O'Neill wrote in 1944:

> The obvious outward characteristic of Tesla's life was his proclivity for feeding pigeons in public places. His friends knew he did it but never knew why. To the pedestrians on Fifth Avenue he was a familiar figure on the plazas of the Public Library at 42nd Street and St. Patrick's Cathedral at 50th Street. When he appeared and sounded a low whistle, the blue- and brown- and white-feathered flocks would appear from all directions, carpet the walks in front of him and even perch upon him while he scattered bird seed or permitted them to feed from his hand.
>
>   During the last three decades of his life, it is probable that not one out of tens of thousands who saw him knew who he was. His fame had died down and the generation that knew him well had passed on. Even when the newspapers, once a year, would break out in headlines about Tesla and his latest predictions concerning scientific wonders to come, no one associated that name with the excessively tall, very lean man, wearing clothes of a bygone era, who almost daily appeared to feed his feathered friends. He was just one of the strange individuals of whom it takes a great many of varying types to make up a complete population of a great metropolis.[37]

⁓

If there was a time that Tesla could use some good news, it was now. And good news came in the disguise of the most famous of all awards, given to a very select few for very specific reasons. It was the Nobel Prize. On November 6, 1915, the *New York Times* reported that both Nikola Tesla and Thomas A. Edison (the president of Columbia University, who nominated Edison, apologized later for having done so) were to be simultaneously awarded the august Nobel Prize for physics. While the Johnsons and many of Tesla's other friends celebrated, Tesla became very circumspect, for he had not heard a word. He did find that his share of the money would be $20,000, a sum he could truly use, but he was still hesitant to react publicly.[38] That said, he did tell the press that he felt the award was deserved because of his discovery of the "transmission of electrical energy without wires."[39]

Although the hard facts of the full story are lost to history, it seems that Edison also said he knew nothing of the award. The original story grew out of a Reuters News Service dispatch from London. As would be expected, rumor of the unsubstantiated story became too noteworthy not to pursue. The thought of two titans of technology, enemies-to-the-bone, would both receive the Nobel Prize, was a speculation with too much momentum to stop. Although the story continued to gain attention, and a great deal of ink splashed across the newspapers and journals worldwide, the Nobel Prize for physics had already been announced on November 14 in Stockholm. The award went to two English scientists, father and son William H. Bragg and W. L. Bragg, for their use of X-rays in discovering the structure of crystals. We can recall that Tesla's work with X-ray technology gave Roentgen pause.

Once the story proved false, Tesla immediately responded by saying that there was a very distinct difference between what he did and what Edison did. While Edison was an inventor of useful appliances, he, Tesla, was the discoverer of new principles—witness the "rotating magnetic field," as an example. Moreover, Tesla questioned what the Nobel Prize was worth if it was given to Marconi. And moreover, then to share one with Edison, well, it was simply too much to ask. It was an insult to him.[40] In a letter to Robert Johnson written November 10, Tesla said the following in his defense for not receiving the award: "My dear Luka, Thank you for your congratulations. . . . To a man of your consuming ambition such a distinction means much. In a thousand years there will be many thousand recipients of the Nobel Prize. But I have not less than four dozen of my creations identified with my name in technical literature. These are honors real and permanent which are bestowed not by a few who are apt to err, but by the whole world which seldom makes a mistake, and for any of these I would give all the Nobel Prizes which will be distributed during the next thousand years."[41] Tesla, in a moment of cutting sarcasm, also suggested that Edison was "worthy of a dozen Nobel Prizes."[42] He was saying that if Edison received such a prize for simply making a better mousetrap, then what was the prize worth to begin with? Interestingly enough, Edison was aware of his fondness for taking the credit due others. One time, when asked about his inventive genius in a discussion with the governor of North Carolina, Edison responded by admitting that his real genius was in his ability to absorb the ideas of others, much like a sponge, and then improve upon them.[43]

~

As aging took its toll, Tesla continued to try and keep ahead of its ravaging of body and mind. Although he had become a scientific outsider to many, he still had ideas worthy of investigation; they were not merely hobbyhorses. The VTOL technology aside, he never fully took advantage of his work on robots. Some suggest that his stubbornness often got in the way. In the case of one member of the wealthy Hammond family, Tesla was convinced that he was stealing his telautomatics ideas. However, it was reported by Jack Hammond in 1903, that Tesla had filed for patent protection for a "prophetic genius patent." It allowed for frequencies of various combinations to send specific commands to machines or lights. Regrettably, Tesla's obstinance pushed the Hammonds away, despite there being an obvious connection between them. John Hays Hammond Jr., an inventor himself, had praised Tesla, defending his scientific methodology—it could neither be replicated by rigid research nor by the simple expenditures of endless dollars. In a direct quote, he said, "It takes a profound personal dedication for a man to achieve high inspiration, and this Nikola Tesla had."[44]

Stubbornness is often an outgrowth of high-level thinking and the successes that result from such thought. However, Tesla's egotism born of great triumphs; his obstinacy the result of believing he was right, were fast becoming observed characteristics by many. Scientists, journalists, and business associates alike all complained about the difficulty of working with him. But one cannot help but think that because of Tesla's many profound successes—albeit now fading in the distance—his unpredictable genius was excusable to some degree.[45] He was firmly convinced that his wireless transmission of power was undoubtedly the most important invention of all time.[46]

In order for Tesla to survive financially during what he believed to be a momentary lapse in income, he took to writing articles for *The Electrical Experimenter*, founded by science editor Hugo Gernsback. As we know, Tesla's close friend Mark Twain was forced to do the same—to keep writing to pay his bills until he wrote another literary classic.[47] So, in 1919, Tesla announced an entirely new concept to a world that had given very little thought to the aging inventor:

> As I stated on a previous occasion, when I was a student at college I conceived a flying machine quite unlike the present ones. The underlying principle was sound but could not be carried into practice for want of a prime-mover of sufficiently great activity. In recent years I have successfully solved this problem and am now planning aerial machines devoid of sustaining planes, ailerons, propellers and other external attachments, which will be capable of immense speeds and are very likely to furnish powerful arguments for peace in the near future. Such a machine, sustained and propelled entirely by reaction . . . is supposed to be controlled either mechani-

cally or by wireless energy. By installing proper plants it will be practicable to project a missile of this kind into the air and drop it almost on the very spot designated, which may be thousands of miles away.[48]

Tesla had a trait not uncommon to great thinkers. He had the uncanny ability to divorce himself from the consequences of being in debt. Thus, in spite of such ideas, by 1928 Tesla had been forced to close his office and furlough his two longtime secretaries Dorothy Skerrit and Muriel Arbus and send some thirty trunks of all manner of documents into storage. However, he did maintain a line of communication with his assistant, George Scherff, who by now, out of necessity, was burdened with being his advisor, accountant, fundraiser, and general ombudsman. In time, even Scherff faced the dilemma of being caught in a financial pinch, informing Tesla that he needed a stable income, and Tesla could not provide it. Tesla did promise him that when his millions came in, he would make it right with him. But the millions never came, again, and after the Wardenclyffe project was no more, Scherff let Tesla know that he could no longer afford to stay, and that he would be seeking employment at the Union Sulphur Company, where Tesla had previously been a consultant on retainer.[49]

As Tesla's age continued to challenge him in ways he would not accept, he became locked into his old ways of thinking, unable to accept new ideas, unless they were his. He took issue with Einstein's work that showed splitting the atom resulted in energy released and said that the theory of relatively was nonsense. But then again, Einstein was no different in his old age, rejecting the new idea of "quantum mechanics." Einstein's midlife transition was characterized by a move from a revolutionary to a conservative point of view, as he became close-minded about quantum theory. He could not square the new quantum mechanics with his search for a unifying theory that would merge with relativity and restore certainty to nature. His efforts to do so would be the focal point of the waning last years of his scientific endeavors.[50] So, in some ways, Tesla's resistance to new ideas was not unusual for great thinkers.

As the slide into the abyss continued, Tesla still managed to have magical thoughts that pierced the growing depression but without the benefit of his old friend, mania.

· 16 ·

Strange Brew

Bizarre, Psychosis, Reality Testing, Magical Thinking

NEW YORK CITY, NEW YORK, 1920

By 1920, Tesla had already licensed the first air-friction speedometer to the Waltham Watch Company. This mechanism was to be installed in various luxury cars, such as the Rolls-Royce, Lincoln, and Pierce-Arrow. The company was so impressed with Tesla's speedometer that it even cited his name in its advertising campaign. Tesla's work on measuring devices for commercial use did not stop there. He extended his technology in the field to metering devices to measure flows and frequencies.[1]

He had also begun, a year earlier, the arduous, sometimes painful, and often joyous recollection of his life from his earliest days as a child in the Austro-Hungarian Empire to the zenith of his inventive powers. His autobiography was published in 1919, in serialized format (six segments) in a new futuristic magazine (journal), *The Electrical Experimenter*. It was a widely read publication, whose editor, writer, and publisher Hugo Gernsback, was excited by the prospect of Tesla's life appearing in his magazine, for which he paid Tesla for his time. Needless to say, given his present financial situation, the monetary reward was not enough for Tesla. Tesla's work for Gernsback and the publication cost him several thousand dollars, inasmuch as he had said a friend had offered him several thousand dollars for a two-hundred-word essay.[2]

He was leery about writing technical articles for Gernsback, because he feared that others would take and infringe on his established patents. He was in constant pitched battles to protect his primacy in the art of radio communications, and any attempt to circumvent his patents would be met with legal actions. By the late summer of 1915, Tesla had filed patent infringement lawsuits against Marconi—challenging the U.S. patent issued to him in 1904. The next year Marconi answered with a lawsuit charging the U.S. govern-

248

ment with an unauthorized use of his wireless during World War I. This charge aggravated then acting secretary of the Navy, Franklin D. Roosevelt, who suggested that the federal government's case against Marconi would be made better by Tesla's correspondence with the Light House Board that clearly showed Tesla's wireless system could be made available—the initial case was adjudicated in France, whose highest court ruled that Tesla's invention had predated Marconi's.[3] Tesla had testified against Marconi's 1896 patent saying that it was but a technical mess, "a mass of imperfection and error." He added: "If anything, it has been the means of misleading many experts and retarding progress in the right direction." The French electrical engineer M. E. Girardeau, who testified on behalf of the company suing Marconi, added his opinion of Tesla's work when he said, "Indeed, one finds in the American patent [Tesla's] extraordinary clearness and precision, surprising even to physicists of today. . . . What a cruel injustice would it be now to try to stifle the pure glory of Tesla in opposing him scornfully."[4] It is most likely that Tesla was financially compensated to some degree.

However, one victory in battle did not make the winning of the war any easier. The government's lawsuit made its way through the U.S. court system, with one decision in favor of Marconi, then another would go in favor of Tesla: it was legal ping-pong. The irony of ironies was that Tesla himself took a rather unexpected, dim view of the radio itself: "The radio, I know I'm its father, but I don't like it. It's a nuisance, I never listen to it. The radio is a distraction and keeps you from concentrating."[5] Today he would most likely hold the same view regarding the television, of which he had a considerable hand in inventing as well.

The court appearances for all sides continued unabated, with experts representing all claimants. Laughably, Michael Pupin, a fellow Serb and one with whom the reader is familiar, surprised everyone when testifying in the 1915 lawsuit of Marconi against Atlantic Communications. The Columbia University professor took credit for inventing the radio! Pupin had several times before metaphorically stabbed Tesla in the back, but now he said that "he had found a wireless wave but had not realized its importance. Tesla had given his discoveries to mankind, and this is one of the points on which the Atlantic Company experts expect to deny claims of Marconi to certain wireless."[6] Years later Pupin brazenly penned a bestselling book that rewrote the history of electrical science, eschewing any attribution of Tesla to the electrical art that he truly had revolutionized. Pupin received major backlash from most leading scientists for having slighted Tesla. Included among Tesla's defenders was one of his most vocal critics, Charles Steinmetz, another Bell Labs scientist.[7] Although certainly and rightfully aggrieved by these false claims to his inventions, the Tesla of this era pursued these interlopers with much of his available energy rather than pursuing new inventions. The Tesla of old, in a grandiose state, may have thought, "You can steal my ideas, but I can have another and you cannot!"—and just moved on to the next invention. Here he was in a more embittered and negative state with these larcenist competitors.

One named expert, John Stone Stone, a research scientist, inventor, and theoretician working at Bell Labs, testified in 1916 that Marconi had missed the mark, saying that "Marconi's view led many to place an altogether too limited scope to the possible range of transmission by the system of grounded, vertical antennae."[8] Stone believed that modern radio technology had circled back to where Tesla had been all the time. He added in closing, "I misunderstood Tesla. I think we all misunderstood Tesla. We thought he was a dreamer and visionary. He did dream and his dreams came true, he did have visions but they were a real future, not an imaginary one."[9]

By 1917, Tesla had managed to empty his pockets and bank account of whatever was left, always believing that his coffers would fill once again; however, he was politely asked to leave the Waldorf-Astoria—his home for twenty years—for a debt of some $18,000 ($400,000 in today's money). With no credit left, he managed to find another luxury suite at the Hotel St. Regis, although only seven years later he found it necessary to leave for unpaid debts. He then moved to The Pennsylvania Hotel and took up feeding his pigeons in a very serious way. In the end, he was also asked to leave because his friends—the unsanitary pigeons—had caused a stink, both in the air and among the other residents.[10]

He spent a good amount of time in parks, rescuing pigeons that had been injured or were sickly. He was so devoted to his feathered friends that it was said he constructed nests on his window ledge and even fashioned a basic shower for them in his room. He had created a bird sanctuary of sorts. If his efforts to help a pigeon failed, then he would take it to a veterinarian, often spending princely sums to save the pigeon. There was one he paid the most attention to. It was a glorious white in color with light gray on its wingtips. Tesla described his unique "romance" with this pigeon to his first biographer, John J. O'Neill: "No matter where I was that pigeon would find me; when I wanted her I had only to wish and call her and she would come flying to me. She understood me and I understood her. . . . I loved that pigeon, I loved her as a man loves a woman, and she loved me. . . . As long as I had her, there was purpose in my life."[11]

Tesla goes on, describing to O'Neill the night when the pigeon came to him:

> "As I looked at her I knew she wanted to tell me—she was dying. And then, as I got her message, there came a light from her eyes—powerful beams of light. . . .

"Yes," he continued, again answering an unasked question, "it was a real light, a powerful, dazzling, blinding light, a light more intense than I had ever produced by the most powerful lamps in my laboratory.

"When the pigeon died, something went out of my life."[12]

Tesla felt that now his life's work would never be finished.

Tesla explained why his devotion to his feathery friends was so necessary to him. In a 1926 magazine interview the stated: "Sometimes I feel that by not marrying, I made too great a sacrifice to my work, so I have decided to lavish all the affection of a man no longer young on the feathery tribe. . . . To care for those homeless, hungry or sick birds is the delight of my life. It is my only means of playing."[13]

The Roaring Twenties was not the best of times for the inventor; he was barely surviving on his moderate royalties remaining from patents that had not yet expired. After having hired attorney Ralph J. Hawkins to handle his ongoing patent matters, he ultimately could not pay the man for his services in the amount $913, thus he found himself in court yet again. His friend Hugo Gernsback (some say it was B. A. Behrend) felt embarrassed for Tesla's lack of funds to live properly, so some years later he managed to have the Westinghouse Company put him on a meager stipend of $125 per month as a consulting engineer and paid for his residency at Hotel New Yorker—suite 3327, to be exact, because it was divisible by three. It comprised two rooms: one for sleeping and the other stuffed with a desk and his ongoing work.[14]

Then in 1925, yet another tragedy struck Tesla, dealing a near death blow to his very weak psyche. His dear Katharine Johnson, who had moved with her husband, Robert, to Italy, in 1920, where he was the new U.S. Ambassador, had been a victim of the great influenza epidemic. Tesla tried to prescribe a special diet, but nothing helped. She lay dying in her Lexington Avenue home, closed off from the world, remembering all the good times with family and friends. She took inventory of her life, now believing that she had lived on the marginalia, engulfed by the successes and failures of others. Her fate inspired Tesla to give an interview on "women," and his thoughts about them.[15] During the interview some months after Katharine's passing, he said this:

.This struggle of the human female toward sex equality will end in a new sex order, with the female as superior. The modern woman, who anticipates in merely superficial phenomena the advancement of her sex, is but a surface symptom of something deeper and more potent fermenting in the bosom of the race.

It is not in the shallow physical imitation of men that women will assert first their equality and later their superiority, but in the awakening of the intellect of women.

Through countless generations, from the very beginning, the social subservience of women resulted naturally in the partial atrophy or at least the hereditary suspension of mental qualities which we now know the female sex to be endowed with no less than men.[16]

Tesla was undoubtedly ahead of his time, even when speaking about the human condition.

Before her passing, Katharine Johnson requested that her husband stay in contact with Tesla. She succumbed to her illness in the autumn of 1925, at the age of sixty-nine. Tesla was devastated; his protector was gone. In October, Luka (Robert) wrote a short missive to Tesla. "Dear Tesla, it was Mrs. Johnson's injunction that last night of her life I should keep in touch with Tesla. This is a pretty hard thing to do, but it will not be my fault if it is not done. Yours faithfully, Luka."[17]

Thankfully, Robert and his daughter Agnes (Holden) continued to keep Tesla in their thoughts, inviting them to holiday dinners and the yearly celebration of Katharine's birthday. All the while, Robert was having his own financial difficulties; Tesla did what he could for him despite his own tenuous financial situation. Once flush with Tesla's monetary donations, he and his daughter made their way to Europe for a vacation.[18] The fact is the relationship was never the same between the two men after Katharine's death.

In a moment of reflection after Robert's passing, Tesla came to the painful realization that he was outliving most of his friends and enemies and even his favorite pigeon. It was all so very bittersweet. He had even outlived another nemesis, Michael Pupin. Tesla, in a moment of great grace, appeared in Pupin's hospital room and held out his hand to greet the dying man. He told Pupin that soon the two would be meeting in the Science Clubrooms and discuss matters of interest as before. After Tesla left, Pupin passed away. Tesla very kindly attended the funeral of the man who spent much of his life attacking his fellow Serb.[19]

In time Tesla gained a new friend, the nineteen-year-old science writer who freelanced for many of the day's most popular magazines from *Popular Science Monthly* to *Life*. His name was Kenneth Swezey, and he made it his personal charge to make certain that Tesla's legacy did not end up in the dustbin of history. Why? After building his own radio at the age of thirteen, he began to learn all he could from scrutinizing information on wireless communications.

It was then that he came to the fast realization that Tesla was its inventor, and he desperately wanted an interview with the mysterious man.

Swezey had a boyish-looking face, even for a teenager, and he and Tesla became fast friends until the inventor's death. Swezey very quickly came to be Tesla's new publicist and over the years their familiarity with each other knew no bounds. When the young man entered Tesla's life, the rumors of homosexuality surfaced, yet Swezey himself said emphatically that Tesla was "an absolute celibate." Swezey made it a point to compile every Tesla manuscript, letter, article that he could get his hands on that would assist John J. O'Neill in writing Tesla's classic biography of the time.[20]

NEW YORK CITY, NEW YORK, 1930

As the Roaring Twenties shut the door on the second decade of the twentieth century, Swezey organized a special seventy-fifth birthday party in 1931 for the mysterious recluse, the Wizard of Electricity, the man who did more to change the world than any mortal who came before or after. Swezey entreated many of the world's most respected scientists to say a few words about how they felt about Tesla. They all were honored to do so, because Tesla had inspired them all, many of whom were Nobel Laureates such as Robert Millikan, Arthur H. Compton, and W. H. Bragg. Bragg said of Tesla, "I shall never forget the effect of your experiments which came to dazzle and amaze us with their beauty and interest."[21] Hugo Gernsback added his thoughts to the memorial book: "If you mean the man who really invented, in other words, *originated* and *discovered*—not merely improved what had already been invented by others—then without a shade of doubt Nikola Tesla is the world's greatest inventor, not only at present but in all of history. . . . His basic as well as revolutionary discoveries, for sheer audacity, have no equal in the annals of the intellectual world."[22]

When complete, the bound memorial volume was presented to Tesla, whose response was rather tepid, as if he had expected it, and well he should have. After all, he cared not for the praises of others who had opposed him all his life. In addition, *Time* magazine did a cover story in July of 1931 on Tesla (he appeared on the front cover painted by Princess Vilna Lwoff-Parlaghy), and myriad other magazines and newspapers the world over also did stories on him. *Time* in particular reported that just finding the recluse proved to be problematic. When they finally found him, he appeared gaunt, a sad pallor had taken over his skeletal-looking face, yet his blue eyes were excited and his voice indicated that he was still thinking at a very high level. In fact, he told the interviewer that he was working new projects, one of which was a new power source and the other was the disproving of Einstein's General Theory of Relativity.[23] He also startled the interviewer when he referred cryptically to his most mysterious invention, the "Teslascope," a device for signaling the nearby stars and went on to say: "I think that nothing can be more important than interplanetary communication. It will certainly come some day, and the certitude that there are other human beings in the universe, working, suffering, struggling like ourselves, will produce a magic effect on mankind, and will form the foundation of a universal brotherhood that will last as long as humanity itself."[24]

Here again Tesla's prediction of interplanetary communication was looked upon as absurd at the time, while the same was said of his belief in communicating with and traveling to Mars. Yet, as we know, on February 18, 2021, the United States' Mars rover named *Perseverance* landed on the Red Planet.

The same year as the *Time* cover story of Tesla was published, his lifelong nemesis Thomas Edison died in October at his home "Glenmont," in West Orange, New Jersey. Very much like Thomas Jefferson and John Adams, who never liked each other, yet in their later years exchanged cordial letters, so too did Tesla and Edison. But the truth be known, Tesla did not hold Edison in high regard, stating that his "method was inefficient in the extreme." Moreover, Tesla would often say of Edison that his rejection of any book learning or knowledge of mathematics caused him to waste a good deal of time using only the empirical methods of investigation.[25] Moreover, Tesla thought Edison to be a mere tinker, while he was a creator, a discoverer.

Perhaps it was Hugo Gernsback who best quantified the differences between Tesla and Edison. "Without wishing to minimize Edison's tremendous amount of work, the fact is well known that he is not so much an original inventor as a genius in perfecting existing inventions. In this respect Tesla has perhaps been the reverse for he has to his credit a number of brilliant as well as original inventions which, however, have not been [as of yet] sufficiently perfected to permit commercial exploitation."[26]

On Tesla's seventy-sixth birthday, the *New York Times* reported on Tesla's latest invention.[27] Tesla related that he was born at the stroke of midnight between July 9 and 10, therefore, he accepted either day as his birthday and went on to say of his new invention: "It will be like the shout with which Joseph's army brought down the walls of Jericho."

Yet, with all the accolades, with all the fuss, the aging genius was still fighting the war within himself. He would have bizarre visions; he could be a curmudgeon at times; and he would on occasion simply be rude to others, particularly women. One of his secretaries was summarily fired for being overweight and sloppy, while another was turned away from the office and sent home because he took issue with the dress she was wearing. Some of his friends thought his love for pigeons was his way of redirecting his affections, since his attitude toward women was so harsh.[28] Moreover, as the inventor aged, close physical contact with others was problematic because his OCD and overt fear of germs caused him to demand that the hotel staff keep its distance. He refused to shake hands with others and washed his hands incessantly. His eating habits eventually led him to a plant-based diet, and he eschewed all stimulants of any kind. He still maintained his immaculate dress but shielded himself from the little beasties that inhabit life with single-use gloves. However, there were exceptions, such as family. He would greet his grandnephew William Terbo with the traditional Serbian hugs and three kisses, alternating cheeks.[29]

Later in life Tesla established a yearly public-relations event to take place on his birthday. Reporters, admirers, and those of celebrity would attend a press conference wherein he would pontificate on his latest work. The 1933 event was what one would expect from the man who lit the world. He stated emphatically—understanding that many had laughed as his predictions before—that his discovery was "so basic that it will undo the Einstein theory of relativity." He added, "They called me crazy in 1896, when I announced the discovery of cosmic rays. Again and again they jeered when I developed something new and then, years later, saw I was right. Now, I suppose, it will be the same old story when I say I have discovered a hitherto unknown source of energy—unlimited energy that can be harnessed."[30]

In spite of receding into his secret world more and more, he always managed to attract the press like no other, no matter what he said. As an example, an August 10, 1932, article in a major New York newspaper's headline read: "Chewing Gum More Fatal Than Rum, Says Tesla."[31] As we now know, it had been his custom for years that on his birthday he would make futuristic predications of what he was about to do next, and his seventy-eighth birthday press conference was no exception. Although some tried to erase him from popular history books, he continued to imagine a future that others could not. In fact, his last years were spent with possibly his most mysterious project yet. It was his "death-beam."

Holding court in the New Yorker Hotel, a day after his birthday, in 1934, Tesla gave a very compelling, if disconcerting, interview with the *New York Times*. In it he let it be known what was on his mind. He had finalized a method and apparatus, which will transmit concentrated beams of particles [tungsten or mercury] through the free air, resulting in an energy force that will neutralize a fleet of ten thousand enemy airplanes at a distance of 250 miles from a defending nation's border and will cause armies of millions to drop dead immediately. It was his "death-beam." Tesla told the reporter: "This 'death-beam' will operate silently but effectively at distances 'as far as a telescope could see an object on the ground and as far as the curvature of the earth would permit it.' It will be invisible and will leave no marks behind it beyond its evidence of destruction." Tesla also told the reporter: "When put in operation, this latest invention of his would make war impossible. This death-beam would surround each country like an invisible Chinese wall, only a million times more impenetrable. It would make every nation impregnable against attack by airplanes or by large invading armies."[32] Could this be the U.S. military's "Star Wars?"

Then again on his seventy-ninth birthday celebration in 1935, before an adoring crowd of well-wishers, eager reporters, and people looking to have their curiosity satisfied, Tesla sat in repose, a stately manner apparent, impeccably dressed, with high concepts flowing from his lips centered within an emaciated face, leaving reporters flummoxed as to just what was the great man saying. One reporter's observation was most telling: "Twenty-odd news-papermen came away from his Hotel New Yorker birthday party yesterday, which lasted six hours, feeling hesitantly that something was wrong either with the old man's mind or else their own, for Dr. Tesla was serene in an old-fashioned Prince Albert and courtly in a way that seems to have gone out of this world."[33]

Although Tesla's killer beam is often referred to as a "death ray," he was insistent that it used particles of mercury that would be accelerated to a speed of at least forty-eight times the speed of sound. As he explained: "I want to state explicitly that this invention of mine does not contemplate the use of any so-called 'death rays.' Rays are not applicable because they cannot be produced in requisite quantities and diminish rapidly in intensity with distance. All the energy of New York City (approximately two million horsepower) transformed into rays and projected twenty miles, could not kill a human being, because, according to a well-known law of physics, it would disperse to such an extent as to be ineffectual.

"My apparatus projects particles which may be relatively large or of microscopic dimensions, enabling us to convey to a small area at a great distance trillions of times more energy than is possible with rays of any kind. Many thousands of horsepower can thus be transmitted by a stream thinner than a hair, so that nothing can resist. This wonderful feature will make it possible, among other things, to achieve undreamed-of results in television, for there will be almost no limit to the intensity of illumination, the size of the picture, or distance of projection."[34]

This begs the question: Did Tesla ever design the death-beam weapon? The short answer is yes. Not much was known as speculation ran rampant, but in 1984, a paper, later confirmed to be genuine by the Tesla Museum in Belgrade, Serbia, surfaced that showed although Tesla still had several problems to work out, his principle behind such a weapon was sound. In 1934, he tried to sell the concept to Jack Morgan, but once again, Morgan said no. Add to that the fact that Tesla attempted to use it as a bargaining chip against his debt of $400 to the Governor Clinton Hotel, and so his desperate financial straits were magnified once again. Even several European countries expressed grave concern that his particle beam could fall into the hands of the enemy and you have the makings of a science-fiction movie.[35]

Needing money, Tesla tried every way he knew to sell the death-beam. The United States and Great Britain were his first thought, while the Amtorg Trading Company, an alleged Soviet firm, had paid him the equivalent of some $450,000 in today's dollars to provide all that was necessary to build its own death-beam—the money figure does not seem plausible, given Tesla's constant need of money. It was also reported that Tesla's laboratory had been broken into, but nothing of value was taken because all the real valuables were stored in his head. There were spies and opportunists who also had a hand in trying to obtain the weapon (if it existed at all). However, his death-beam does bear a striking similarity to a particle-beam accelerator developed during the Cold War by the United States, and the Soviet Union.[36]

Consider this. During the Cold War, the United States and its military were frantically searching for an advantage against the Soviet regime. The name Nikola Tesla still held great fascination, so the military was convinced that the "death ray" was real, and that the United States needed to have it; after all, it was invented in their country by one of its citizens. So the powers that be set to work on what was called "Project Nick."[37]

Today, rumors of "Project Nick" still circulate in military circles and among conspiracy theorists.

~

Tesla's endless efforts to seek money to live and funds to invest in his latest concepts usually ended in great disappointment, yet in 1936, he celebrated his eightieth birthday, with well-wishes sent to the physically frail yet vitally thinking inventor from as far away as Russia (Soviet Union), France, Japan, and Great Britain. He still had endless ideas to present to the public at large, but most people had moved on.

By 1937, life was slowly leaving his physical being—one day, he had been hit by a city taxi and slammed to the ground while crossing a street in the city. Adamantly rejecting any thought of seeing a doctor, he gradually made his way home, only to be bedridden for some six months, broken bones and all. In spite of his injuries, he managed to send a message for someone to bring pigeon feed to his hotel room. This accident and constant financial problems set in motion his final decline into the dark abyss—he walked the rest of his days aided by a cane.[38]

During that same year, another tragedy blindsided the sickly inventor. Robert Johnson died at the age of eighty-four. Now Tesla had lost his other "protector," who had kept his promise to his wife Katharine to look after Tesla.[39] There would be no more dinners, soirees, deep salon conversations, and holiday celebrations. It was to be no more. Tesla continued to outlive all his friends and enemies, but he knew death was soon to arrive for him as well.

At this uncertain time Tesla became deeply involved in the nexus between religion and science. He said of both:

> There is no conflict between the ideal of religion and the ideal of science, but science is opposed to theological dogmas because science is founded on fact. To me, the universe is simply a great machine which never came into being and never will end. The human being is no exception to the natural order. Man, like the universe, is a machine. Nothing enters our minds or determines our actions which is not directly or indirectly a response to stimuli beating upon our sense organs from without. Owing to the similarity of our construction and the sameness of our environment, we respond in like manner to similar stimuli, and from the concordance of our reactions, understanding is born. In the course of ages, mechanisms of infinite complexity are developed, but what we call "soul" or "spirit," is nothing more than the sum of the functionings of the body. When this functioning ceases, the "soul" or the "spirit" ceases likewise.
>
> I expressed these ideas long before the behaviorists, led by Pavlov in Russia and by Watson in the United States, proclaimed their new psychology. This apparently mechanistic conception is not antagonistic to an ethical conception of life. The acceptance by mankind at large of these tenets will not destroy religious ideals. Today Buddhism and Christianity are the greatest religions both in number of disciples and in importance. I believe that the essence of both will be the religion of the human race in the twenty-first century.[40]

By 1942, the deleterious effects of aging had assigned Tesla to his bed most of the time. He permitted no one to enter his room except for the hotel staff to see to his personal needs. He talked to no one and kept secluded. To pay some of his living expenses and other requirements, the Nikola Institute in Yugoslavia arranged for a monthly stipend of $600 to be paid to its most famous "son" beginning in 1939.[41] This helped reduce and in some cases, defer the mounting debts.

His ongoing convalescence increased his irrationality at times and even exacerbated the onset of early-stage senility, yet his moments of absolute lucidity outnumbered the negative aspects of his present situation. In July, he requested that his favorite Western Union messenger boy William Kerrigan deliver an envelope containing $100 to his dear friend Mark Twain (he died in 1913) at 35 South Fifth Avenue, which happened to be the address of his old laboratory, before its name was changed to West Broadway. When Kerrigan returned to Tesla's hotel room, he told the famous inventor that he could not deliver the message because there was no such address and because Mark Twain had died many years ago. Tesla rejected the idea that his friend was dead, and he told the messenger boy in a forceful tone to either deliver the envelope to his friend Mark Twain because he needed the money or to keep it himself.[42]

~

On a snowy Serbian Christmas, January 7, 1943, room 3327 on the thirty-third floor of the Hotel New Yorker suddenly became famous and that fame would live on to this very day. It was the very room where Nikola Tesla died in his sleep at the age of eighty-six. The next day, Alice Monaghan, the floor maid, remembered she had put the "Do Not Disturb" sign on the door the day before at Tesla's request, but for some inexplicable reason, she decided to enter the room. There she discovered his silent body in sublime repose with a gaunt face looking satisfied (a death mask was made later in the day by Hugo Gernsback). The assistant medical examiner, H. W. Wembley, had noted the time of death to be 10:30 p.m., January 7, 1943. He added that the cause of death (under no suspicious circumstances) was coronary thrombosis.[43]

Nikola Tesla was born at the stroke of midnight during a powerful thunderstorm. It would be his lifelong dream to "lift the burdens from the shoulders of mankind." And to that end, his dreams did come true, except for one, for he was now gone forever . . . into the ether. But what he would *never* know, was that his dream of being rightfully known as the "father of the radio" would indeed come true.

The U.S. federal court case (No. 369) known as *Marconi Wireless Telegraph Company of America v. the United States*, took several decades of legal wrangling to make its way through the complicated federal court system. However, a 1935 legal ruling by the United States Court of Federal Claims invalidated the fundamental Marconi patent because it was "anticipated" by Tesla. . . . On June 21, 1943, the United States Supreme Court upheld the ruling. At the very least, Tesla finally gained a legal victory over Marconi, and his dream was realized. He was and still is the father of the radio![44]

There would be a postmortem.

IV

DEEP IN THE ABYSS

• *17* •

Into the Ether

Postmortem

NEW YORK CITY, NEW YORK, 1943

$\mathcal{T}$he morning of January 12 was marked by more than two thousand mourn-
ers crowding into the massive Cathedral of St. John the Divine. For the first
time in some twenty years the great golden doors of the city's cathedral were
unlocked in an effort to accommodate the throngs that wished entrance.
Noting the ongoing contretemps between Serbians and Croatians, Episcopal
Bishop William Manning said that no political speeches were allowed. Tesla's
coffin appeared in the colossal cathedral as if it were the center jewel in a king's
crown. It was draped in both American and Yugoslav national flags. The long,
traditional funeral service was conducted in Serbian by Orthodox Christian
priests led by the Very Reverend Dusan Sukletović. Because Tesla's country
of birth deemed the funeral an official state function, the Yugoslav government encircled the casket with a dozen soldiers and was officially represented
by Constantine Fotitch, Yugoslav Ambassador to the United States.[1] Despite
Serbian Orthodox Christian church tradition and dogma, Tesla's nephew Sava
Kosanović made the very monumental decision to have his uncle's remains
cremated.[2] Subsequently, in February of 1957, the cremated remains traveled
with Charlotte Muzar, Kosanović's secretary, to the Nikola Tesla Museum
(Belgrade, Serbia), and they are presently on display in a brilliant, gold-plated
orb set atop a marble pedestal, all of which is exquisitely backlit.

$\sim$

The immediate days following Tesla's passing were filled with endless praise
for the inventor and condolences, beginning with President Franklin D.
Roosevelt's wife, Eleanor, who said, "The President and I are deeply sorry to
hear of the death of Mr. Nikola Tesla. We are grateful for his contribution to

science and industry and to this country."[3] Vice President Henry A. Wallace said of Tesla that "in Nikola Tesla's death the common man lost one of his best friends."[4] A gaggle of Noble Prize winners, including Millikan, Compton, and Franck joined in a eulogy that called Tesla "one of the outstanding intellects of the world who paved the way for many of the important technological developments of modern times."[5] The city's mayor, Fiorello La Guardia, gave a resounding tribute in a radio broadcast, and David Sarnoff, the president of RCA said of Tesla, "Nikola Tesla's achievements in electrical science are monuments that symbolize America as a land of freedom and opportunity. . . . Tesla's mind was a human dynamo that whirled to benefit mankind."[6] Even Sarnoff understood what drove Tesla to succeed beyond all measure: his desire to benefit mankind.

The list of honorary pallbearers included many of the leading scientists, luminaries, and captains of industry: Dr. E. F. W. Alexanderson of the General Electric Company, Professor E. H. Armstrong of Columbia University, Consul General D. M. Stanoyavitch of Yugoslavia, and Dr. H. C. Rentachler of Westinghouse Electric & Manufacturing Company.

The obituary that ran in the *New York Times* described Tesla as the "father of radio and of modern electrical generation and transmission systems."[7] The City of New York, where Tesla lived for some sixty years, was represented by Newbold Morris, president of the City Council.[8]

One very insightful science writer said with great prescience: "He was so far ahead of his contemporaries that his patents often expired before they could be put to practical use."[9] This was a central problem inherent in Tesla's inventing style. He was never driven by how his inventions' commercial use would make him money, while Edison and virtually all other inventors were driven by financial gain.

It could be said that the *New York Sun* just might have had the most on-target evaluation of Tesla when it printed the following after his death:

> He was a nonconformist, possibly. At any rate, he would leave his experiments and go for a time to feed the silly and inconsequential pigeons in Herald Square. . . . Granting he was a difficult man to deal with . . . here, still, was an extraordinary man of genius. He must have been. He was seeing a glimpse into that confused and mysterious frontier which divides the known from the unknown. . . . But . . . at times he was trying with superb intelligence to find the answers. His guesses were right so often that he would be frightening. Probably we shall appreciate him better a few million years from now.[10]

One will never know, the extent of his genius, because he took so much with him on his endless journey across the heavens that began in 1856.

   Tesla's remains were sent to the Ferncliffe Cemetery at Ardsley-on-the-Hudson on that horrifically cold winter morning. The hearse was followed by a car containing Swezey and Kosanović.[11]

   But prior to the funeral and the weeks that followed, something very secretive happened in room 3327 at the Hotel New Yorker. What was left behind?

   Immediately after Tesla's death, the FBI, the Office of Alien Property Custodian (OAPC), and various principals of the War Department all gathered to take control of and protect the materials pertaining to Tesla's secret-weaponry, which he had conceived of as a way to stop all wars and to ensure that they never happened.[12] More to the point, the U.S. government became rather concerned that the documents related to Tesla's particle beam (also called the death ray) and his personal papers, especially his technical papers and schematics, contained information related to national security. This concern and the subsequent release by the FBI under the Freedom of Information Act of several hundred of Tesla's papers only served to excite conspirators and opportunists. When one reads through the heavily "redacted documents," it cannot be said with certainty that they contain no information that should be protected.[13]

   When his nephew Sava Kosanović arrived the morning after his uncle's death at the Hotel New Yorker, he fully expected to find a will. No will. He then assumed, because of his relationship to Tesla, that he would be the recipient of his uncle's estate, which he said was of "value unknown."[14] He also thought that, since he held a high position as director of the Eastern and Central European Planning Board for the Balkans, there would be no problem. So Kosanović called in a locksmith to open the safe. Here it must be said that the chain of possession is still questionable. Some sources state that Kosanović ordered the safe opened, and no one truly knows if he took anything of importance—the FBI had its own suspicions. Even the quantity of items taken varies from source to source. However, some sources state without reservation that around January 9 officials of the OAPC, led by Walter C. Gorsuch, seized all of the inventor's belongings, including instruments, plans, and other paraphernalia, and had them sent under seal directly to Manhattan Storage and Warehouse Company. The OAPC also caused two truckloads of additional papers, artifacts, and other materials, as well as a very large number of other documents, to be joined with what had already been seized from Tesla's room and placed under OAPC seal.[15]

   From this point on, conspiracy theories appeared from all ends—the FBI, other security agencies, and various factions within the U.S. military, as well as from Kosanović himself. To hopefully quell any suspicions and get the facts of what was in the seized property, the OAPC appointed John G. Trump, a scientist and expert on high-voltage generators to sift through the documents to see if there was anything that would be vital to the war effort. Trump was assisted by a naval intelligence officer and two other enlisted men whose charge it was to microfilm the papers that were of interest to them. No copies exist within any federal agencies.[16] This begs the question: What papers were of interest to the Navy? After two days (January 26–27) and page after page examined, they decided there was nothing to report that was of interest, and subsequently the papers were released to his nephew.[17] However, Tesla still had financial troubles that followed him even after death, because the papers released to his nephew remained under lock and key with the New York Department of Taxation until his back taxes were paid, which took seven years to resolve.[18]

Thinking he had completed his review of the Tesla papers, Trump did call attention to several items that appeared to be scientific equipment. One in particular was thought to be the Holy Grail of weaponry. Possibly the death-beam itself. The apparatus was located in a depository at the Governor Clinton Hotel, where several years before, Tesla had stored it in a box. It was said to have a curse-like quality to it: if you were not authorized to open the wooden box, it would explode. Managers at the hotel brought the box to Trump, then promptly left the storeroom. It was wrapped in brown shipping paper and secured with a string. He held his breath, as he reluctantly and slowly opened the polished wooden box. Once opened, it revealed an old box used to measure resistance. Did Tesla have the last laugh?[19]

∽

In June of 1946, Tesla's nephew Kosanović returned to the United States as his country's ambassador. He brought with him from Yugoslavia sufficient funds to pay Tesla's back taxes and other debts, and he also made the necessary preparations to have Tesla's papers and all other belongings shipped to Belgrade.

Finally, in the middle of 1952, Tesla's original papers (those that were released), including whatever models of his experiments had survived over the years as he moved from laboratory to laboratory, as well as a variety of his Tesla coils, induction motors, fluorescent lamps, turbines, robots, vessels, photographs, plans for various inventions that included his VTOL plane, his Colorado Springs diary, and even exhibits demonstrated at the 1893 Chicago World's Fair that featured his Egg of Columbus, were all shipped to the Nikola Tesla Museum in Belgrade, Yugoslavia (now Serbia), which opened to the public in 1955. The total shipment amounted to some eighty trunks. His ashes followed five years later.[20] The Edison Medal went missing.

It would seem logical that the story of the Tesla papers would have ended then; after all, they were at the time secure in a museum thousands of miles away. However, the search continued under the aegis of various agencies from Naval Intelligence to the Office of Strategic Services (predecessor of the CIA). The FBI would regularly say that the papers did not exist or they were not in its possession. Suddenly the papers were nowhere to be found. They had disappeared! To add to the secrecy of it all, during the 1970s the Defense Advanced Research Projects Agency (DARPA) began a project code-named ALPHA (a chemical laser) and TALON GOLD, a project that focused on target systems. Numerous other such projects were conducted well into the 1980s, all of which led to the Strategic Defense Initiative (SDI) that President Ronald Reagan announced to the public in 1983. It has often been stated that SDI was instrumental in the eventual collapse of the U.S.S.R.

Moreover, there exists a particular treatise written by Tesla in 1937 entitled *The New Art of Projecting Concentrated Non-dispersive Energy through the Natural Media*. Specifics of the treatise have never been released, but what is known is that it contains numerous drawings and mathematical calculations by Tesla—its detail was said to be astonishing. This treatise is to this very day classified "top secret" by the U.S. military, and only a few federal government agencies have access to it.

So a few questions arise: One, did Tesla build his death-beam? Back in 1934, Tesla's yearly birthday announcement ginned up the possibility in a New York City newspaper at the time.[21] And Tesla said he was well on the way to completing the other two elements, but we will never know. Also, is the United States still engaged in Tesla's "death-beam" technology and other inventions and discoveries? Yes, because Nikola Tesla's box of riddles has been opened.

· 18 ·

The Brand

Legacy

The year 1956, marked the centenary of Nikola Tesla's birth in Smiljan, Province of Lika, Austrian Empire. A great celebration was organized. Niels Bohr, a Nobel Laureate and father of quantum mechanics, highlighted the event as speaker during the centennial congress held in Tesla's honor. In the same spirit, other honors were underway. The Yugoslav government issued a 100-dinar note (equivalent to one U.S. dollar with Washington on its obverse) with Tesla's likeness on its obverse, and its national postal service issued a commemorative Tesla stamp. In 1983, the U.S. Post Office did the same. Schools across the continents erected statues in his memory, and "Tesla days" were set aside to honor the great man. Today you walk along the paths next to Niagara Falls and see bronze statues depicting Tesla, one standing as he looks down upon his creation of harnessing its endless power; the other of Tesla seated while in deep contemplation. New York hotels that were honored with his presence erected commemorative plaques. Examples of his world-famous Tesla coil can be seen from sea to shining sea, at the Griffith Park Observatory in Los Angeles and the Boston Museum of Science and many places in between. There are also societies that serve to preserve his legacy.[1] And Tesla would be very honored by the International Electrotechnical Commission in Geneva that agreed to adopt the name "tesla" (symbol "T") as the unit of "magnetic flux" or the strength of a magnetic field—it is used in MRI equipment and atom-smashing accelerators. Just a handful of other Americans have had that distinct honor, and Edison is not among them. Tesla has become part of the scientific language.[2] Moreover, it is interesting to note that there is no scientific unit of measurement named after Marconi.

To honor Nikola Tesla, countries have even been able to cross the great religious divide, as when Croatia rebuilt Tesla's birthplace in Smiljan,

268

Province of Lika, as well as the church where his father preached the Serbian Orthodox gospel. It has become a tourist destination and even offers a cultural multimedia center. *Encyclopaedia Britannica* has listed him as one of the "ten most interesting historical figures" and *Life* magazine as one of the one hundred most famous people of the last one-thousand years.[3] And as we recall, Tesla graced the cover of *Time* magazine back in 1931. His adopted country gave him a special honor when on September 25, 1943, during the height of World War II, when the U.S. Navy christened ("I baptize you and give you the name Nikola Tesla. Bring happiness and liberty!") a Liberty ship, the SS *Nikola Tesla.*[4]

At the time of the energy crisis in the 1970s, the inventor was hailed as a hero because of his belief in free energy—when Morgan found out that was Tesla's ultimate plan, it was one of the reasons he stopped investing in Tesla's projects. Inspired by Tesla's free-energy work, from 1970 to 1988, a group of scientists led by physicist Robert K. Golka tried to re-create Tesla's famous Colorado Springs experiments in an effort to produce artificial lightning. This project has spawned numerous Tesla acolytes who come together to meet in societies dedicated to his name and work.[5] The work goes on.

As one would expect, Tesla has been the subject of numerous movies throughout the years, and it has been said that at any moment in Hollywood there are several movie scripts circulating among directors who are pursuing the opportunity to make a movie about the mystical man as producers understand that Tesla's name could translate into big box office; and actors are desirous to play this most complex of characters. His image has also been used in video games and comic strips. But perhaps today one of the most noticeable tips of the hat to Tesla's memory is the name of one of the world's most famous automobiles: The Tesla's power source is a version of the AC motor of Nikola Tesla's invention—using his discovery of the "rotating magnetic field."[6]

But what of Wardenclyffe, his unfulfilled dream lying fallow on Long Island? After its use for several years as a manufacturing facility for photograph supplies by both the Peerless Photo Company and later Agfa, all production ceased in 1987. A massive cleanup of the polluted site was completed in 2012. A few years later the Tesla Science Center at Wardenclyffe grew out of the generosity of thousands of crowdfunding participants, and a 2015 donation of $1 million by none other than Elon Musk gave the project real life.[7]

Look on any university's website and type the name Nikola Tesla in its search bar and up will appear page after page of articles, classes, research projects, and so on. Google (the founders of Google were greatly influenced by Tesla) his name and millions of websites pop up. Drive on any interstate highway in virtually any country in the world and you will see electrical towers along the road, and it should remind you of the great man who made them possible. The power he invented is "green" power. You get the point. Today every wireless communication owes its capability to Tesla.

Tesla's brand is growing at an exponential rate. He came to America as a poor man; he became a rich man; he died a poor man; and now he has become a rich man once again, because his name and accomplishments have become the richest of worldwide brands.

However, some of the blame for Tesla's legacy having to wait until the dawn of the twenty-first century must be laid at his own feet, because he left no presumptive heir, no progeny, no acolytes. He never sought to be the center of attention—yes, his demonstrations were phantasmagorical to be sure, but they served a purpose—or to monetize his work like Edison, Ford, or Marconi. It has been said that "the key to Tesla's character is his indifference to the commercial side of his work."[8]

As we now understand, his work was never about the marketplace. It was always about his inventions and discoveries and how they would "lift the burdens from the shoulders of mankind!"

In truth, a biographer can fill page after page, chapter after chapter with Tesla's legacy, but it can be quantified into a single, simple sentence: He gave us the modern world.

Epilogue

*W*hen an author sets to writing the epilogue of a life of great accomplishments and significance, a problem arises: What to say that has not already been said in the body of the text? As such, we faced the same vexing problem. What can we say but offer our deepest appreciation and gratitude to Nikola Tesla, for he gave us the richest of subjects to research and write about. He was a man who occupied multiple aspects of his time. He was an "electric" personality.

He has been called a genius, a visionary, an inventor, a discoverer, a creator, a wizard, a dreamer, a polymath, a Prometheus redivivus, an outlier, a futurist, an interloper, a Beau Brummel, an electric sorcerer, a cult figure, the first geek, an enigma, a scientific saint, a mysterious magician, the consummate iconoclast, psychologically unstable, a maverick, a mystic, the father of green energy, and the ultimate loner. He was all of these and more. He was the "supernova scientist" who brooked no opposition to his way of thinking. He was sui generis. He rejected received wisdom out of hand. He always looked for another way to do something. He did not need to be confirmed by a coterie of scientists to know he was right. His belief in himself could not be destroyed. His life was guided by "principle over profit." And he devoted his long life to an unquenchable passion, to "lift the burdens from the shoulders of mankind."

For so very long he had been nothing more than a forgotten footnote in most history books of the last half of the twentieth century or used as a historical antagonist to illuminate the lives of Thomas A. Edison and others. Moreover, he transcended the methodology known as "Yankee ingenuity" or the "American way," exemplified by such personages as Edison or the Wright Brothers.[1] Edison and his "improvement" of the lightbulb as well as the Wright Brothers and their "first in flight" image have served through the decades to cement their place in the history books.[2] Tesla did not take the road most traveled, for he had always rejected received wisdom in all its manifestations, believing he knew better—and most of the time he did—and so he was purposely redacted from history. Although he became a naturalized citizen soon after his arrival in the United States in 1884, that was never enough for the naysayers, be they historians, scientists, or merely observers.

What Tesla did from early childhood on was to look at the way Nature worked and knock on the front door for the answers. If no reply, he would go to the back door, where most scientists would never want to be seen because it was unscientific to them; it lacked methodology. Once he had visualized a problem or an idea for something entirely new, having crystalized it in his mind, and having built and tested it through a unique mental process, he then went to work making it work. Yes, he had his endless thought experiments, much like Einstein, but unlike Einstein, he was actually able to take those thought experiments from the mental to the physical stage of development. Thinking about an induction motor and then being able to build it and have it function was a gift few have ever possessed. He sought to disrupt the general way scientists thought, working mostly on instinct and unbridled ingenuity.

He said his life was guided by peerless instinct, about which he remarked, "I could not demonstrate my belief at that time, but it came to me through what I call instinct, for lack of a better name. But instinct is something which transcends knowledge. We undoubtedly have in our brains some finer fibers which enable us to perceive truths which we could not attain through logical deductions, and which it would be futile to attempt to achieve through any willful effort of thinking."[3] Unlike intuition, which processes are learned, instinct is something that is "hardwired" from birth—it is innate, it is mapped in one's DNA.

Tesla's May 16, 1888, lecture before the American Institute of Electrical Engineers was groundbreaking, for it established a new era in the electrical sciences. His lecture topic, titled: "A New System of Alternate Current Motors and Transformers," presented the theory and practical application of alternating current to power engineering. Thus AC became, and still is, the foundation upon which has been built the entire electrical system of the world today.[4] With all eyes upon him, the aristocratically styled Tesla began his lecture by starting and stopping his miraculous motors and showing drawings and diagrams to prove his points. Simply put, with the flip of a switch a new era in technology began.[5] The highly respected scientist B. A. Behrend said of the lecture: "He left nothing to be done by those who followed him. His paper contained the skeleton even of the mathematical theory."[6]

What Tesla had done that night before many of the world's most respected scientists and electrical engineers was to challenge them to break the bonds of ossified scientific thinking, to be bolder and to take chances and not accept the narrative of received wisdom. In effect, his approach was a scientific version of the modern Overton Window. His Overton Window can be defined in scientific terms as a method of identifying the ideas that define the range of acceptability within the scientific community.[7] Much like politicians, most scientists generally only act within the acceptable range inside the Overton Window, particularly if they receive government funding. What Tesla managed to do, at great psychological and financial costs to him, was to reject what was acceptable and pursue avenues of thought and experimentation that he was convinced would lead him to the answers found on the left side of the Overton Window.

At the time, much of what Tesla said and did was considered out of the mainstream of contemporary thought and challenged standard industries. From his discovery of the rotating magnetic field and invention of alternating current to wireless communications and his death-beam, Tesla demonstrated his ability to always think in a nonlinear fashion. While scientists of the day worked in a very linear way, not wanting to veer too far from the collective consciousness for fear of not being accepted by the majority of the scientific community, Tesla did not fear living in exile; after all, he chose to live some six decades in New York City completely alone.

As we now know, his mental health struggles with bipolar disorder as revealed in this biography served to endow him with a laser focus, a tenacious attitude, and the will to continue on. When his Fifth Avenue laboratory burned to the ground, leaving only "four cracked and blackened walls," he soldiered on, despite being psychologically crushed.[8]

Nikola Tesla's love of invention was everlasting. He said, "Archimedes was my ideal. I admired the works of artists, but to my mind they were on shadows and semblances. The inventor, I thought, gives the world creations which are palpable, which live and work."[9]

A question that will be asked, for which there is no certain answer, is whether Tesla could have made his discoveries and inventions without being bipolar. Looking at the timeline of his life, there can be little doubt that Tesla had what psychiatrists today call "bipolar I disorder" (BD-I) and what was called "manic depressive disorder" in the past. He experienced "highs" with full-blown mania sometimes accompanied by psychotic features such as visions and sounds that likely were hallucinations; he also experienced long periods of sustained "hypomania," during which he got most of his work done, his inventions made, and his patents filed. As anyone with bipolar disorder experiences, his moods were not stable and even during his productive years he experienced not only depressive episodes, but periods when his mood cycled rapidly between mania and depression, and even hypomania "mixed" with some features of depression. This is very typical for patients with bipolar I disorder.

During his lifespan, Tesla had what psychiatrists call a "comorbid" condition, obsessive compulsive disorder (OCD), which added color and flavor to his persona but was not likely to have been disabling. However, during the last half of his life, his mania and hypomania largely dissipated with only some short periods of infectious highs, and these were usually tinged with mixed features of depression as well. Mostly he was depressed, irascible, irritable, and stubborn, and although he continued to express novel ideas in this state, he was not successful in getting the finances to develop them. What went wrong? Did Tesla need his hypomania in order to invent? Although the answer to this cannot be known with any certainty, it seems that Tesla needed his hypomania to convince other scientists of the truth of his findings and to convince those with money to invest in them, since they could not understand his ideas. In the absence of hypomania, ideas flowed, but they were not accepted nearly as well coming from an unpleasant crank as from a charismatic, self-assured genius. Overall, it is not possible to separate bipolar disorder from the productive genius, and the life of Nicola Tesla will continue the debate of the role of bipolar disorder in those who have the most stunning creativity.

In the final analysis, history has at last come to its senses and is catching up to Nikola Tesla, for he was the man who, "in his time," changed the world for "all time." So now it is the twenty-first century, and "*it is his time*," as the world enters his domain of ever-expanding "electric power and wireless communications." He did indeed "lift the burdens from the shoulders of mankind" in his own time. Hence, Nikola Tesla has become the synecdoche for the modern world.

Acknowledgments

$\mathscr{A}$ special acknowledgment to my parents, Darinka and Mike Perko, who always allowed me to find my own way and forever encouraged my curiosity.

To Charles D. Holland, Esquire, for your wise counsel.

To Franklin M. Chu, MD; Hrayr Karnig Shahinian, MD; Natalija Nogulich; Liz Chu, NP; Jessica E. Reynolds; Shelly J. Preston; James E. Cheeley, DC; and Darrell E. Parks for your endless support. You are all deserving of my deepest respect.

To William Terbo, Nikola Tesla's grandnephew, who many years ago accommodated my questions about the great man during a pleasant Sunday afternoon we spent together. It was most illuminating.

And to those who knowingly and unknowingly inspired me to write this book, I humbly thank you.—*MP*

∼

To Professor Hagop Akiskal, my former colleague at the University of California, San Diego, for many useful discussions about the expression of bipolar disorder in creative geniuses, and for his help in interpreting the evidence in the record of Nikola Tesla that documents the ups and downs of Tesla's mood over his long lifetime.

To the numerous other psychiatrists, mental health professionals, and experts in bipolar disorder who have written on the topic and opened my eyes to the age-old question of the relationship of bipolar disorder and creativity.

To the Nikola Tesla Museum in Belgrade, Serbia, for facilitating my visit there, and for providing me insights into the man Nikola Tesla and his genius. —*SMS*

∽

We, the authors, have many people to thank, without whom this book would not have been possible. These talented people, from beginning to end, were part of the journey from concept to the first book sale and beyond. The list begins with our literary agent extraordinaire, Mark Gottlieb, of the Trident Media Group. From a late-night exchange of emails a few years ago to this very moment, he has been the one individual whose conviction that Tesla's unique story was worth telling "our way" has never wavered. To our exceptional (executive) editor, Jonathan Kurtz, of Prometheus Books, for his belief in the project, and his generosity in giving us the freedom to write *Tesla: His Tremendous and Troubled Life.* To Jessica McCleary, Gary Hamel, and the entirety of the production staff at Prometheus Books for their assiduous work and support in making the vision of our book become reality. And to Tara C. Craig for granting our requests at the Rare Book & Manuscript Library, Columbia University.

And lastly, writing as a two-author team would have been virtually impossible without Nikola Tesla's invention of wireless communication. —*MP & SMS*

Appendix A: Patents

Inventions and Discoveries

$\mathscr{D}$irect Current. Motor and generator controls, arc lights, etc.:

334,823	335,786	335,787	336,961	336,962,	350,954
382,845					

Polyphase Currents. Electric transmissions of power, dynamos, motors, transformers, electrical distribution:

395,748	381,968	381,969	381,970	382,279	382,280
382,281	382,282	390,413	390,414	390,415	390,721
390,820	401,520	405,858	405,859	406,968	413,353
413,702	413,703	416,191	416,192	416,193	416,194
416,195	417,794	418,248	424,036	433,700	433,700
445,207	455,067	459,772	464,666	487,796	511,559
511,560	511,915	512,340	524,426	555,190	

Currents of High Frequency and High Potential. Apparatuses for generating, controls, circuits and systems, insulation, and applications:

314,167	314,168	447,920	447,921	454,622	455,069
462,418	464,667	511,916	514,170	567,818	568,176
568,177	568,178	568,179	568,180	568,671	577,670
583,953	609,246	609,247	609,248	609,229	609,250
609,251	611,719	613,735			

Wireless Systems. Radio telegraphy, radio mechanics, methods of tuning and selection, detectors, etc.:

568,178	593,138	613,809	645,576	649,621	655,838
685,012	685,953	685,594	685,955	685,956	685,957
685,958	723,188	725,605	787,412	1,119,73[2]	

Various Patents. Steam turbines, pumps, speedometers, airplanes, mechanical oscillators, thermo-magnetic motors, etc.:

396,121	428,057	455,068	514,169	517,900	524,972
524,973	1,061,142	1,113,716	1,209,359	1,266,175	1,274,816
1,314,718	1,329,559	1,365,547	1,402,025	1,655,114	

From John J. O'Neill, *Prodigal Genius: The Life of Nikola Tesla* (Albuquerque, NM: Brotherhood of Life, 1994), 321–22. Some patents cannot be accounted for and are secreted away in patent archives or never registered. The total number is estimated to exceed three hundred patents.

The complete list of Tesla's United States and International Patents is available in detail at: https://en.wikipedia.org/wiki/List_of_Nikola_Tesla_patents#:~:text=Nikola%20 Tesla%20was%20an%20inventor,patents%20worldwide%20for%20his%20 inventions.&text=There%20are%20a%20minimum%20of,that%20have%20been%20 accounted%20for

http://web.mit.edu/most/Public/Tesla1/alpha_tesla.html

www.uspto.gov

www.TeslaBiography.com

Appendix B: Bibliographical Essay

The authors found that among the myriad books and other materials that they consulted on Tesla's life and work, several were referenced on a regular basis and deserve special mention. It is hoped that they may assist future biographers in their worthy work in chronicling the tremendous and troubled life of Nikola Tesla, much of which is still a mystery.

Let us first begin with books that were of immediate value. Marc J. Seifer's *Wizard: The Life of Nikola Tesla*. His work was both comprehensive and contains interesting "reconstructions" of Tesla's relationships with prominent people. W. Bernard Carlson published *Tesla: Inventor of the Electrical Age*, a book the authors found to be a reliable technical source for the complex scientific topics. *Tesla: Inventor of the Modern* by Richard Munson proved be a very worthwhile source. Then there is Margaret Cheney's biography *Tesla: Man Out of Time* and her follow-up with Robert Uth, *Tesla: Master of Lightning*, both of which are laudable reads.

Reaching back to a 1944 source, the authors found *Prodigal Genius: The Life of Nikola Tesla*—written by Pulitzer Prize–winning John J. O'Neill, who knew Nikola Tesla personally over many years—often recognized as the second biography of the inventor, to be of value, however, it lacks footnotes or endnotes. Reaching even further back to 1894 in our research, *The Inventions, Researches and Writings of Nikola Tesla*, prepared by Thomas C. Martin, his editor and friend, was an invaluable source in areas dealing with Tesla's scientific experiments, theories, and eventual inventions. It is considered to be the first biography of Nikola Tesla. Then there is *Nikola Tesla: Colorado Springs Notes 1899–1900*, a primary source, written by Nikola Tesla himself. Also of assistance was *Dr. Nikola Tesla Bibliography* edited by John T. Ratzlaff and Leland I. Anderson and *Tesla Said* compiled by John T. Ratzlaff. The authors also mined the "Nikola Tesla Papers" at Columbia University Library's Rare Books and Manuscript Library in their research. The papers contain copious letters, documents, and other significant materials pertaining to Tesla.

The only primary source of consequence regarding Tesla's early life is his autobiography *My Inventions: The Autobiography of Nikola Tesla.* This autobiography first appeared in serialized form in *The Electrical Experimenter* magazine in 1919. Subsequent to its first appearance, the autobiography has since been reprinted in book form by numerous publishers over many decades. Each publisher's format is somewhat different; thus, the authors chose to draw from several publishers of the autobiography in an effort to get as close to what Tesla actually said and did as possible.

Another invaluable source was the HathiTrust Digital Library. It was both a powerful search engine and source of historical newspapers, journals, and so on.

Moreover, as is all too often the case, the researching and writing of a biography about an individual who was so productive and so complex exposes inconsistencies in the facts relied upon in previous biographies. Once again, the authors made the prudent decision to multi-source such occurrences where possible.

See also the full bibliography, including archival collections, periodicals, newspapers, and websites that the authors used as fruitful references in the researching and writing of *Tesla: His Tremendous and Troubled Life.*

Notes

PROLOGUE

1. John Brockman, ed., *The Greatest Inventions of the Past 2,000 Years: Today's Leading Thinkers Choose the Creations That Shaped Our World* (New York: Simon & Schuster, 2000), 44–46.

2. Eminent psychobiographer/psychologist Alan C. Elms states in *Uncovering Lives: The Uneasy Alliance of Biography and Psychology* that "the psychobiographer's empathy for the subject—an important research tool—is often helped along if he or she has experienced similarities in life history or at least comes from a similar cultural background" (248). Elms further emphasizes his point when he states: "I'd like to see a Russian psychobiography of Vladimir Nabokov, for example, . . . it should add insights that I and other American psychobiographers wouldn't think of" (250).

CHAPTER 1

1. W. Bernard Carlson, *Tesla: Inventor of the Electrical Age* (Princeton, NJ: Princeton University Press, 2013), 18.

2. The Serbian Orthodox Church adheres to the Julian calendar (proposed by Julius Caesar in 46 BC), as do Eastern Orthodox countries across the globe. The Eastern Orthodox Church and its many adherents insist on still using the Julian calendar to calculate Easter and all other moveable feasts. Hence, all other dates are also affected. Most countries today utilize the Gregorian calendar, an issue that is still very explosive among Eastern Orthodox Churches of today. Using the Gregorian calendar to calculate dates would put Nikola Tesla's birthday between July 20–21, 1856.

3. "The theme of this year's American Institute of Electrical Engineers 'Fall General Meeting' (October 1–5, 1956, Chicago, Illinois) will be the Nikola Tesla Centennial. At the meeting, Dr. Samuel G. Hibben will present a demonstration-type lecture com-

memorating Dr. Tesla's outstanding contributions to the electrical industry and to the world." "The Nikola Tesla Papers": Columbia University, Rare Book and Manuscript Library, New York.

4. Carlson, *Tesla*, 18.

5. Margaret Cheney, *Tesla: Man Out of Time* (New York: Barnes & Noble, 1981/1993), 6. Factually, the Austro-Hungarian Empire did not begin until 1867, the result of the Austro-Hungarian Compromise, which established the dual monarch of Austria-Hungary. The correct name for this area prior to 1867 is the Austrian Empire. Andrew Wheatcroft, *The Habsburg: Embodying Empire* (New York: Viking, 1995), 246–47, 278–79; Richard Munson, *Tesla: Inventor of the Modern* (New York: W.W. Norton, 2018), 9.

6. Serbian (Eastern) Orthodox Church vs. Roman Catholic Church: The Catholic Church believes the pope to be "infallible" in matters of doctrine. The (Serbian) Orthodox Church rejects the infallibility of the pope and considers its own patriarchs, too, as human and, hence, subject to error. Most Orthodox Churches permit both ordained married priests and celibate monastics, so celibacy, to a certain level in the Church hierarchy, is an option.

7. The predicate for what would evolve into a worldwide religious schism had its roots in a battle on October 28, 312, at the Battle of the Milvian Bridge. It proved to be the most momentous event in Roman Emperor Constantine I's (born in Niš, present-day *Serbia*) life. Moreover, the battle between Roman Emperors Constantine I and Maxentius brought about a sea-change that has influenced world events from that day to this day. It was a vicious confrontation that took place at the Milvian Bridge, which even today spans the Tiber River in northern Rome. Legend has it that on or about the day before or maybe even during the battle itself, Emperor Constantine I had his famous vision. As Christian historian Eusebius of Caesarea portrays it: "A marvelous sign appeared to him from heaven. . . . He said that about midday, when the sun was beginning to decline, he saw with his own eyes the trophy of a cross of light in the heavens, above the sun, and bearing the inscription Conquer by This (*Hoc Vince*). He himself was struck with amazement, and his whole army also." Constantine I, also known as Constantine the Great, successfully routed Maxentius and became master of all he surveyed. He also became the first Christian emperor of the Roman Empire and was well on his way to converting to Christianity. By AD 330, Emperor Constantine I renamed Byzantium after himself, hence, Constantinople, a city that historians often describe as "the World's Desire." Constantinople (present-day Istanbul) would prove to be for more than a millennium the bastion and defender of Eastern Orthodox Christianity. By the late ninth century, wandering Slavic tribes of the Balkans were full-fledged acolytes of Orthodox Christianity. Nonetheless, Roman (Catholic) hegemony continued to hold sway over Croatia, the northern reaches of the Dalmatian coast and Moravia, yet in Serbia, Macedonia, and Greece the Orthodox faith and the dominance of Constantinople were willingly adopted. John Julius Norwich, *A Short History of Byzantium* (New York: Alfred A. Knopf, 1997), 5, 155.

8. Carlson, *Tesla*, 10.

9. Marc J. Seifer, *Wizard: The Life and Times of Nikola Tesla—Biography of a Genius* (New York: Citadel Press/Kensington Publishing, 1998), 5.

10. John J. O'Neill, *Prodigal Genius: The Life of Nikola Tesla* (Albuquerque, New Mexico: Brotherhood of Life, 1994), 20–21.

11. Carlson, *Tesla*, 13.

12. Ibid., 13–14.

13. The coauthor Perko can tell you from personal experience that the many trodden paths that carried everything from old Babas (grandmothers) and donkeys to milk-laden cows and packhorses through the ages are still peppered with the very same rocks.

14. Nikola Tesla, "Zmai Iovan Iovanovich–The Servian Poet," *The Century Magazine*, May 1, 1894, 130.

15. It is June 28 by the Gregorian Calendar.

16. Carlson, *Tesla*, 13.

17. Seifer, *Wizard*, 5–6.

18. "Although conventional wisdom deems Waterloo as the site of Napoleon's defeat . . . it was not. The actual battle, the last of the Napoleonic Wars, was fought approximately four miles south at a site somewhere between the two villages of Plancenoit and Mont-Saint-Jean." Marko Perko, *Did You Know That . . . ?: "Revised and Expanded" Edition: Surprising-But-True Facts about History, Science, Inventions, Geography, Origins, Art, Music and More* (New York: Open Road Distribution, 2017), 12.

19. Nikola Tesla, *My Inventions: Nikola Tesla's Autobiography* (Las Vegas: Lits, 2011), 10. Tesla's autobiography remains the single greatest primary source on the history of his childhood. (Several different published versions of his autobiography were cited.)

20. Seifer, *Wizard*, 6.

21. Ibid.; O'Neill, *Prodigal Genius*, 19.

22. Carlson, *Tesla*, 16.

23. Ibid., 14.

24. Ibid., 14–15; Seifer, *Wizard*, 7.

25. Tesla, *My Inventions*, 11; Nikola Tesla, "My Inventions: 1. My Early Life," *The Electrical Experimenter*, February 1919, 696.

26. Senj is a northern coastal fortification situated some seventy-five miles from the Italian seaport of Trieste, and the former home of the coauthor Perko's grandfather and namesake.

27. Munson, *Tesla: Inventor of the Modern*, 10.

28. Norwich, *A Short History of Byzantium*, 146.

29. Coauthor Perko spent a very illuminating afternoon in a comprehensive discussion of Tesla's life with William Terbo several years ago.

30. Carlson, *Tesla*, 15.

31. Tesla, *My Inventions* (Lits), 11.

32. Thomas C. Martin, *The Inventions, Researches, and Writings of Nikola Tesla* (CreateSpace/Skytower Press, 2013), 6.

33. Nikola Tesla, *My Inventions: Nikola Tesla's Autobiography*, ed. Ben Johnston (Austin, TX: Hart Brothers, 1982), 30.

34. Ibid., 30–31.

35. Tesla, *My Inventions* (Lits), 12; O'Neill, *Prodigal Genius*, 19.

36. Seifer, *Wizard*, 6–7; Tesla, *My Inventions* (Lits), 12.

37. Carlson, *Tesla*, 16.

38. Ibid., 17; Seifer, *Wizard*, 10.

39. Walter Isaacson, *The Innovators: How a Group of Hackers, Geniuses, and Geeks Created the Digital Revolution* (New York: Simon & Schuster, 2014), 9. Augusta Ada King (1815–1852), Countess of Lovelace, the only legitimate daughter of the celebrated poet Lord Byron, was considered by many to be the world's first computer programmer and the first to understand the power of computers.

40. Tesla, "My Inventions: 1," 746.

41. Seifer, *Wizard*, 7–8.

42. Tesla, "My Inventions: 1," 841.

43. Seifer, *Wizard*, 8.

44. Nikola Tesla, "Story of Youth Told by Age" (dedicated to Miss Pola Fotitch by its Author), Hotel New Yorker, Rastko Project, 1939; Munson, *Tesla: Inventor of the Modern*, 15.

45. Miss Pola Fotitch (Triandis) was the eight-year-old daughter of Konstantin Fotić, ambassador of Yugoslavia to the United States at the time of Peter II, King of Yugoslavia's reign. The coauthor Perko, as a teenage boy, had the occasion to meet and spend time with the former King Peter in Los Angeles, California. Margaret Cheney and Robert Uth, *Tesla: Master of Lightning* (New York: Barnes & Noble, 1999), 4.

46. Tesla, "Story of Youth Told by Age."

47. Ibid.

48. Tesla, "My Inventions: 1," 696.

49. Tesla, "Story of Youth Told by Age."

50. Ibid.

51. Tesla, "My Inventions: 1," 841.

52. Nikola Tesla, *My Inventions* (Lits), 31; Carlson, *Tesla*, 19. Other sources indicate sixteen May–(June)–bugs were used. See: Note to the Reader: Numbers.

53. Tesla, *My Inventions* (Lits), 32.

54. Tesla, "My Inventions: 1," 841.

55. Ibid.

56. Seifer, *Wizard*, 12.

57. Tesla, *My Inventions* (Lits), 12–13.

58. Ibid., 13.

59. Tesla (ed. Johnston), *My Inventions*, 32–33.

60. Tesla, *My Inventions* (Lits), 14.

61. Ibid., 18.

62. Ibid.

63. See: Note to the Reader: Dates.

64. See: Note to the Reader: Places.

65. Seifer, *Wizard*, 12.

66. Carlson, *Tesla*, 21.

67. Tesla, *My Inventions* (Lits), 8.

68. O'Neill, *Prodigal Genius*, 19.

69. Carlson, *Tesla*, 16.

70. Munson, *Tesla: Inventor of the Modern*, 16; Seifer, *Wizard*, 10, 488.

71. Margaret Cheney, *Tesla: Man Out of Time* (New York: Barnes & Noble, 1981/1993), 9.

72. Munson, *Tesla: Inventor of the Modern*, 16.

73. Carlson, *Tesla*, 21; Seifer, *Wizard*, 9.

74. Carlson, *Tesla*, 21; Tesla, "My Inventions: 1," 697.

75. Seifer, *Wizard*, 10.

76. Munson, *Tesla: Inventor of the Modern*, 16.

77. Tesla, *My Inventions* (Lits), 33.

78. Ibid.

79. Ibid., 33–34.

80. Ibid., 9; Munson, *Tesla: Inventor of the Modern*, 16.

81. Seifer, *Wizard*, 11; Tesla (ed. Johnston), *My Inventions*, 46.

82. Munson, *Tesla: Inventor of the Modern*, 18.

CHAPTER 2

1. Nikola Tesla, *My Inventions: Nikola Tesla's Autobiography* (Las Vegas: Lits, 2011), 19.

2. Marc J. Seifer, *Wizard: The Life and Times of Nikola Tesla—Biography of a Genius* (New York: Citadel Press/Kensington Publishing, 1996/1998), 11.

3. W. Bernard Carlson, *Tesla: Inventor of the Electrical Age* (Princeton, NJ: Princeton University Press, 2013), 23; *A History of Hungarian Literature: From the Earliest Times to the mid-1970s*, chapter 10, "Social Criticism and the Novel in the Age of Reform," http://mek.niif.hu/02000/02042/html/23.html.

4. Richard Munson, *Tesla: Inventor of the Modern* (New York: W.W. Norton, 2018), 21; *A History of Hungarian Literature*, chapter 10.

5. Tesla, *My Inventions* (Lits), 19.

6. Ibid., 19–20.

7. Carlson, *Tesla*, 24.

8. John J. O'Neill, *Prodigal Genius: The Life of Nikola Tesla* (Albuquerque, NM: Brotherhood of Life, 1994), 22.

9. In *Thus Spoke Zarathustra: A Book for All and None* (1883–1885), author Friedrich Nietzsche imagines an *Übermensch* or "Superman."

10. Michael Strevens, *The Knowledge Machine: How Irrationality Created Modern Science* (New York: Liveright, 2020), 106–7.

11. O'Neill, *Prodigal Genius*, 17.

12. Ibid., 22.

13. Munson, *Tesla: Inventor of the Modern*, 22.

14. Ibid.

15. Strevens, *The Knowledge Machine*, 135.

16. Tesla, *My Inventions* (Lits), 37; Carlson, *Tesla*, 26.

17. Carlson, *Tesla*, 26. Today one can walk along the pathway that parallels Niagara Falls on the Canadian side and see a giant-size bronze statue of Nikola Tesla, wearing a tuxedo and long coat, as he stands proudly, looking down upon his great invention—that signaled the beginning of the electrification of the world—erected in 2006. Another statue is located on the United States (New York) side of the falls that depicts him seated, reading from a book of his notes. The statue was donated in 1976 by the then-country of Yugoslavia.

18. Tesla, *My Inventions* (Lits), 37.

19. Seifer, *Wizard*, 13.

20. Tesla's ability with high-level mathematics stood in stark contrast to Albert Einstein's inability with high-level mathematics, yet both were world-class physicists.

21. Tesla, *My Inventions* (Lits), 37.

22. Ibid.

23. Ibid., 38.

24. Ibid.

25. Carlson, *Tesla*, 27.

26. Tesla, *My Inventions* (Lits), 38.

27. Ibid., 39.

28. Carlson, *Tesla*, 28.

29. Seifer, *Wizard*, 12.

30. Munson, *Tesla: Inventor of the Modern*, 22.

31. Tesla, *My Inventions* (Lits), 39. It should be noted that once Nikola had achieved great prominence in America, Mark Twain, as well as scores of other luminaries, sought him out, hoping to be his friend.

32. Seifer, *Wizard*, 13. Today the City of Karlovac is home to the Nikola Tesla Foundation for gifted students.

33. Margaret Cheney, *Tesla: Man Out of Time* (New York: Barnes & Noble, 1981/1993), 17; Tesla, *My Inventions* (Lits), 40.

34. Seifer, *Wizard*, 13.

35. Munson, *Tesla: Inventor of the Modern*, 24.

36. Although the numbers vary from one source to the next, Tesla was said to have been fluent in anywhere from eight to twelve languages.

37. Munson, *Tesla: Inventor of the Modern*, 23.

38. Tesla, *My Inventions* (Lits), 40.

39. Seifer, *Wizard*, 13.

40. Tesla, *My Inventions* (Lits), 40.

41. Seifer, *Wizard*, 13.

42. Tesla, *My Inventions* (Lits), 40.

43. Nikola Tesla, "My Inventions: 1. My Early Life," *The Electrical Experimenter*, April 1919, 865.

44. Tesla, *My Inventions* (Lits), 41.

45. Munson, *Tesla: Inventor of the Modern*, 25.

46. Seifer, *Wizard*, 14.

47. Tesla, *My Inventions* (Lits), 42.

48. Ibid.

49. Munson, *Tesla*, 25–26.

50. Ibid., 26.

51. Daniel Mrkich, *Nikola Tesla: The European Years* (Part One) (Ottawa, ON: Commoners' Publishing, 2002), 34.

52. Munson, *Tesla: Inventor of the Modern*, 26.

53. Carlson, *Tesla*, 30.

54. Ibid. Tesla's original concept became reality years later where cylindrical containers were propelled through networks of tubes using compressed air or by partial vacuum. They were used during the late 1800s to the mid-1950s in New York City and elsewhere. The concept is still in use in various areas of industry today.

55. Carlson, *Tesla*, 30; O'Neill, *Prodigal Genius*, 28.

56. Carlson, *Tesla*, 30.

57. Mrkich, *Nikola Tesla: The European Years* (Part One), 26.

58. Seifer, *Wizard*, 14.

59. Carlson, *Tesla*, 31.

60. Ibid. HSAM's cause is not known, but the obvious places to look are psychological and biological. Within the biological aspect lives the genetic component. It is worth noting that young Tesla's mother was gifted with a superior memory, as she would often regale her children with stories of Serbian heroes.

61. Carlson, *Tesla*, 31–32.

62. Mrkich, *Nikola Tesla: The European Years* (Part One), 8.

63. Tesla, *My Inventions* (Lits), 44.

64. Seifer, *Wizard*, 15.

65. Ibid.; Charles M. Meister, "Tesla Nearly Missed His Calling," *New York Sun*, August 27, 1931.

66. Tesla, *My Inventions*, Eastford, CT: Martino Fine Books, 2018, 25.

67. Ibid., 45.

68. Seifer, *Wizard*, 17.

69. Ibid., 15; Munson, *Tesla: Inventor of the Modern*, 27.

70. Seifer, *Wizard*, 15–16; Munson, *Tesla: Inventor of the Modern*, 27.

71. Tesla, *My Inventions* (Lits), 44; Munson, *Tesla: Inventor of the Modern*, 27.

72. Tesla, *My Inventions* (Lits), 45.

73. Carlson, *Tesla*, 35.

74. Seifer, *Wizard*, 16.

75. Tesla, *My Inventions* (Lits), 44.

76. Munson, *Tesla: Inventor of the Modern*, 29.

77. Meister, "Tesla Nearly Missed His Calling"; Mrkich, *Nikola Tesla: The European Years* (Part One), 18.

78. Mrkich, *Nikola Tesla: The European Years* (Part One), 17–19.

79. Ibid.

80. The Gramme dynamo (Gramme ring or Gramme magneto) is an electrical generator of "direct current." It was developed by Zénobe Théophile Gramme, a Belgian inventor and first demonstrated in 1871. It was inspired by an earlier machine invented by Antonio Pacinotti in 1860. Although the Gramme dynamo is no longer in use because of many technical limitations identified by Tesla, it was the first generator

(if supplied with a constant voltage, it becomes an electric motor as well) to produce electric power for commercial use.

81. Tesla, *My Inventions* (Lits), 45–46. A commutator is a rotary attachment or electrical switch in specific types of electric motors and electrical generators that regularly reverses the current direction between the rotor and the external circuit. It is comprised of a cylinder of multiple metal contact segments on the rotating armature of the machine. Two or more electrical contacts are referred to as brushes, which are made of a soft conductive material such as carbon press against the commutator, thus making a sliding contact with successive segments of the commutator as it rotates. The windings—coils of wire—on the armature are connected to the commutator segments.

82. Ibid., 45–46.

83. Munson, *Tesla: Inventor of the Modern*, 28.

84. Walter Isaacson, *The Innovators: How a Group of Hackers, Geniuses, and Geeks Created the Digital Revolution* (New York: Simon & Schuster, 2014), 457.

85. Ibid., 277.

86. Tesla, *My Inventions* (Lits), 45–46; O'Neill, *Prodigal Genius*, 48; Cheney, *Tesla: Man Out of Time*, 18–19; Munson, *Tesla: Inventor of the Modern*, 28.

87. Carlson, *Tesla*, 43; Munson, *Tesla: Inventor of the Modern*, 28.

88. Ibid.; Carlson, *Tesla*, 43.

89. Tesla, *My Inventions* (Lits), 46.

90. Nikola Tesla, *My Inventions: The Autobiography of Nikola Tesla*, ed. Ben Johnston (Austin, TX: Hart Brothers, 1982), 37.

91. Ibid.

92. Seifer, *Wizard*, 17.

93. Carlson, *Tesla*, 47.

94. Munson, *Tesla: Inventor of the Modern*, 31.

95. Ibid.; Seifer, *Wizard*, 18.

96. Munson, *Tesla: Inventor of the Modern*, 32.

97. Ibid.

98. Seifer, *Wizard*, 18.

99. Carlson, *Tesla*, 48.

CHAPTER 3

1. Daniel Mrkich, *Nikola Tesla: The European Years* (Part Three) (Ottawa, ON: Commoners' Publishing, 2002), 16.

2. Margaret Cheney, *Tesla: Man Out of Time* (New York: Barnes & Noble, 1993), 78.

3. Alan C. Elms, *Uncovering Lives: The Uneasy Alliance of Biography and Psychology* (New York: Oxford University Press, 1994), 83.

4. Nikola Tesla, "My Inventions: IV. The Discovery of the Tesla Coil and Transformer," *The Electrical Experimenter*, May 1919, 16.

5. Richard Munson, *Tesla: Inventor of the Modern* (New York: W.W. Norton, 2018), 36.

6. Mrkich, *Nikola Tesla: The European Years* (Part Three), 16.

7. Jenny Uglow, *The Lunar Men: Five Friends Whose Curiosity Changed the World* (New York: Farrar, Straus and Giroux, 2002), xv.

8. Normal household electricity that comes from a wall outlet is alternating current (AC), which is delivered over powerlines. Alternating current is also utilized throughout the world in all types of industry. From its outset, it became the world standard of electrical power and remains so to this very day. More specifically, alternating current (AC) is an electric current that periodically reverses direction or polarity, as opposed to direct current (DC), which flows only in one direction. Alternating current is the most efficient form in which electric power is delivered for commercial and industrial applications, as well as for residential purposes. It is also the preferred form of electric power that drives virtually all regularly used appliances, electric lights, televisions, desktop computers, and so on. Moreover, it is the worldwide standard for electric power.

9. Essentially, Tesla's AC polyphase system provides a more economical and superior means of distributing electrical power (over great distances with no power loss) and utilizes three or more energized electrical conductors that carry alternating currents.

10. Nikola Tesla, *My Inventions: Nikola Tesla's Autobiography* (Las Vegas: Lits, 2011), 27.

11. Munson, *Tesla: Inventor of the Modern*, 37.

12. Mrkich, *Nikola Tesla: The European Years* (Part Three), 12–13.

13. W. Bernard Carlson, *Tesla: Inventor of the Electrical Age* (Princeton, NJ: Princeton University Press, 2013), 50. Nikola Tesla, "My Inventions: III. My Later Endeavors: The Discovery of the Rotating Magnetic Field," *The Electrical Experimenter*, April 1919, 907.

14. Munson, *Tesla: Inventor of the Modern*, 36.

15. Margaret Cheney, *Tesla: Man Out of Time*, 21.

16. Tesla, "My Inventions: III," 907.

17. Cheney, *Tesla: Man Out of Time*, 22.

18. Ibid.

19. Walter Isaacson, *Einstein: His Life and Universe* (New York: Simon & Schuster, 2007), 113.

20. Oliver Sacks, *The Man Who Mistook His Wife for a Hat and Other Clinical Tales* (New York: Touchstone, 1985), 87.

21. Tesla, *My Inventions* (Lits), 49.

22. John J. O'Neill, *Prodigal Genius: The Life of Nikola Tesla* (Albuquerque, NM: Brotherhood of Life, 1994), 49.

23. Direct current (DC) is electrical current that flows steadily in one direction. It is the same current that powers a flashlight or another device requiring batteries. Also, unlike alternating current (AC), the loss of energy when direct current is carried over long distances is significant.

24. Tesla, *My Inventions* (Lits), 49.

25. Margaret Cheney and Robert Uth, *Tesla: Master of Lightning* (New York: Barnes & Noble, 1999), 11.

26. O'Neill, *Prodigal Genius*, 49.

27. Carlson, *Tesla*, 55.

28. Isaacson, *Einstein*, 79.

29. O'Neill, *Prodigal Genius*, 237.

30. Ibid.

31. Carlson, *Tesla*, 51–52, 54.

32. Tesla, *My Inventions* (Lits), 50.

33. Carlson, *Tesla*, 51; Cheney, *Tesla: Man Out of Time*, 23.

34. Tesla, *My Inventions* (Lits), 50.

35. Isaacson, *Einstein*, 122–24.

36. Carlson, *Tesla*, 57.

37. Nikola Tesla, "My Inventions: I. My Early Life," *The Electrical Experimenter*, February 1919), 696.

38. O'Neill, *Prodigal Genius*, 237.

39. Sacks, *The Man Who Mistook*, 168.

40. Craig Wright, *The Hidden Habits of Genius: Beyond Talent, IQ, and Grit—Unlocking the Secrets of Greatness* (New York: Dey St./William Morrow, 2020), 42.

41. Sacks, *The Man Who Mistook*, 168. Hildegard left detailed accounts of other visions that we can draw on in both *Scivias* and *Liber divinorum operum*.

42. Wright, *The Hidden Habits of Genius*, 115.

43. Carlson, *Tesla*, 63.

44. Ibid.

45. Tesla, *My Inventions* (Lits), 52.

46. The Second Industrial Revolution, also referred to as the Technological Revolution, was a period of very rapid standardization and industrialization from the mid-nineteenth century into the mid-twentieth century (1850–1970). It saw the burgeoning field of "electricity" become a major industry, while oil and steel experienced greater growth.

47. Mrkich, *Nikola Tesla: The European Years* (Part Three), 11.

48. Carlson, *Tesla*, 63–64.

49. Tesla, "My Inventions: IV, 17; "The Nikola Tesla Papers": Columbia University Rare Book and Manuscript Library, New York.

50. Mrkich, *Nikola Tesla: The European Years* (Part Three), 26.

51. Inez Hunt and Wanetta W. Draper, *Lightning in His Hand: The Life Story of Nikola Tesla* (Hawthorne, CA: Omni Publications, 1977), 38; Carlson, *Tesla*, 68.

52. Hunt and Draper, *Lightning in His Hand*, 38.

53. B. A. Behrend, "Dynamo-Electric Machinery and Its Evolution," *Western Electric*, September 1907, 503 ("The Nikola Tesla Papers": Columbia University Rare Book and Manuscript Library, New York).

54. Walter Isaacson, *Leonardo da Vinci* (New York: Simon & Schuster, 2017), 90–91.

55. Carlson, *Tesla*, 69; Mrkich, *Nikola Tesla: The European Years* (Part Three), 33–34.

56. Cheney, *Tesla: Man Out of Time*, 9.

57. Dan P. McAdams, *George W. Bush and the Redemptive Dream: A Psychological Portrait* (New York: Oxford University Press, 2011), 11.

58. Hunt and Draper, *Lightning in His Hand*, 39; Cheney and Uth, *Tesla: Master of Lightning*, 17; Ronald W. Clark, *Edison: The Man Who Made the Future* (New York: G.P. Putnam's Sons, 1977), 158.

59. Cheney, *Tesla: Man Out of Time*, 26.

CHAPTER 4

1. Richard Munson, *Tesla: Inventor of the Modern* (New York: W.W. Norton, 2018), 47–48; Margaret Cheney, *Tesla: Man Out of Time* (New York: Barnes & Noble, 1993), 27.

2. Munson, *Tesla: Inventor of the Modern*, 47–48; Nikola Tesla, *My Inventions: The Autobiography of Nikola Tesla*, ed. Ben Johnston (Austin, TX: Hart Brothers, 1982), 71.

3. Nikola Tesla, "My Inventions: IV. The Discovery of the Tesla Coil and Transformer," *The Electrical Experimenter*, May 1919, 64.

4. Ibid.

5. Inez Hunt and Wanetta W. Draper, *Lightning in His Hand: The Life Story of Nikola Tesla* (Hawthorne, CA: Omni Publications, 1977), 39; Margaret Cheney and Robert Uth, *Tesla: Master of Lightning* (New York: Barnes & Noble, 1999), 17.

6. David Berlinski, *Newton's Gift: How Sir Isaac Newton Unlocked the System of the World* (New York: Free Press, 2000), 18–19.

7. Munson, *Tesla: Inventor of the Modern*, 47; Marc J. Seifer, *Wizard: The Life and Times of Nikola Tesla—Biography of a Genius* (New York: Citadel Press/Kensington Publishing, 1998), 33.

8. Nigel Cawthorne, *Tesla: The Life and Times of an Electric Messiah* (New York: Chartwell Books, 2014), 29.

9. Jill Jonnes, *Empires of Light: Edison, Tesla, Westinghouse, and the Race to Electrify the World* (New York: Random House, 2004), 106.

10. Margaret Cheney and Robert Uth, *Tesla: Master of Lightning* (New York: Barnes & Noble, 1999), 13.

11. Jonnes, *Empires of Light*, 104.

12. Munson, *Tesla: Inventor of the Modern*, 48.

13. Letter from Nikola Tesla to the National Institute of Immigrant Welfare (on his being unable to attend the award ceremony in his honor, and that of Felix Frankfurter, and Giovanni Martinelli, all of whom received awards from the NIMF), May 11, 1938, in *Tesla Said*, compiled by John T. Ratzlaff (Millbrae, CA: Tesla Book Company, 1984), 280. The celebrated event was reported in the *New York Times*, May 12, 1938.

14. Munson, *Tesla: Inventor of the Modern*, 48.

15. Seifer, *Wizard*, 33.

16. Ibid., 33–34.

17. Munson, *Tesla: Inventor of the Modern*, 48. Tesla lived eighty-six-plus years while Edison lived to the age of eighty-four.

18. Ibid.

19. Seifer, *Wizard*, 35.

20. Jonnes, *Empires of Light*, 110.

21. Seifer, *Wizard*, 35.

22. Ibid.

23. Mike Winchell, *The Electric War: Edison, Tesla, Westinghouse, and the Race to Light the World* (New York: Henry Holt and Company, 2019), 25.

24. Munson, *Tesla: Inventor of the Modern*, 49.

25. Ibid., 49–50.

26. Seifer, *Wizard*, 35.

27. Ibid., 34.

28. "Tesla Says Edison Was an Empiricist," *New York Times*, October 19, 1931, p. 25, https://todayinsci.com/T/Tesla_Nikola/TeslaNikola-Quotations.htm.

29. Ohm's law is named after German physicist Georg Ohm (1789–1854). The law states that the voltage or potential difference between two points is directly proportional to the electric current passing through the resistance, and directly proportional to the resistance of the circuit.

30. Munson, *Tesla: Inventor of the Modern*, 50.

31. Ibid., 50–51.

32. It has been a long-held misconception that Thomas Edison invented the "light bulb." He did not. In 1802, British chemist Sir Humphry Davy developed the first incandescent light, and then he produced the first practical electric arc light in 1806. Marko Perko, *Did You Know That . . . ?: "Revised and Expanded" Edition: Surprising-But-True Facts about History, Science, Inventions, Geography, Origins, Art, Music and More* (New York: Open Road Distribution, 2017), 76–77.

33. Paul Israel, *Edison: A Life of Invention* (New York: John Wiley & Sons, 1998), 95; Frank Lewis Dyer and Thomas Commerford Martin, *Edison, His Life and Inventions* (New York: Harper & Brothers, 1910), available from Project Gutenberg (EBook #820), 411, https://www.gutenberg.org/files/820/820-h/820-h.htm.

34. Ronald W. Clark, *Edison: The Man Who Made the Future* (New York: G.P. Putnam's Sons, 1977), 10.

35. Jon Gertner, *The Idea Factory: Bell Labs and the Great Age of American Innovation* (New York: Penguin, 2012), 12.

36. Clark, *Edison*, 158.

37. Michael Strevens, *The Knowledge Machine: How Irrationality Created Modern Science* (New York: Liveright, 2020), 7, 293.

38. Walter Isaacson, *The Innovators: How a Group of Hackers, Geniuses, and Geeks Created the Digital Revolution* (New York: Simon & Schuster, 2014), 408.

39. Nikola Tesla: *My Inventions: Nikola Tesla's Autobiography* (Eastford, CT: Martino Fine Books, 2018), 34.

40. W. Bernard Carlson, *Tesla: Inventor of the Electrical Age* (Princeton, NJ: Princeton University Press, 2013), 70.

41. Ibid., 71.

42. Munson, *Tesla: Inventor of the Modern*, 53.

43. Ibid.

44. Carlson, *Tesla*, 72–73.

45. Cheney, *Tesla: Man Out of Time*, 34.

46. Munson, *Tesla: Inventor of the Modern*, 54.

47. Ibid., 53–54.

48. When listing Tesla's patents chronologically by "number," his actual first patent is: U.S. Patent 0,334,823—Commutator for Dynamo Electric Machines issued January 26, 1886. Its purpose: elements to prevent sparking on dynamo-electric machines, drum-style with brushes.

49. Seifer, *Wizard*, 40.

50. Ibid., 41; Richard Munson, *Tesla: Inventor of the Modern*, 54; Charles M. Meister, "Tesla Nearly Missed His Calling," *New York Sun*, August 27, 1931.

51. Seifer, *Wizard*, 41.

52. Ibid.; *Electrical Review*, August 14, 1886.

53. Carlson, *Tesla*, 74.

54. Ibid., 75.

55. Letter from Nikola Tesla to the National Institute of Immigrant Welfare, May 11, 1938, in *Tesla Said*, 280; Seifer, *Wizard*, 41.

56. Letter from Nikola Tesla to the National Institute of Immigrant Welfare, May 11, 1938, 280.

57. Ibid.

58. Munson, *Tesla: Inventor of the Modern*, 55.

59. Hunt and Draper, *Lightning in His Hand*, 44.

60. Munson, *Tesla: Inventor of the Modern*, 55.

61. Ibid.

62. Ibid., 56.

63. Carlson, *Tesla*, 82.

64. Seifer, *Wizard*, 42.

65. Cheney, *Tesla: Man Out of Time*, 37.

66. Munson, *Tesla: Inventor of the Modern*, 56.

67. Seifer, *Wizard*, 42–43.

68. See chapter 3.

69. Seifer, *Wizard*, 43; Cheney and Uth, *Tesla: Master of Lightning*, 21; Cawthorne, *Tesla: The Life and Times of an Electric Messiah*, 33.

70. Munson, *Tesla: Inventor of the Modern*, 56.

71. Cheney and Uth, *Tesla: Master of Lightning*, 21; Nikola Tesla, *My Inventions: Nikola Tesla's Autobiography* (Las Vegas: Lits, 2011), 59–60.

72. Seifer, *Wizard*, 43.

73. Munson, *Tesla: Inventor of the Modern*, 61.

74. John J. O'Neill, *Prodigal Genius: The Life of Nikola Tesla* (Albuquerque, NM: Brotherhood of Life, 1994), 73.

75. Ibid.

76. Munson, *Tesla: Inventor of the Modern*, 59.

CHAPTER 5

1. Marc J. Seifer, *Wizard: The Life and Times of Nikola Tesla—Biography of a Genius* (New York: Citadel Press/Kensington Publishing, 1998), 42.

2. Nikola Tesla, "Tesla's Egg of Columbus: How Tesla Performed the Feat of Columbus without Cracking the Egg," *The Electrical Experimenter*, March 1919, 775.

3. Seifer, *Wizard*, 42.

4. Tesla, "Tesla's Egg of Columbus," 775.

5. Christopher Columbus did not discover America. Marko Perko, *Did You Know That . . . ?: "Revised and Expanded" Edition: Surprising-But-True Facts about History, Science, Inventions, Geography, Origins, Art, Music and More* (New York: Open Road Distribution, 2017), 32–35.

6. Tesla, "Tesla's Egg of Columbus," 775; W. Bernard Carlson, *Tesla: Inventor of the Electrical Age* (Princeton, NJ: Princeton University Press, 2013), 91.

7. Richard Munson, *Tesla: Inventor of the Modern* (New York: W.W. Norton, 2018), 59; Seifer, *Wizard*, 42; Tesla, "Tesla's Egg of Columbus," 775.

8. Munson, *Tesla: Inventor of the Modern*, 56.

9. Seifer, *Wizard*, 43.

10. Jill Jonnes, *Empires of Light: Edison, Tesla, Westinghouse, and the Race to Electrify the World* (New York: Random House, 2004), 154. An "induction motor" is an AC electric motor in which the electric current in the rotor required to produce torque is obtained by electromagnetic induction from the magnetic field of the stator winding. Hence, an induction motor can be built without electrical connections to the rotor. An induction motor's rotor can be either a wound type or squirrel-cage type. A "polyphase system" is a means of distributing alternating-current electrical power where the power transfer remains constant during each electrical cycle. Polyphase systems are especially useful for transmitting power to electric motors which rely on alternating current to rotate. The most common example is the three-phase power system used for industrial applications and for power transmission. When compared to a single-phase, two-wire system, a three-phase three-wire system transmits three times as much power for the same conductor size and voltage.

11. Thomas Commerford Martin, "Nikola Tesla," *The Century Illustrated Monthly Magazine*, November 1893–April 1894, vol. 47, New Series vol. 25: 582, https://babel.hathitrust.org/cgi/pt?id=mdp.39015025903512&view=1up&seq=596.

12. Seifer, *Wizard*, 44.

13. Jonnes, *Empires of Light*, 154.

14. Munson, *Tesla: Inventor of the Modern*, 61.

15. Carlson, *Tesla*, 104; W. A. Anthony to D. C. Jackson, March 11, 1888, cited in Kenneth M. Swezey, "Nikola Tesla," *Science*, no. 127 (May 16, 1958), 1149.

16. Seifer, *Wizard*, 44–45; Nigel Cawthorne, *Tesla: The Life and Times of an Electric Messiah* (New York: Chartwell Books, 2014), 38.

17. Jonnes, *Empires of Light*, 154. The cache of patents for Tesla's complete AC system was so all-encompassing that the patent office required Tesla's submission to be broken down into subsets.

18. Jonnes, *Empires of Light*, 156.

19. See Thomas C. Martin, *The Inventions, Researches, and Writings of Nikola Tesla* (CreateSpace/Skytower Press, 2013), 9.

20. John J. O'Neill, *Prodigal Genius: The Life of Nikola Tesla* (Albuquerque, NM: Brotherhood of Life, 1994), 74–75; Margaret Cheney, *Tesla: Man Out of Time* (New York: Barnes & Noble, 1993), 39.

21. Munson, *Tesla: Inventor of the Modern*, 62.

22. Margaret Cheney and Robert Uth, *Tesla: Master of Lightning* (New York: Barnes & Noble, 1999), 23.

23. Martin, *The Inventions*, 9–10; Jonnes, *Empires of Light*, 156–57.

24. Ibid., 157.

25. Discussion of Tesla's paper before the AIEE, *AIEE Transactions* 5 (September 1887–October 1888), 324, https://play.google.com/books/reader?id=_l9QRiE89 pkC&hl=en&pg=GBS.PA324.

26. Jonnes, *Empires of Light*, 157.

27. Carlson, *Tesla*, 106–7.

28. Jonnes, *Empires of Light*, 158; discussion of Tesla's paper before the AIEE, 325–26.

29. Carlson, *Tesla*, 107.

30. Munson, *Tesla: Inventor of the Modern*, 63.

31. Carlson, *Tesla*, 107.

32. Ibid., 103.

33. Seifer, *Wizard*, 44.

34. Ibid., 45.

35. Jonnes, *Empires of Light*, 159.

36. Ibid.

37. Cheney, *Tesla: Man Out of Time*, 39.

38. Carlson, *Tesla*, 107.

39. Cawthorne, *Tesla: The Life and Times of an Electric Messiah*, 39.

40. Munson, *Tesla: Inventor of the Modern*, 69–70.

41. Nikola Tesla, "Tributes of Former Associates," *Electrical World*, March 21, 1914.

42. Munson, *Tesla: Inventor of the Modern*, 64.

43. Carlson, *Tesla*, 111; Seifer, *Wizard*, 48–49.

44. Carlson, *Tesla*, 111; Munson, *Tesla: Inventor of the Modern*, 64.

45. Carlson, *Tesla*, 111.

46. Munson, *Tesla: Inventor of the Modern*, 71.

47. Carlson, *Tesla*, 111.

48. Ibid.

49. Jonnes, *Empires of Light*, 160.

50. Carlson, *Tesla*, 111–12.

51. Ibid., 112.

52. Munson, *Tesla: Inventor of the Modern*, 65.

53. Iwan Rhys Morus, *Nikola Tesla and the Electrical Future* (London: Icon Books, 2019), 83–84.

54. Carlson, *Tesla*, 113.

55. Munson, *Tesla: Inventor of the Modern*, 64. (It must be noted that the precise numbers involved in Westinghouse's purchase of the Tesla patents vary somewhat from chronicler to chronicler.)

56. Morus, *Nikola Tesla and the Electrical Future*, 84.

57. Munson, *Tesla: Inventor of the Modern*, 65.

58. Carlson, *Tesla*, 113.

59. Seifer, *Wizard*, 53.

60. Cheney and Uth, *Tesla: Master of Lightning*, 24.

61. Ibid.

62. Jonnes, *Empires of Light*, 163.

63. Seifer, *Wizard*, 53.

64. Munson, *Tesla: Inventor of the Modern*, 71.

65. Ibid., 68–69; Jonnes, *Empires of Light*, 180–81.

66. Munson, *Tesla: Inventor of the Modern*, 71.

67. Carlson, *Tesla*, 117.

68. John T. Ratzlaff, ed., *Tesla Said* (Millbrae, CA: Tesla Book Company, 1984), 272 ("Prepared [Press] Statement by Nikola Tesla dated July 10, 1937, on the occasion of his eighty-first birthday observance").

69. Carlson, *Tesla*, 117.

70. Munson, *Tesla: Inventor of the Modern*, 71; Carlson, *Tesla*, 117.

71. Carlson, *Tesla*, 117.

72. Ibid., 118.

73. Munson, *Tesla: Inventor of the Modern*, 72.

74. Ibid.

75. O'Neill, *Prodigal Genius*, 77.

CHAPTER 6

1. W. Bernard Carlson, *Tesla: Inventor of the Electrical Age* (Princeton, NJ: Princeton University Press, 2013), 117.

2. Nigel Cawthorne, *Tesla: The Life and Times of an Electric Messiah* (New York: Chartwell Books, 2014), 44.

3. John J. O'Neill, *Prodigal Genius: The Life of Nikola Tesla* (Albuquerque, NM: Brotherhood of Life, 1994), 302.

4. Richard Munson, *Tesla: Inventor of the Modern* (New York: W.W. Norton, 2018), 43.

5. Cawthorne, *Tesla: The Life and Times of an Electric Messiah*, 44; Nikola Tesla, *Electrical Review* 11 (August 14, 1896), 193; Margaret Cheney and Robert Uth, *Tesla: Master of Lightning* (New York: Barnes & Noble, 1999), 86–87.

6. O'Neill, *Prodigal Genius*, 302–3. Even Socrates thought marriage was a great distraction.

7. David J. Kent, *Tesla: The Wizard of Electricity* (New York: Fall River Press, 2013), 58.

8. Marc J. Seifer, *Wizard: The Life and Times of Nikola Tesla—Biography of a Genius* (New York: Citadel Press/Kensington Publishing, 1998), 207.

9. O'Neill, *Prodigal Genius*, 303.

10. Kent, *Tesla: The Wizard of Electricity*, 58.

11. Anthony Storr, *Churchill's Black Dog, Kafka's Mice and Other Phenomena of the Human Mind* (New York: Grove Press, 1988), 5–6.

12. Ronald W. Clark, *Edison: The Man Who Made the Future* (New York: G.P. Putnam's Sons, 1977), 159.

13. Ibid.

14. Kent, *Tesla: The Wizard of Electricity*, 110.

15. John F. Wasik, *Lightning Strikes: Timeless Lessons in Creativity from the Life and Work of Nikola Tesla* (New York: Sterling, 2016), 86.

16. Munson, *Tesla: Inventor of the Modern*, 80.

17. Thomas A. Edison, "The Dangers of Electric Lighting," *The North American Review* (November 1889), 632.

18. Ibid.

19. Clark, *Edison*, 159; Harrold C. Passer, *The Electrical Manufacturers 1875–1900: A Study in Competition, Entrepreneurship, Technical Change, and Economic Growth* (New York: Arno Press, 1972), 168.

20. Munson, *Tesla: Inventor of the Modern*, 80.

21. Jill Jonnes, *Empires of Light: Edison, Tesla, Westinghouse, and the Race to Electrify the World* (New York: Random House, 2004), 168.

22. Ibid.

23. Richard Holmes, *The Age of Wonder: How the Romantic Generation Discovered the Beauty and Terror of Science* (New York: Pantheon Books, 2008), 444.

24. Sir John F. W. Herschel, *Preliminary Discourse on the Study of Natural Philosophy* (London: Longman Press, 1851), 329–30.

25. Static electricity is a stationary electric charge, usually produced by friction, which creates sparks or crackling or the attraction of dust or hair.

26. Carlson, *Tesla*, 118.

27. Seifer, *Wizard*, 61.

28. Carlson, *Tesla*, 117; Seifer, *Wizard*, 41.

29. Carlson, *Tesla*, 118.

30. Munson, *Tesla: Inventor of the Modern*, 73; Nikola Tesla, *Nikola Tesla on His Work with Alternating Currents and Their Application to Wireless Telegraphy, Telephony and Transmission of Power: An Extended Interview*, Leland I. Anderson, ed. (Breckenridge, CO: Twenty-First Century Books, 2002), http://www.tfcbooks.com/tesla/nt_on_ac.htm#Section_2.

31. Carlson, *Tesla*, 118.

32. Ibid., 118–19.

33. Munson, *Tesla: Inventor of the Modern*, 80.

34. Kent, *Tesla: The Wizard of Electricity*, 112.

35. Ibid., 110, 112.

36. Iwan Rhys Morus, *Nikola Tesla and the Electrical Future* (London: Icon Books Ltd., 2019), 39.

37. Jon Gertner, *The Idea Factory: Bell Labs and the Great Age of American Innovation* (New York: Penguin Press, 2012), 12.

38. Morus, *Nikola Tesla and the Electrical Future*, 40; Wasik, *Lightning Strikes*, 92.

39. Seifer, *Wizard*, 55–56.

40. Cawthorne, *Tesla: The Life and Times of an Electric Messiah*, 40.

41. Ibid.; Munson, *Tesla: Inventor of the Modern*, 81.

42. Munson, *Tesla: Inventor of the Modern*, 81.

43. Cawthorne, *Tesla: The Life and Times of an Electric Messiah*, 40.

44. David J. Kent, *Tesla: The Wizard of Electricity* (New York: Fall River Press, 2013), 110.

45. Wasik, *Lightning Strikes*, 86; Seifer, *Wizard*, 55.

46. Wasik, *Lightning Strikes*, 87; Munson, *Tesla: Inventor of the Modern*, 81.

47. Munson, *Tesla: Inventor of the Modern*, 81.

48. Terry S. Reynolds and Theodore Bernstein, "Edison and 'The Chair,'" *IEEE Technology & Society Magazine*, March 1989, 19.

49. Seifer, *Wizard*, 56.

50. Jonnes, *Empires of Light*, 177.

51. Ibid., 178.

52. Clark, *Edison*, 160.

53. Reynolds and Bernstein, "Edison and 'The Chair,'" 19, 21.

54. Ibid., 21.

55. Randall Stross, *The Wizard of Menlo Park: How Thomas Alva Edison Invented the Modern World* (New York: Crown Publishers, 2007), 184.

56. Munson, *Tesla: Inventor of the Modern*, 81–82; Mike Winchell, *The Electric War: Edison, Tesla, Westinghouse, and the Race to Light the World* (New York: Henry Holt, 2019), 5–6.

57. Seifer, *Wizard*, 56.

58. Winchell, *The Electric War*, 151.

59. Tom McNichol, *AC/DC: The Savage Tale of the First Standards War* (San Francisco: Jossey-Bass, 2006), 121.

60. Winchell, *The Electric War*, 153.

61. Ibid., 154.

62. Stross, *The Wizard of Menlo Park*, 184.

63. Nigel Cawthorne, *Tesla vs Edison: The Life-Long Feud That Fueled the World* (New York: Chartwell Books, 2016), 79.

64. Winchell, *The Electric War*, 239.

65. Ibid., 155; see "William Kemmler," Wikipedia, https://en.wikipedia.org/wiki/William_Kemmler#cite_note-execution1-8. The chair's electrified system had been tested the day before when a horse was put to death by electrocution.

66. Winchell, *The Electric War*, 156.

67. Ibid., 157.

68. Ibid., 160.

69. Ibid., 160–61; Munson, *Tesla: Inventor of the Modern*, 82.

70. Ibid., 82.

71. Michael Strevens, *The Knowledge Machine: How Irrationality Created Modern Science* (New York: Liveright Publishing Corporation, 2020), 58.

72. Ibid., 61.

73. Ibid., 24, 28.

74. Ibid., 28.

CHAPTER 7

1. Mike Winchell, *The Electric War: Edison, Tesla, Westinghouse, and the Race to Light the World* (New York: Henry Holt and Company, 2019), 239.

2. Jill Jonnes, *Empires of Light: Edison, Tesla, Westinghouse, and the Race to Electrify the World* (New York: Random House, 2004), 212–13.

3. Michael Strevens, *The Knowledge Machine: How Irrationality Created Modern Science* (New York: Liveright, 2020), 59.

4. Jonnes, *Empires of Light*, 163.

5. Ibid., 213.

6. Ibid.

7. Richard Munson, *Tesla: Inventor of the Modern* (New York: W.W. Norton, 2018), 62; Jonnes, *Empires of Light*, 226.

8. W. Bernard Carlson, *Tesla: Inventor of the Electrical Age* (Princeton, NJ: Princeton University Press, 2013), 130–31; Harrold C. Passer, *The Electrical Manufacturers 1875–1900: A Study in Competition, Entrepreneurship, Technical Change, and Economic Growth* (New York: Arno Press, 1972), 278. The amount of $2.50 per horsepower is the most accepted figure among biographers.

9. Jonnes, *Empires of Light*, 229.

10. John J. O'Neill, *Prodigal Genius: The Life of Nikola Tesla* (Albuquerque, NM: Brotherhood of Life, 1994), 87.

11. Jonnes, *Empires of Light*, 228.

12. O'Neill, *Prodigal Genius*, 87–88; Jonnes, *Empires of Light*, 228–29; Carlson, *Tesla*, 131.

13. Munson, *Tesla: Inventor of the Modern*, 79–80; Carlson, *Tesla*, 132–33.

14. Carlson, *Tesla*, 133.

15. Ibid., 132.

16. Munson, *Tesla: Inventor of the Modern*, 79.

17. Marc J. Seifer, *Wizard: The Life and Times of Nikola Tesla—Biography of a Genius* (New York: Citadel Press/Kensington Publishing, 1998), 66.

18. Munson, *Tesla: Inventor of the Modern*, 79; Carlson, *Tesla*, 131–32.

19. When Tesla inscribed, for the ages, the schematic of his "rotating magnetic field" in the wet earth of a Budapest park.

20. Munson, *Tesla: Inventor of the Modern*, 79.

21. Ibid., 78; Carlson, *Tesla*, 119.

22. Seifer, *Wizard*, 66.

23. Jonnes, *Empires of Light*, 229.

24. Ibid.

25. Munson, *Tesla: Inventor of the Modern*, 82.

26. Carlson, *Tesla*, 133.

27. Seifer, *Wizard*, 70.

28. Ibid.

29. Carlson, *Tesla*, 134.

30. Ibid.

31. Ibid.

32. Seifer, *Wizard*, 70.

33. Nikola Tesla, *My Inventions: The Autobiography of Nikola Tesla*, ed. Ben Johnston (Austin, TX: Hart Brothers, 1982), 87–88.

34. Joseph Wetzler, "Electric Lamps Fed from Space, and Flames That Do Not Consume," *Harper's Weekly*, July 11, 1891, 524.

35. Margaret Cheney and Robert Uth, *Tesla: Master of Lightning* (New York: Barnes & Noble, 1999), 38.

36. Seifer, *Wizard*, 71.

37. Ibid.

38. Munson, *Tesla: Inventor of the Modern*, 83.

39. Jonnes, *Empires of Light*, 231.

40. Tesla (Johnston, ed.), *My Inventions*, 81–82.

41. Cheney and Uth, *Tesla: Master of Lightning*, 38.

42. Thomas Commerford Martin, *The Inventions, Researches, and Writings of Nikola Tesla with Special Reference to His Work in Polyphase Currents and High Potential Lighting* (New York: D. Van Nostrand Company, 1894), 196–97; Seifer, *Wizard*, 71; Munson, *Tesla: Inventor of the Modern*, 83.

43. Tesla (Johnston, ed.), *My Inventions*, 87.

44. Nikola Tesla, "My Inventions: IV. The Discovery of the Tesla Coil and Transformer," *The Electrical Experimenter*, June 1919, 178.

45. Charles Culver Johnson, "Nikola Tesla's Revolution in War Telegraphy," *Philadelphia Press*, May 1, 1898.

46. Munson, *Tesla: Inventor of the Modern*, 85; Carlson, *Tesla*, 155–56.

CHAPTER 8

1. W. Bernard Carlson, *Tesla: Inventor of the Electrical Age* (Princeton, NJ: Princeton University Press, 2013), 143.

2. "Progress of Mr. Tesla's High Frequency Work," *The Electrical World* 19 (January 9, 1892), 20.

3. Ben Wilson, *Metropolis: A History of the City, Humankind's Greatest Invention* (New York: Doubleday, 2020), 135, 198.

4. Nikola Tesla, *My Inventions: The Autobiography of Nikola Tesla* (Eastford, CT: Marino Fine Books, 2018), 39–40; Michael Faraday (1791–1867) was an English scientist of great repute who added significantly to the study of electromagnetism and electrochemistry.

5. Carlson, *Tesla*, 143–44.

6. Marc J. Seifer, *Wizard: The Life and Times of Nikola Tesla—Biography of a Genius* (New York: Citadel Press/Kensington Publishing, 1998), 73; Carlson, *Tesla*, 144.

7. Carlson, *Tesla*, 144; Carl Hering, "Electrical Practice in Europe as Seen by an American—IV," *The Electrical World* (September 19, 1891), 193–96.

8. Carlson, *Tesla*, 144.

9. Seifer, *Wizard*, 73–74.

10. Carlson, *Tesla*, 146; Nigel Cawthorne, *Tesla: The Life and Times of an Electric Messiah* (New York: Chartwell Books, 2014), 46.

11. Richard Munson, *Tesla: Inventor of the Modern* (New York: W.W. Norton, 2018), 88; Seifer, *Wizard*, 85.

12. Nikola Tesla, "Experiments with Alternate Currents of High Potential and High Frequency," delivered before the Institution of Electrical Engineers/Royal Institution, London, February 1892; Nikola Tesla, *The Nikola Tesla Treasury* (Radford, VA: Wilder Publications, 2007), 121; Carlson, *Tesla*, 442.

13. Thomas Commerford Martin, *The Inventions, Researches, and Writings of Nikola Tesla with Special Reference to His Work in Polyphase Currents and High Potential Lighting* (New York: D. Van Nostrand Company, 1894), 235; Margaret Cheney and Robert Uth, *Tesla: Master of Lightning* (New York: Barnes & Noble, 1999), 40.

14. "Mr. Tesla's Lecture," Our London Correspondence, *Electrical Review* 20 (March 5, 1892), 20.

15. Martin, *The Inventions*, 201.

16. Ibid., 200–201.

17. "Mr. Tesla before the London Institution of Electrical Engineers and the Royal Institution," (Special Cablegram to *The Electrical Engineer*) *The Electrical Engineer* (February 10, 1892), 139, https://teslauniverse.com/nikola-tesla/articles/mr-tesla-london-institution-electical-engineers-and-royal-institution.

18. Michael Strevens, *The Knowledge Machine: How Irrationality Created Modern Science* (New York: Liveright, 2020), 190.

19. Martin, *The Inventions*, 200.

20. Ibid.

21. Albert Schmid, "Mr. Tesla before the Royal Institution, London," *Electrical Review* (March 19, 1892), 57. http://www.teslacollection.com/tesla_articles/1892/electrical_review_ny/albert_schmid/mr_tesla_before_the_royal_institution_london.

22. James Croft, "Performing Invention: On the Revelation of Technology," *Liminalities: A Journal of Performance Studies* 6, no. 2 (October 2010), 2, 5, http://liminalities.net/6-2/croft.pdf.

23. Ibid., 5.

24. Strevens, *The Knowledge Machine*, 187, 212, 227.

25. "Mr. Tesla's Lecture," *The Electrical Review* (February 12, 1892), 192.

26. Carlson, *Tesla*, 151–52.

27. Seifer, *Wizard*, 90–91; John Gribbin, *The Scientists: A History of Science Told through the Lives of Its Greatest Inventors* (New York: Random House, 2002), 489.

28. Iwan Rhys Morus, *Nikola Tesla and the Electrical Future* (London: Icon Books, 2019), 197–98.

29. Seifer, *Wizard*, 92.

30. Seeing great works of art can drive you crazy: Since 1978, specialists and psychiatrists have counseled over 110 foreign visitors to Florence, Italy, for what has come to be known as Stendhal's syndrome. These experts believe that exposure to great works of art in the city leads those with the propensity for psychological problems to be overwhelmed with bouts of mental turmoil such as suicidal urges, confusion, and panic. Apparently, the reaction is not modern, for this Renaissance city has overwhelmed travelers for centuries with its great works of art. French novelist Marie Henri Beyle (1783–1842), whose pseudonym was Stendhal, succumbed to the beauty of Florence almost two hundred years ago. He chronicled his emotional reaction in his book *Rome, Naples, and Florence*. The syndrome seems to occur in other cities that arouse one's emotions: Paris, Venice, Jerusalem, and Athens, to name a few. It appears that it's not a city such as Florence that causes the problem, but rather the exposure to great works of art that can trigger dramatic reactions in an already unstable individual. Even authors Henry James and Marcel Proust, along with psychoanalyst Sigmund Freud, admitted to being under the spell of Stendhal's syndrome. Marko Perko, *Did You Know That . . . ?: "Revised and Expanded" Edition: Surprising-But-True Facts about History, Science, Inventions, Geography, Origins, Art, Music and More* (New York: Open Road Distribution, 2017), 195; Wilson, *Metropolis*, 241.

31. Nikola Tesla, "Mechanical Therapy" (undated, four-page essay/letter on personal stationery), "The Nikola Tesla Papers": Reel # 3842, Columbia University, Rare Book and Manuscript Library, New York; Nikola Tesla, *Tesla Said* (Millbrae, CA: Tesla Book Company, 1984), 286–87.

32. Édouard Hospitalier, "Mr. Tesla's Experiments of Alternating Currents of Great Frequency" (translated from a report in *La Nature*), *Scientific American* (March 26, 1892), 195–96; Carlson, *Tesla*, 153.

33. Carlson, *Tesla*, 153.

34. Munson, *Tesla: Inventor of the Modern*, 92.

35. Ibid.

36. Nikola Tesla, *My Inventions* (Martino Fine), 48.

37. Munson, *Tesla: Inventor of the Modern*, 92–93.

38. Carlson, *Tesla*, 155.

39. Nikola Tesla, *My Inventions: The Autobiography of Nikola Tesla*, ed. Ben Johnston (Austin, TX: Hart Brothers, 1982), 104–5.

40. See: Note to the Reader: Dates.

41. Carlson, *Tesla*, 155.

42. Munson, *Tesla: Inventor of the Modern*, 93.

43. Ibid., 94.

44. Seifer, *Wizard*, 95.

45. Ibid., 96–97.

46. Carlson, *Tesla*, 156.

47. Nikola Tesla, *My Inventions: Nikola Tesla's Autobiography* (Las Vegas: Lits, 2011), 67–68; Carlson, *Tesla*, 157.

CHAPTER 9

1. "The Nikola Tesla Papers": Letterhead—Letter to R.U. Johnson, Columbia University, Rare Book and Manuscript Library, New York, March 31, 1894, Reel #3840.

2. Richard Munson, *Tesla: Inventor of the Modern* (New York: W.W. Norton, 2018), 97; Marc J. Seifer, *Wizard: The Life and Times of Nikola Tesla—Biography of a Genius* (New York: Citadel Press/Kensington Publishing, 1998), 98; W. Bernard Carlson, "Places of Invention: Nikola Tesla's Life in New York," Gotham Center for New York City History, June 14, 2013, https://www.gothamcenter.org/blog/places-of-invention-nikola-teslas-life-in-new-york; Moses King, *King's Photographic Views of New York: A Souvenir Companion to King's Handbook of New York City* (Boston: Moses King, 1895), 598–99.

3. Seifer, *Wizard*, 99.

4. Walter T. Stephenson, "Nikola Tesla and the Electric Light of the Future," *Scientific American Supplement* 39, no. 1004 (March 30, 1895), 16048–49.

5. Ibid.

6. Ibid., 16049.

7. Ibid.

8. Ibid.

9. H. Gernsback, "Nikola Tesla and His Inventions," an Announcement, *The Electrical Experimenter*, January 1919, 614.

10. Munson, *Tesla: Inventor of the Modern*, 98.

11. Ibid.

12. Seifer, *Wizard*, 99; Nikola Tesla, "On the Dissipation of the Electrical Energy of the Hertz Resonator," *The Electrical Engineer*, 15 (December 21, 1892), 587–88.

13. Thomas Commerford Martin, *The Inventions, Researches, and Writings of Nikola Tesla with Special Reference to His Work in Polyphase Currents and High Potential Lighting*, (New York: D. Van Nostrand Company, 1894), 348.

14. Seifer, *Wizard*, 99; Martin, *The Inventions*, 349.

15. W. Bernard Carlson, *Tesla: Inventor of the Electrical Age* (Princeton, NJ: Princeton University Press, 2013), 161.

16. Harrold C. Passer, *The Electrical Manufacturers 1875–1900: A Study in Competition, Entrepreneurship, Technical Change, and Economic Growth* (New York: Arno Press, 1972), 331.

17. Margaret Cheney and Robert Uth, *Tesla: Master of Lightning* (New York: Barnes & Noble, 999), 28.

18. Passer, *The Electrical Manufacturers*, 142.

19. Ibid.; see: Note to the Reader: Numbers.

20. John F. Wasik, *Lightning Strikes: Timeless Lessons in Creativity from the Life and Work of Nikola Tesla* (New York: Sterling, 2016), 103.

21. Christopher Columbus did not discover America. Marko Perko, *Did You Know That . . . ?: "Revised and Expanded" Edition: Surprising-But-True Facts about History, Science, Inventions, Geography, Origins, Art, Music and More* (New York: Open Road Distribution, 2017), 32–35.

22. Seifer, *Wizard*, 100.

23. Carlson, *Tesla*, 161.

24. Passer, *The Electrical Manufacturers*, 143.

25. Seifer, *Wizard*, 117–19.

26. Cheney and Uth, *Tesla: Master of Lightning*, 28–29, 33; Passer, *The Electrical Manufacturers*, 281: Munson, *Tesla: Inventor of the Modern*, 110.

27. Seifer, *Wizard*, 119–20.

28. Cheney and Uth, *Tesla: Master of Lightning*, 33; Passer, *The Electrical Manufacturers*, 281.

29. Nikola Tesla, "On Light and Other High Frequency Phenomena," *Nature*, 48, no. 1232 (June 8, 1893), 137.

30. "Electricians Listen in Wonder to the 'Wizard of Physics,'" *Chicago Tribune*, August 26, 1893; Munson, *Tesla*, 110.

31. Seifer, *Wizard*, 120–21; Martin, *The Inventions*, 486; "Electricians Listen in Wonder."

32. Martin, *The Inventions*, 477, 486–88.

33. "Electricians Listen in Wonder."

34. Carlson, *Tesla*, 163; Passer, *The Electrical Manufacturers*, 282.

35. Passer, *The Electrical Manufacturers*, 282.

36. Wasik, *Lightning Strikes*, 105.

37. Ibid.

38. Passer, *The Electrical Manufacturers*, 283–84.

39. Carlson, *Tesla*, 165.

40. Passer, *The Electrical Manufacturers*, 285.

41. Ibid.; Carlson, *Tesla*, 165.

42. Passer, *The Electrical Manufacturers*, 287, 289–90.

43. Carlson, *Tesla*, 168.

44. Passer, *The Electrical Manufacturers*, 290; Carlson, *Tesla*, 166.

45. Passer, *The Electrical Manufacturers*, 291.

46. Carlson, *Tesla*, 168.

47. Ibid., 173.

48. "Tesla's Work at Niagara," *New York Times*, July 16, 1895.

49. Carlson, *Tesla*, 173–74.

50. Ibid., 174; Munson, *Tesla: Inventor of the Modern*, 116–17.

51. Perko, *Did You Know*, 77; Munson, *Tesla: Inventor of the Modern*, 116.

52. Munson, *Tesla: Inventor of the Modern*, 117.

53. "Nikola Tesla at Niagara Falls," *The Western Electrician*, 19, no. 5 (August 1, 1896), 5; Munson, *Tesla: Inventor of the Modern*, 117.

54. Jill Jonnes, *Empires of Light: Edison, Tesla, Westinghouse, and the Race to Electrify the World* (New York: Random House, 2004), 354.

CHAPTER 10

1. W. Bernard Carlson, *Tesla: Inventor of the Electrical Age* (Princeton, NJ: Princeton University Press, 2013), 198.

2. Lieutenant F. Jarvis Patten, "Nikola Tesla and His Work," *The Electrical World* 23, no. 15 (April 14, 1894), 496.

3. Richard Munson, *Tesla: Inventor of the Modern* (New York: W.W. Norton, 2018), 122. As an aside, Tesla and Martin did have a disagreement that festered over time regarding royalties from the book *The Inventions, Researches and Writings of Nikola Tesla*, which was selling quite well.

4. Marc J. Seifer, *Wizard: The Life and Times of Nikola Tesla—Biography of a Genius* (New York: Citadel Press/Kensington Publishing, 1998), 122–23.

5. Ibid., 123.

6. Munson, *Tesla: Inventor of the Modern*, 128, 132.

7. Ibid., 133–34.

8. Margaret Cheney and Robert Uth, *Tesla: Master of Lightning* (New York: Barnes & Noble, 1999), 47; Nikola Tesla, "My Inventions: III. My Later Endeavors, The Discovery of the Rotating Magnetic Field," *The Electrical Experimenter*, April 1919, 864–65.

9. Nikola Tesla, "Mechanical Therapy," https://teslauniverse.com/nikola-tesla/articles/mechanical-therapy; Nikola Tesla, "Mechanical Therapy," in *Tesla Said* (Millbrae, CA: Tesla Book Company, 1984), 286–87; Munson, *Tesla: Inventor of the Modern*, 124.

10. Nigel Cawthorne, *Tesla: The Life and Times of an Electric Messiah* (New York: Chartwell Books, 2014), 64; John J. O'Neill, *Prodigal Genius: The Life of Nikola Tesla* (Albuquerque, NM: Brotherhood of Life, 1994), 160.

11. John F. Wasik, *Lightning Strikes: Timeless Lessons in Creativity from the Life and Work of Nikola Tesla* (New York: Sterling, 2016), 121.

12. O'Neill, *Prodigal Genius*, 159–61; Cawthorne, *Tesla: The Life and Times of an Electric Messiah*, 64.

13. Munson, *Tesla: Inventor of the Modern*, 124.

14. Randal K. Buddington, Thomas Wong, and Scott C. Howard, "Paracellular Filtration Secretion Driven by Mechanical Force Contributes to Small Intestinal Fluid Dynamics," *Med Sci* (Basel) 9, no. 1 (March 2021), 19.

15. Munson, *Tesla: Inventor of the Modern*, 124.

16. Carlson, *Tesla*, 198. See: Notes on Dates. Following the Julian calendar as well, many Orthodox Christians celebrate their Christmas on January 7.

17. Carlson, *Tesla*, 199.

18. Many of these letters and other communiqués are available to be viewed at the Butler Library, Manuscript Division, Columbia University, New York City.

19. "The Nikola Tesla Papers": 21380E. Columbia University, Rare Book and Manuscript Library, New York.

20. O'Neill, *Prodigal Genius*, 289.

21. Ibid.

22. Cawthorne, *Tesla: The Life and Times of an Electric Messiah*, 64.

23. Munson, *Tesla: Inventor of the Modern*, 129.

24. Ibid.

25. Ibid., 130.

26. Ibid.

27. Ibid., 130–31.

28. Margaret Cheney, *Tesla: Man Out of Time* (New York: Barnes & Noble, 1993), 107.

29. Robert Underwood Johnson, *Remembered Yesterdays* (Boston: Little, Brown, and Company, 1923), 401; Wasik, *Lightning Strikes*, 109–10.

30. Johnson, *Remembered Yesterdays*, 401.

31. Robert Underwood Johnson, *Songs of Liberty and Other Poems: Including Paraphrases from the Servian after Translations by Nikola Tesla, with a Prefatory Note by Him on Servian Poetry* (New York: Century Co., 1897), 54–57; Seifer, *Wizard*, 124–25.

32. Carlson, *Tesla*, 199; Seifer, *Wizard*, 125; Cheney, *Tesla: Man Out of Time*, 83. Cheney states that Tesla called the Johnsons: Luka Filipov and Madame Filipov.

33. Munson, *Tesla: Inventor of the Modern*, 129; Seifer, *Wizard*, 125.

34. Munson, *Tesla: Inventor of the Modern*, 129.

35. Ibid.

36. Seifer, *Wizard*, 123–24.

37. Ibid., 125–26.

38. "The Nikola Tesla Papers": Letter to Katharine Johnson, November 3, 1898, Columbia University, Rare Book and Manuscript Library, New York.

39. Petar Ivić, "Nikola Tesla and Women—The Only Love of a Great Scientist," *Bazar* (*Harper's Bazaar*), 1974.

40. Seifer, *Wizard*, 126.

41. Johnson, *Remembered Yesterdays*, 400.

42. Ibid.

43. Seifer, *Wizard*, 127; Johnson, *Remembered Yesterdays*, 400.

44. Johnson, *Poems*, Second Edition (New York: Century Co., 1908), 116.

45. Munson, *Tesla: Inventor of the Modern*, 136; Curtis Brown, "A Man of the Future," *Savannah Morning News*, October 21, 1894.

46. Carlson, *Tesla*, 203–4; Seifer, *Wizard*, 131.

47. Brown, "A Man of the Future."

48. Ibid.

49. Carlson, *Tesla*, 201–3.

CHAPTER 11

1. Walter T. Stephenson, "Fruits of Genius Were Swept Away," *New York Herald*, March 14, 1895.

2. Ibid.

3. Margaret Cheney and Robert Uth, *Tesla: Master of Lightning* (New York: Barnes & Noble, 1999), 53.

4. Stephenson, "Fruits of Genius Were Swept Away."

5. Richard Munson, *Tesla: Inventor of the Modern* (New York: W.W. Norton, 2018), 136.

6. Stephenson, "Fruits of Genius Were Swept Away."

7. Thomas Commerford Martin, "The Burning of Tesla's Laboratory," *The Engineering Magazine* (April 1, 1895), 104.

8. "Mr. Tesla's Great Loss," *New York Times*, March 14, 1895.

9. "Destruction of the Tesla Laboratory by Fire," *The Electrical Engineer* (March 20, 1895), 275.

10. Martin, "The Burning of Tesla's Laboratory," 101–4.

11. Munson, *Tesla: Inventor of the Modern*, 139.

12. Vitold Kreutzer, "Nikola Tesla," Beverley Viljakainen, editor, Institute for Consciousness Research, 2019, https://www.icrcanada.org/research/literaryresearch/tesla.

13. Munson, *Tesla: Inventor of the Modern*, 139; Kreutzer, "Nikola Tesla."

14. Marc J. Seifer, *Wizard: The Life and Times of Nikola Tesla—Biography of a Genius* (New York: Citadel Press/Kensington Publishing, 1998), 158.

15. Nikola Tesla, "Mechanical Therapy," in *Tesla Said* (Millbrae, CA: Tesla Book Company, 1984), 286; Munson, *Tesla: Inventor of the Modern*, 140.

16. "The Nikola Tesla Company," Trade Notes and Novelties and Mechanical Department, *The Electrical Engineer* 19, no. 534 (February 13, 1895), 149; Munson, *Tesla: Inventor of the Modern*, 136–37; Carlson, *Tesla*, 205–7.

17. Marc J. Seifer, *Wizard*, 148–49; Munson, *Tesla: Inventor of the Modern*, 140.

18. Arthur Brisbane, "Our Foremost Electrician," *New York World*, July 1894.

19. Carlson, *Tesla*, 1–3.

20. Brisbane, "Our Foremost Electrician."

21. Ibid.

22. Ibid.; Carlson, *Tesla*, 3.

23. Ibid., 5.

24. George Heli Guy, "Tesla: Man and Inventor," *New York Times*, March 31, 1895.

25. Brisbane, "Our Foremost Electrician."

26. A theory of everything (TOE) is also referred to as the ultimate theory, final theory, or master theory. It is a hypothetical single, all-encompassing, theoretical matrix of physics that completely explains and joins together all physical attributes of the universe.

27. Harrold C. Passer, *The Electrical Manufacturers 1875–1900: A Study in Competition, Entrepreneurship, Technical Change, and Economic Growth* (New York: Arno Press, 1972, 1953), 328.

28. Munson, *Tesla: Inventor of the Modern*, 141; Seifer, *Wizard*, 158.

29. Carlson, *Tesla*, 207–9.

30. Ibid., 211.

31. Thomas Commerford Martin, "Tesla's Oscillator and Other Inventions: An Authoritative Account of Some of His Recent Electrical Work," *The Century Magazine* (April 1895), 933; Carlson, *Tesla*, 213.

CHAPTER 12

1. Marc J. Seifer, *Wizard: The Life and Times of Nikola Tesla—Biography of a Genius* (New York: Citadel Press/Kensington Publishing, 1998), 156–57.
2. W. C. Peake, "Is Tesla to Signal the Stars?" *The Electrical World*, April 4, 1896, 369.
3. Ibid.
4. Nigel Cawthorne, *Tesla: The Life and Times of an Electric Messiah* (New York: Chartwell Books, 2014), 71.
5. Ibid.
6. Christopher Cooper, *The Truth about Tesla: The Myth of the Lone Genius in the History of Innovation* (New York: Race Point Publishing, 2018), 44–45.
7. Richard Munson, *Tesla: Inventor of the Modern* (New York: W.W. Norton, 2018), 124.
8. Earl Sparling, "Nikola Tesla, at 79, Uses Earth to Transmit Signals; Expects to Have $100,000,000 within Two Years: Could Destroy Empire State Building with Five Pounds of Air Pressure, He Says," *New York World-Telegram*, July 11, 1935.
9. Margaret Cheney, *Tesla: Man Out of Time* (New York: Barnes & Noble, 1993), 115.
10. Sparling, "Nikola Tesla, at 79."
11. Margaret Cheney and Robert Uth, *Tesla: Master of Lightning* (New York: Barnes & Noble, 1999), 75.
12. Seifer, *Wizard*, 68–70.
13. Ibid., 167.
14. Cheney, *Tesla: Man Out of Time*, 100–101.
15. "The X Rays and the New Photography," *The Electrical Review*, 38, no. 959, 470.
16. Nikola Tesla, "The Streams of Lenard and Roentgen and Novel Apparatus for Their Production," in *Lecture Before the New York Academy of Sciences, April 6, 1897*, ed. Leland Anderson (New York: Twenty-First Century Books, 1994); Munson, *Tesla: Inventor of the Modern*, 141.
17. Tesla, "The Streams of Lenard and Roentgen."
18. Cheney, *Tesla: Man Out of Time*, 101.
19. Ibid.
20. Nikola Tesla, "Tesla's Latest Results—He Now Produces Radiographs at a Distance of More Than Forty Feet," *The Electrical Review*, March 18, 1896, 147.
21. Cheney, *Tesla: Man Out of Time*, 103–4.
22. Ibid., 104.
23. Munson, *Tesla: Inventor of the Modern*, 142.

24. Nikola Tesla, "Tesla on the Roentgen Streams," *The Electrical Review*, December 2, 1896; Munson, *Tesla: Inventor of the Modern*, 142.

25. Munson, *Tesla: Inventor of the Modern*, 143.

26. Ibid., 142–43.

27. "Tesla's 'Power Center' Speech," January 12, 2019, https://teslasciencecenter .org/announcements/teslas-power-banquet-speech/.

28. Ibid.

29. Ibid.

30. Ibid.

31. Seifer, *Wizard*, 173.

32. Munson, *Tesla: Inventor of the Modern*, 143.

33. Seifer, *Wizard*, 183.

34. Ibid., 193.

35. Ibid.; Isaac Asimov coined the term "robotics" in 1941.

36. Nikola Tesla, "Tesla's Latest Invention: Details of an Invention Which May Assure the Peace of the World," *The Electrical Review*, November 16, 1898, 305–12; Seifer, *Wizard*, 193.

37. Nikola Tesla, *My Inventions: The Autobiography of Nikola Tesla*, ed. Ben Johnston (Austin, TX: Hart Brothers, 1982), 105–6.

38. Ibid., 107.

39. Nikola Tesla, "Tesla Describes His Efforts in Various Fields of Work" (Reprint from the *New York Sun*, November 21, 1898), *The Electrical Review*, 344–45.

40. W. Bernard Carlson, *Tesla: Inventor of the Electrical Age* (Princeton, NJ: Princeton University Press, 2013), 227.

41. Seifer, *Wizard*, 194.

42. Munson, *Tesla: Inventor of the Modern*, 143.

43. Cheney and Uth, *Tesla: Master of Lightning*, 79.

44. John F. Wasik, *Lightning Strikes: Timeless Lessons in Creativity from the Life and Work of Nikola Tesla* (New York: Sterling, 2016), 113–14.

45. Tesla, *My Inventions*, ed. Johnston, 107.

46. Carlson, *Tesla*, 229.

47. Cheney and Uth, *Tesla: Master of Lightning*, 82.

48. F. L. Christman, "Tesla Declares He Will Abolish War," *New York Herald*, November 8, 1898.

49. Ibid.

50. Seifer, *Wizard*, 196.

51. Munson, *Tesla: Inventor of the Modern*, 145.

52. Nikola Tesla, "Torpedo Boat without a Crew," *New York Journal*, February 1, 1899, 136.

53. Seifer, *Wizard*, 196–97; Carlson, *Tesla*, 231–32.

54. Nikola Tesla, "Mr. Tesla to His Friends," *The Electrical Engineer* 26, no. 551, November 18, 1898, 514.

55. Ibid.

56. Carlson, *Tesla*, 232.

57. Ibid., 234.

58. Seifer, *Wizard*, 203.

59. Nikola Tesla, *The Problem of Increasing Human Energy: With Special Reference to the Harnessing of the Sun's Energy*. As printed in *Century Illustrated Magazine*, June 1900, (Merchant Books, 2019), 26–27.

60. Cheney and Uth, *Tesla: Master of Lightning*, 82.

61. Ibid.

62. Orrin E. Dunlap, "Nikola Tesla at Niagara Falls," *The Western Electrician* 19, no. 5, August 1, 1896, 55.

63. Cheney and Uth, *Tesla: Master of Lightning*, 51.

64. Carlson, *Tesla*, 240.

65. Munson, *Tesla: Inventor of the Modern*, 148.

66. Carlson, *Tesla*, 241–43.

67. John J. O'Neill, *Prodigal Genius: The Life of Nikola Tesla* (Albuquerque, NM: Brotherhood of Life, 1994), 4.

CHAPTER 13

1. Nikola Tesla, *My Inventions: Nikola Tesla's Autobiography* (Las Vegas: Lits, 2011), 72.

2. Nikola Tesla, "The True Wireless," *The Electrical Experimenter*, May 1919, 61.

3. Richard Munson, *Tesla: Inventor of the Modern* (New York: W.W. Norton, 2018), 149.

4. *Rough Riders* is the celebrated sobriquet given to the First Regiment of U.S. Cavalry Volunteers in the Spanish-American War of 1898. It is commonly believed that Teddy Roosevelt (1858–1919) was their commander and led this regiment up hill and down dale in various flourishes of bravado against the enemy. As assistant secretary of the Navy, Roosevelt did acquire Congressional approval to recruit a volunteer regiment of cavalry. He personally selected several hundred men, mostly Harvard blue bloods, horsemen, cowboys, and famous athletes. In effect, Roosevelt did organize the Rough Riders, but command of them fell to Colonel Leonard Wood, who had legitimate military credentials and was also a physician. Lieutenant Colonel Teddy Roosevelt did not lead the charge up San Juan Hill, but he did lead the charge up nearby Kettle Hill on his horse, Texas. His success helped the Americans win the battle of San Juan Hill, and therein lies the confusion. Roosevelt's ride became mistakenly associated with the San Juan Hill assault. He was second in command to Wood, whose Rough Riders soon called themselves "Wood's Weary Walkers," because they fought much of the war entirely on foot. What happened to the horses? They had to be left in Florida because the ships that were transporting the cavalry soldiers to Cuba had no room for the horses the soldiers were supposed to have ridden. See Marko Perko, *Did You Know That . . . ?: "Revised and Expanded" Edition: Surprising-But-True Facts about History, Science, Inventions, Geography, Origins, Art, Music and More* (New York: Open Road Distribution, 2017), 11.

5. W. Bernard Carlson, *Tesla: Inventor of the Electrical Age* (Princeton, NJ: Princeton University Press, 2013), 256.

6. Marc J. Seifer, *Wizard: The Life and Times of Nikola Tesla—Biography of a Genius* (New York: Citadel Press/Kensington Publishing, 1998), 210.

7. Ibid.

8. Carlson, *Tesla*, 258; Seifer, *Wizard*, 211.

9. Munson, *Tesla: Inventor of the Modern*, 152–53; Seifer, *Wizard*, 211.

10. Carlson, *Tesla*, 259.

11. Seifer, *Wizard*, 213.

12. "Tesla's Visit to Chicago," *The Western Electrician*, May 20, 1899, 285.

13. Ibid.

14. Ibid.

15. Ibid.

16. Carlson, *Tesla*, 259–61.

17. "Tesla's Visit to Chicago," 285.

18. Chauncy M. McGovern, "Nikola Tesla Will 'Wire' to France," *Colorado Springs Evening Telegraph*, May 17, 1899; Munson, *Tesla: Inventor of the Modern*, 153.

19. McGovern, "Nikola Tesla Will 'Wire' to France."

20. Ibid.

21. Munson, *Tesla: Inventor of the Modern*, 154.

22. Ibid., 153–54.

23. Seifer, *Wizard*, 214.

24. Ibid.; Carlson, *Tesla*, 220–21.

25. Harry Goldman, "Nikola Tesla's Bold Adventure," *The American West*, March 1971, 4–9.

26. Carlson, *Tesla*, 266–67.

27. Munson, *Tesla: Inventor of the Modern*, 151.

28. Margaret Cheney and Robert Uth, *Tesla: Master of Lightning* (New York: Barnes & Noble, 1999), 85.

29. Carlson, *Tesla*, 265.

30. Nikola Tesla, "The Transmission of Electric Energy without Wires," *Electrical World and Engineer*, March 5, 1904.

31. Ibid.

32. Cheney and Uth, *Tesla: Master of Lightning*, 87; John J. O'Neill, *Prodigal Genius: The Life of Nikola Tesla* (Albuquerque, NM: Brotherhood of Life, 994), 183.

33. Tesla, "The Transmission of Electric Energy without Wires."

34. Ibid.

35. Ibid.

36. Munson, *Tesla: Inventor of the Modern*, 157–58.

37. Tesla, "The Transmission of Electric Energy without Wires."

38. Carlson, *Tesla*, 211; Munson, *Tesla: Inventor of the Modern*, 158.

39. Carlson, *Tesla*, 293.

40. Nikola Tesla, *Nikola Tesla: Colorado Springs Notes, 1899–1900* (BN Publishing, 2007), 18. Scientific commentaries by Aleksander Marinčić, D.Sc.

41. Ibid., 18; Munson, *Tesla: Inventor of the Modern*, 155.

42. Tesla, *Nikola Tesla: Colorado Springs Notes*, 12.

43. Munson, *Tesla: Inventor of the Modern*, 159.

44. O'Neill, *Prodigal Genius*, 187.

45. Munson, *Tesla: Inventor of the Modern*, 159; O'Neill, *Prodigal Genius*, 187; Cheney and Uth, *Tesla: Master of Lightning*, 89.

46. O'Neill, *Prodigal Genius*, 187.

47. Munson, *Tesla: Inventor of the Modern*, 159; O'Neill, *Prodigal Genius*, 189.

48. O'Neill, *Prodigal Genius*, 190–91.

49. Tesla, *Nikola Tesla: Colorado Springs Notes*, 100.

50. Tesla, *Nikola Tesla: Colorado Springs Notes*, 137–38.

51. Munson, *Tesla: Inventor of the Modern*, 160.

52. Nikola Tesla, *The Problem of Increasing Human Energy: With Special Reference to the Harnessing of the Sun's Energy* (as printed in *Century Illustrated Magazine*, June 1900), unabridged ed. (Merchant Books, 2019), 69.

53. Cheney and Uth, *Tesla: Master of Lightning*, 94; also see Wikipedia, "Search for extraterrestrial intelligence," https://en.wikipedia.org/wiki/Search_for_extraterrestrial_intelligence#History.

54. Cheney and Uth, *Tesla: Master of Lightning*, 94–95; Wikipedia, "Search for extraterrestrial intelligence."

55. Carlson, *Tesla*, 256; Munson, *Tesla: Inventor of the Modern*, 160.

56. Nikola Tesla, "The True Wireless," *The Electrical Experimenter*, May 1919, 30.

57. Munson, *Tesla: Inventor of the Modern*, 161.

58. *Tesla: Master of Lightning*, PBS, https://www.pbs.org/tesla/ll/ll_whoradio.html.

59. https://patents.google.com/patent/US645576A/en; https://patentimages.storage.googleapis.com/62/90/92/45a5932052a940/US645576.pdf; https://pdfpiw.uspto.gov/.piw?Docid=00645576&homeurl=http%3A%2F%2Fpatft.uspto.gov%2Fnetacgi%2Fnph-Parser%3FSect1%3DPTO1%2526Sect2%3DHITOFF%2526d%3DPALL%2526p%3D1%2526u%3D%25252Fnetahtml%25252FPTO%25252Fsrchnum.htm%2526r%3D1%2526f%3DG%2526l%3D50%2526s1%3D0645576.PN.%2526OS%3DPN%2F0645576%2526RS%3DPN%2F0645576&PageNum=&Rtype=&SectionNum=&idkey=NONE&Input=View+first+page.

60. Munson, *Tesla: Inventor of the Modern*, 161–62.

61. Perko, *Did You Know That . . .*, 238–39.

62. Munson, *Tesla: Inventor of the Modern*, 162.

63. Seifer, *Wizard*, 436; Munson, *Tesla: Inventor of the Modern*, 162.

64. Munson, *Tesla: Inventor of the Modern*, 163.

65. Nikola Tesla, "Talking with the Planets," *Collier's Weekly: An Illustrated Journal*, February 9, 1901, 4–5.

66. Munson, *Tesla: Inventor of the Modern*, 164.

67. Tesla, *Nikola Tesla: Colorado Springs Notes*, 367.

68. Carlson, *Tesla*, 294–95.

69. Tesla, *Nikola Tesla: Colorado Springs Notes*, 433; Margaret Cheney, *Tesla: Man Out of Time* (New York: Barnes & Noble), 151.

70. Seifer, *Wizard*, 234–35; "The Nikola Tesla Papers": 40032E, etc., Columbia University, Rare Book and Manuscript Library, New York.

CHAPTER 14

1. "Issued Nikola Tesla," *Colorado Springs Gazette*, April 6, 1804, 3; "Left Property Here; Skips; Sheriff's Sale," *Colorado Springs Gazette*, March 22, 1906, 3; "Nikola Tesla Says He Is Not Indebted to Duffner," *Colorado Springs Gazette*, September 6, 1905, 5; "Nikola Tesla Must Pay Up," *Colorado Springs Gazette*, November 19, 1905, 5.

2. "Notes," *The Electrician*, January 19, 1900, 423.

3. Ibid.

4. Leland I. Anderson, "Wardenclyffe—A Forfeited Dream," Long Island Forum, August 1968.

5. Richard Munson, *Tesla: Inventor of the Modern* (New York: W.W. Norton, 2018), 167; Marc J. Seifer, *Wizard: The Life and Times of Nikola Tesla—Biography of a Genius* (New York: Citadel Press/Kensington Publishing, 1998), 205.

6. Anderson, "Wardenclyffe."

7. John J. O'Neill, *Prodigal Genius: The Life of Nikola Tesla* (Albuquerque, NM: Brotherhood of Life, 1994), 98.

8. Nikola Tesla, comments written on page 774 of *The Century Magazine*, vol. 63, March 1902, "The Nikola Tesla Papers": Columbia University, Rare Book and Manuscript Library, New York.

9. Seifer, *Wizard*, 236.

10. Ibid., 246.

11. Ibid., 239–40.

12. W. Bernard Carlson, *Tesla: Inventor of the Electrical Age* (Princeton, NJ: Princeton University Press, 2013), 307.

13. "Decision in Favor of the Tesla Rotating Magnetic Field Patents," *Electrical World and Engineer* 36, no. 10 (September 8, 1900), 394–95.

14. Margaret Cheney and Robert Uth, *Tesla: Master of Lightning* (New York: Barnes & Noble, 1999), 98.

15. Seifer, *Wizard*, 246; Cheney and Uth, *Tesla: Master of Lightning*, 99.

16. Jill Jonnes, *Empires of Light: Edison, Tesla, Westinghouse, and the Race to Electrify the World* (New York: Random House, 2004), 360–61.

17. Seifer, *Wizard*, 252–53.

18. Munson, *Tesla: Inventor of the Modern*, 174.

19. Theodore Roosevelt Center, https://www.theodorerooseveltcenter.org/Research/Digital-Library/Record.aspx?libID=o286435.

20. Seifer, *Wizard*, 255.

21. Carlson, *Tesla*, 318–19; Munson, *Tesla: Inventor of the Modern*, 174–75.

22. Munson, *Tesla: Inventor of the Modern*, 175.

23. Ibid., 177.

24. Cheney and Uth, *Tesla: Master of Lightning*, 100–101.

25. Tesla's description of the Long Island plant and inventory of the installation as reported in 1922 foreclosure appeal proceedings in the Supreme Court–Appellate Division of the State of New York, p. 177.

26. Cheney and Uth, *Tesla: Master of Lightning*, 100.

27. Carlson, *Tesla*, 327.

28. Ibid., 320, 322.

29. Seifer, *Wizard*, 284.

30. Cheney and Uth, *Tesla: Master of Lightning*, 100.

31. O'Neill, *Prodigal Genius*, 207–8.

32. Natalie Aurucci Stiefel, "Nikola Tesla at Wardenclyffe," in *Looking Back at Rocky Point: In the Shadow of the Radio Towers, Vol. 1* (self-published).

33. Munson, *Tesla: Inventor of the Modern*, 176; Cheney and Uth, *Tesla: Master of Lightning*, 101; Seifer, *Wizard*, 266; O'Neill, *Prodigal Genius*, 207.

34. "Wireless Signals Across the Ocean," *New York Times*, December 15, 1901, 1.

35. Aaron Antonio Toscano, "Positioning Guglielmo Marconi's Wireless: A Rhetorical Analysis of an Early Twentieth-Century Technology," PhD diss., University of Louisville, 2006, 1.

36. Carlson, *Tesla*, 334–36; Munson, *Tesla: Inventor of the Modern*, 179.

37. Lawrence W. Lichty, and Malachi C. Topping, *American Broadcasting: A Source Book on the History of Radio and Television* (New York: Hastings House, 1975), 11.

38. Munson, *Tesla: Inventor of the Modern*, 179–80.

39. Seifer, *Wizard*, 295; John F. Wasik, *Lightning Strikes: Timeless Lessons in Creativity from the Life and Work of Nikola Tesla* (New York: Sterling, 2016), 131; Carlson, *Tesla*, 339–40.

40. Munson, *Tesla: Inventor of the Modern*, 173–74.

41. Seifer, *Wizard*, 296.

42. "Edison Becomes Marconi's Ally," *New York World*, May 28, 1903, 16.

43. Iwan Rhys Morus, *Nikola Tesla and the Electrical Future* (London: Icon Books, 2019), 172.

44. Cheney and Uth, *Tesla: Master of Lightning*, 105; Seifer, *Wizard*, 320.

45. Munson, *Tesla: Inventor of the Modern*, 181.

46. Nikola Tesla, "The Transmission of Electrical Energy without Wires as a Means for Furthering Peace," *Electrical World and Engineer*, January 7, 1905, 21–24.

47. Munson, *Tesla: Inventor of the Modern*, 182.

48. Ibid., 183.

49. "The Nikola Tesla Papers": Columbia University, Rare Book and Manuscript Library, New York.

50. Nikola Tesla, *My Inventions: The Autobiography of Nikola Tesla*, ed. Ben Johnston (Austin, TX: Hart Brothers, 1982), 93.

51. Munson, *Tesla: Inventor of the Modern*, 184.

52. Seifer, *Wizard*, 300; Munson, *Tesla: Inventor of the Modern*, 184.

53. Carlson, *Tesla*, 347.

54. Ibid.; Munson, *Tesla: Inventor of the Modern*, 185.

55. Munson, *Tesla: Inventor of the Modern*, 185.

56. Carlson, *Tesla*, 352.

57. Ibid., 352–53.

58. Munson, *Tesla: Inventor of the Modern*, 186.

59. Ibid.

60. O'Neill, *Prodigal Genius*, 292.

61. Ibid., 286.

62. Munson, *Tesla: Inventor of the Modern*, 187.

63. As an example, see letter of April 14, 1903(?) to George Scherff (21380E) in "The Nikola Tesla Papers": Columbia University, Rare Book and Manuscript Library, New York.

64. Munson, *Tesla: Inventor of the Modern*, 187–88.

65. Ibid., 188.

66. Ibid., 189.

67. "Wardenclyffe Tower," Wikipedia, https://en.wikipedia.org/wiki/Wardenclyffe_Tower.

68. "Tesla Science Center at Wardenclyffe," Wikipedia, https://en.wikipedia.org/wiki/Tesla_Science_Center_at_Wardenclyffe.

69. Carlson, *Tesla*, 367.

CHAPTER 15

1. Some people with remarkable memories were Nikola Tesla, Charles Darwin, and Teddy Roosevelt. Photographic memory has been directly linked to high intelligence in such individuals. Tesla's IQ was said to be between 160 and 310. See https://www.betterhelp.com; https://www.businessinsider.com/the-40-smartest-people-of-all-time-2015-2#14-nikola-tesla-27.

2. Frank G. Carpenter, "Wonderful Discoveries in Electricity: Electrical Force without Wires," *Pittsburgh Dispatch*, December 18, 1904.

3. W. Bernard Carlson, *Tesla: Inventor of the Electrical Age* (Princeton, NJ: Princeton University Press, 2013), 364.

4. Marc J. Seifer, *Wizard: The Life and Times of Nikola Tesla—Biography of a Genius* (New York: Citadel Press/Kensington Publishing, 1998), 309.

5. Nikola Tesla, "The People's Forum: 'Mr. Tesla on the Wireless Transmission of Power,'" *New York World*, May 19, 1907.

6. Nikola Tesla, *My Inventions: The Autobiography of Nikola Tesla*, ed. Ben Johnston (Austin, TX: Hart Brothers, 1982), 91–92.

7. Margaret Cheney and Robert Uth, *Tesla: Master of Lightning* (New York: Barnes & Noble, 1999), 109.

8. Seifer, *Wizard*, 339; "Boundary layer," Wikipedia, https://en.wikipedia.org/wiki/Boundary_layer.

9. Seifer, *Wizard*, 339.

10. Carlson, *Tesla*, 371.

11. Ibid., 341.

12. Cheney and Uth, *Tesla: Master of Lightning*, 110–15.

13. Seifer, *Wizard*, 347.

14. Nikola Tesla, "Mr. Tesla Speaks Out," *New York World*, November 29, 1929, 10.

15. James Shapiro, *Contested Will: Who Wrote Shakespeare?* (New York: Simon & Schuster, 2010), 111.

16. Cheney and Uth, *Tesla: Master of Lightning*, 118–19.

17. Allan L. Benson, "Nikola Tesla, Dreamer," *World To-day (Heart's Magazine)*, February 1, 1912, 1763.

18. Ibid.

19. Carlson, *Tesla*, 372–73.

20. Cheney, *Tesla: Man Out of Time* (New York: Barnes & Noble, 1993), 166.

21. Frank Parker Stockridge, "Tesla's New Monarch of Machines," *New York Herald (Tribune)*, October 15, 1991, 1.

22. H. Winfield Secor, "Tesla Has Only Credit," *New York Times*, March 18, 1916.

23. Cheney, *Tesla: Man Out of Time*, 215.

24. John J. O'Neill, *Prodigal Genius: The Life of Nikola Tesla* (Albuquerque, NM: Brotherhood of Life, 1994), 232–33; Cheney and Uth, *Tesla: Master of Lightning*, 119–20.

25. Ibid.

26. O'Neill, *Prodigal Genius*, 234–36.

27. Ibid., 237–38.

28. "Nikola Tesla's Acceptance Speech on Receiving the Edison Medal," Minutes of the Annual Meeting of the American Institute of Electrical Engineers, May 18, 1917, http://rastko.rs/rastko/delo/10842.

29. Cheney and Uth, *Tesla: Master of Lightning*, 123; O'Neill, *Prodigal Genius*, 238.

30. Cheney and Uth, *Tesla: Master of Lightning*, 123.

31. H. Winfield Secor, "Tesla's Views on Electricity and the War," *The Electrical Experimenter*, August 1917, 270.

32. Cheney, *Tesla: Man Out of Time*, 213.

33. Seifer, *Wizard*, 334.

34. Richard Munson, *Tesla: Inventor of the Modern*, 198–99.

35. Cheney and Uth, *Tesla: Master of Lightning*, 129.

36. Ibid., 119–20.

37. John J. O'Neill, *Prodigal Genius*, 308–9.

38. Cheney and Uth, *Tesla: Master of Lightning*, 120; Munson, *Tesla: Inventor of the Modern*, 207.

39. Munson, *Tesla: Inventor of the Modern*, 207.

40. O'Neill, *Prodigal Genius*, 230–31.

41. Seifer, *Wizard*, 380.

42. Ibid.

43. Marko Perko, *Did You Know That . . . ?: "Revised and Expanded" Edition: Surprising-But-True Facts about History, Science, Inventions, Geography, Origins, Art, Music and More* (New York: Open Road Distribution, 2017), 77.

44. Munson, *Tesla: Inventor of the Modern*, 199.

45. Ibid.

46. Ibid.

47. Shapiro, *Contested Will*, 111.

48. Nikola Tesla, "My Inventions: VI. The Art of Telautomatics," *The Electrical Experimenter*, October 1919, 601; Cheney and Uth, *Tesla: Master of Lightning*, 130.

49. O'Neill, *Prodigal Genius*, 218; Munson, *Tesla: Inventor of the Modern*, 202.

50. Walter Isaacson, *Einstein: His Life and Universe* (New York: Simon & Schuster, 2007), 320–21.

CHAPTER 16

1. Richard Munson, *Tesla: Inventor of the Modern* (New York: W.W. Norton, 2018), 210; W. Bernard Carlson, *Tesla: Inventor of the Electrical Age* (Princeton, NJ: Princeton University Press, 2013), 378.

2. Munson, *Tesla: Inventor of the Modern*, 211.

3. Ibid., 211–12.

4. Marc J. Seifer, *Wizard: The Life and Times of Nikola Tesla—Biography of a Genius* (New York: Citadel Press/Kensington Publishing, 1998), 359; Munson, *Tesla: Inventor of the Modern*, 212.

5. Munson, *Tesla: Inventor of the Modern*, 212.

6. Waldemar Kaempffert, "Prof. Pupin Now Claims Wireless His Invention," *Los Angeles Examiner*, May 13, 1915.

7. Munson, *Tesla: Inventor of the Modern*, 214.

8. Leland I. Anderson, "John Stone Stone on Nikola Tesla's Priority in Radio and Continuous-Wave Radiofrequency Apparatus," *The A.W.A. Review-First Edition*, 1986, 39.

9. Ibid.

10. Munson, *Tesla: Inventor of the Modern*, 216–17; Margaret Cheney and Robert Uth, *Tesla: Master of Lightning* (New York: Barnes & Noble, 1999), 133.

11. John J. O'Neill, *Prodigal Genius: The Life of Nikola Tesla* (Albuquerque, NM: Brotherhood of Life, 1994), 317.

12. Ibid., 318.

13. Seifer, *Wizard*, 414; http://www.icrcanada.org/research/literaryresearch/tesla.

14. Carlson, *Tesla*, 379; Munson, *Tesla: Inventor of the Modern*, 218.

15. Margaret Cheney, *Tesla: Man Out of Time* (New York: Barnes & Noble, 1993), 232.

16. "When Woman Is Boss," an Interview with Nikola Tesla by John B. Kennedy, *Colliers*, January 30, 1926.

17. Seifer, *Wizard*, 410.

18. Cheney, *Tesla: Man Out of Time*, 234.

19. Seifer, *Wizard*, 437.

20. Ibid., 412–13.

21. Cheney, *Tesla: Man Out of Time*, 239.

22. Ibid.

23. Ibid., 239–40.

24. Seifer, *Wizard*, 421.

25. Munson, *Tesla: Inventor of the Modern*, 226.

26. Hugo Gernsback, "Edison and Tesla," *The Electrical Experimenter*, December 1915, 379.

27. "Tesla, 76, Reports His Talents at Peak—Says His New Invention, Almost Done, Will Come as '100,000 Trumpets of Apocalypse,'" *New York Times*, July 10, 1932.

28. Cheney and Uth, *Tesla: Master of Lightning*, 135.

29. Coauthor Marko Perko had the pleasure of spending a very informative day interviewing William Terbo several years ago, as he reminisced about his *veliki ujak Nikola*. Munson, *Tesla: Inventor of the Modern*, 218.

30. "Tesla Certain of His Power—Inventor Says Only Details Remain to Be Checked," *New York Sun*, July 10, 1933.

31. Nikola Tesla, "Chewing Gum More Fatal Than Rum," *New York World-Telegram*, August 10, 1932.

32. "Tesla, at 78, Bares New Death-Beam," *New York Times*, July 11, 1934.

33. Cheney and Uth, *Tesla: Master of Lightning*, 151–52; Carlson, *Tesla*, 380.

34. Nikola Tesla as told to George Sylvester Viereck, "A Machine to End War: A Famous Inventor, Picturing Life 100 Years from Now, Reveals an Astounding Scientific Venture Which He Believes Will Change the Course of History," *Liberty*, February 1937, https://www.pbs.org/tesla/res/res_art11.html.

35. Carlson, *Tesla*, 386–87.

36. Munson, *Tesla: Inventor of the Modern*, 230.

37. Sam Kean, "The Underlying Appeal of Nikola Tesla's Death Ray," Science History, October 26, 2020, https://www.sciencehistory.org/distillations/the-undying-appeal-of-nikola-teslas-death-ray.

38. Munson, *Tesla: Inventor of the Modern*, 232–33.

39. Ibid., 233.

40. Nikola Tesla, as told to George Sylvester Viereck, "A Machine to End War."

41. Carlson, *Tesla*, 389.

42. O'Neill, *Prodigal Genius*, 274–75. Note: During one of coauthor Perko's popular lectures on the life of Nikola Tesla, an elderly man approached him regarding this event with the messenger boy. He told Perko that he was one of those messenger boys in New York City. He also said that every messenger boy wanted to deliver a message to Tesla or to deliver one for him, because he was always a big tipper. Perko also inquired about what has often been said regarding Tesla's presence—that he was magnetic, mysterious, with piercing blue eyes that caused one to stare up at the tall man, both slack-jawed and mesmerized. The elderly man confirmed the description.

43. Cheney, *Tesla: Man Out of Time*, 265.

44. Carlson, *Tesla*, 378; Cheney, *Tesla: Man Out of Time*, 267; Eric Wenass, "An Examination of Nikola Tesla's Priority in Discovering Radio," *The A.W.A. Review* 32 (2019): 215–97, http://richardn1.sg-host.com/wp-content/uploads/AWA32_Wenaas_19-05-07-Tesla-revised-new-title.pdf; Marconi Wireless Telegraph Company of America vs. United States, https://tile.loc.gov/storage-services/service/ll/usrep/usrep320/usrep320001/usrep320001.pdf; Antique Wireless Association, "Nicola Tesla's Priority in the Discovery of Radio," https://www.antiquewireless.org/homepage/nikola-tesla/; Munson, *Tesla: Inventor of the Modern*, 236; Cheney and Uth, *Tesla: Master of Lightning*, 160.

CHAPTER 17

1. Richard Munson, *Tesla: Inventor of the Modern* (New York: W.W. Norton, 2018), 234; Margaret Cheney, *Tesla: Man Out of Time* (New York: Barnes & Noble, 1993), 266; "2,000 Are Present at Tesla Funeral," *New York Times*, January 13, 1943.

2. W. Bernard Carlson, *Tesla: Inventor of the Electrical Age* (Princeton, NJ: Princeton University Press, 2013), 390; Margaret Cheney and Robert Uth, *Tesla: Master of Lightning* (New York: Barnes & Noble, 1999), 161.

3. "2,000 Are Present at Tesla Funeral."

4. Munson, *Tesla: Inventor of the Modern*, 234; Cheney, *Tesla: Man Out of Time*, 267.

5. Cheney, *Tesla: Man Out of Time*, 266–67.

6. Nigel Cawthorne, *Tesla: The Life and Times of an Electric Messiah* (New York: Chartwell Books, 2014), 174.

7. "2,000 Are Present at Tesla Funeral."

8. Ibid.

9. Munson, *Tesla: Inventor of the Modern*, 235.

10. Ibid., 236.

11. Cheney, *Tesla: Man Out of Time*, 267.

12. Marc J. Seifer, *Wizard: The Life and Times of Nikola Tesla—Biography of a Genius* (New York: Citadel Press/Kensington Publishing, 1998), 446.

13. Carlson, *Tesla*, 390.

14. "Tesla Left No Will—Nephew Says Estate Consists of Research Data, Models," *New York Times*, January 22, 1943.

15. Cheney and Uth, *Tesla: Master of Lightning*, 158; Carlson, *Tesla*, 392–93. Note: the OAPC is sometimes referred to as the OAP.

16. Cheney and Uth, *Tesla: Master of Lightning*, 163.

17. Carlson, *Tesla*, 392.

18. Ibid., 393.

19. Cheney and Uth, *Tesla: Master of Lightning*, 160; Carlson, *Tesla*, 392.

20. Cheney, *Tesla: Man Out of Time*, 279; Seifer, *Wizard*, 460; Zorica Civric, "Nikola Tesla Legacy," in *Nikola Tesla Museum 1952–2003* (Belgrade, Serbia: Museum of Science and Technology, 2004), 10.

21. "Death-Ray Machine Described—Dr. Tesla Says Two of Four Necessary Pieces of Apparatus Have Been Built," *New York Sun*, July 11, 1934.

CHAPTER 18

1. Marc J. Seifer, *Wizard: The Life and Times of Nikola Tesla—Biography of a Genius* (New York: Citadel Press/Kensington Publishing, 1998), 464. See chapter 1, endnote 3.

2. Richard Munson, *Tesla: Inventor of the Modern* (New York: W.W. Norton, 2018), 250; Margaret Cheney and Robert Uth, *Tesla: Master of Lightning* (New York:

Barnes & Noble, 1999), 172; "List of scientific units named after people," Wikipedia, https://en.wikipedia.org/wiki/List_of_scientific_units_named_after_people.

3. Ibid., 249.

4. Cheney and Uth, *Tesla: Master of Lightning*, 161; Zorica Civric, "Nikola Tesla Legacy," in *Nikola Tesla Museum 1952–2003* (Belgrade, Serbia: Museum of Science and Technology, 2004), 4.

5. W. Bernard Carlson, *Tesla: Inventor of the Electrical Age* (Princeton, NJ: Princeton University Press, 2013), 398–99.

6. Munson, *Tesla: Inventor of the Modern*, 250.

7. Ibid.; Tesla Science Center, https://teslasciencecenter.org/.

8. W. K. Brasher, "Nikola Tesla," *Journal of the Institution of Electrical Engineers* 90, no. 35 (November 1943), 467.

EPILOGUE

1. W. Bernard Carlson, *Tesla: Inventor of the Electrical Age* (Princeton, NJ: Princeton University Press, 2013), 397.

2. Edison did not invent the lightbulb and the Wright Brothers were not the first to fly. Marko Perko, *Did You Know That . . . ?: "Revised and Expanded" Edition: Surprising-But-True Facts about History, Science, Inventions, Geography, Origins, Art, Music and More* (New York: Open Road Distribution, 2017), 76–77, 90–92.

3. John J. O'Neill, *Prodigal Genius: The Life of Nikola Tesla* (Albuquerque, NM: Brotherhood of Life, 1994), 49.

4. Ibid., 74–75; Margaret Cheney, *Tesla: Man Out of Time* (New York: Barnes & Noble, 1993), 39.

5. Margaret Cheney and Robert Uth, *Tesla: Master of Lightning* (New York: Barnes & Noble, 1999), 23.

6. Cheney, *Tesla: Man Out of Time*, 23.

7. "Overton window," Wikipedia, https://en.wikipedia.org/wiki/Overton_window; Mackinac Center for Public Policy, "Overton Window," https://www.mackinac.org/OvertonWindow; Maggie Astor, "How the Politically Unthinkable Can Become Mainstream," *New York Times*, February 26, 2019.

8. Walter T. Stephenson, "Fruits of Genius Were Swept Away," *New York Herald*, March 14, 1895.

9. Cheney, *Tesla: Man Out of Time*, 23.

Bibliography

Anderson, Leland I. *Bibliography Dr. Nikola Tesla (1856–1943), Second Enlarged Edition.* Compiled and edited by Leland I. Anderson. Minneapolis: The Tesla Society, 1956.

Aurucci Stiefel, Natalie. *Looking Back at Rocky Point: In the Shadow of the Radio Towers, Vol. 1.* Self-published.

Bodanis, David. *Electric Universe: The Shocking True Story of Electricity.* New York: Crown Publishers, 2005.

Brockman, John., editor. *The Greatest Inventions of the Past 2,000 Years: Todays' Leading Thinkers Chose the Creations That Shaped Our World.* New York: Simon & Schuster, 2000.

Carlson, W. Bernard. *Tesla: Inventor of the Electrical Age.* Princeton, NJ: Princeton University Press, 2013.

Cawthorne, Nigel. *Tesla: The Life and Times of an Electric Messiah.* New York: Chartwell Books, 2014.

———. *Tesla vs Edison: The Life-Long Feud That Fueled the World.* New York: Chartwell Books, 2016.

Cheney, Margaret. *Tesla: Man Out of Time.* New York: Barnes & Noble, 1993.

Cheney, Margaret, and Robert Uth. *Tesla: Master of Lightning.* New York: Barnes & Noble, 1999.

Clark, Ronald W. *Edison: The Man Who Made the Future.* New York: G.P. Putnam's Sons, 1977.

Cline, Adam. *The Current War: A Battle Story between Two Electrical Titans, Thomas Edison and George Westinghouse.* Second edition. Self-published, 2017.

Cooper, Christopher. *The Truth about Tesla: The Myth of the Lone Genius in the History of Innovation.* New York: Race Point Publishing, 2018.

Dolnick, Edward. *The Clockwork Universe: Isaac Newton, the Royal Society, and the Birth of the Modern World.* New York: Harper, 2011.

Dyer, Frank Lewis, and Thomas Commerford Martin. *Edison, His Life and Inventions,* Salt Lake City, UT: Project Gutenberg, January 21, 2006 [eBook #820].

Elms, Alan C. *Uncovering Lives: The Uneasy Alliance of Biography and Psychology*. New York: Oxford University Press, 1994.

Gertner, Jon. *The Idea Factory: Bell Labs and the Great Age of American Innovation*. New York: Penguin, 2012.

Ghiselin, Brewster (ed.). *The Creative Process: A Symposium*. New York: A Mentor Book, 1952.

Goldstone, Lawrence, and Nancy Goldstone. *The Friar and Cipher: Roger Bacon and the Unsolved Mystery of the Most Unusual Manuscript in the World*. New York: Doubleday, 2005.

Govorchin, Gerald Gilbert. *Americans from Yugoslavia*. Gainesville: University of Florida Press, 1961.

Greenblatt, Stephen. *The Swerve: How the World Became Modern*. New York: W.W. Norton & Company, 2011.

Gribbin, John. *The Scientists: A History of Science Told Through the Lives of Its Greatest Inventors*. New York: Random House, 2002.

Hart, B. H. Liddell. *Why Don't We Learn from History?* Edited and with an Introduction by Giles Laurén. Self-published, 2012.

Herschel, Sir John F. W., and K. H. Bart. *Preliminary Discourse on the Study of Natural Philosophy*. London: Longman Press, 1851. (https://ia800504.us.archive.org/16/items/preliminarydisco00hersiala/preliminarydisco00hersiala.pdf)

Holmes, Richard. *The Age of Wonder: How the Romantic Generation Discovered the Beauty and Terror of Science*. New York: Pantheon Books, 2008.

Hunt, Inez, and Wanetta W. Draper. *Lightning in His Hand: The Life Story of Nikola Tesla*. Hawthorne, CA: Omni Publications, 1977.

Isaacson, Walter. *Einstein: His Life and Universe*. New York: Simon & Schuster, 2007.

———. *Leonardo da Vinci*. New York: Simon & Schuster, 2017.

———. *The Innovators: How a Group of Hackers, Geniuses, and Geeks Created the Digital Revolution*. New York: Simon & Schuster, 2014.

Israel, Paul. *Edison: A Life of Invention*. New York: John Wiley & Sons, 1998.

Johnson, Robert Underwood. *Poems: Second Edition, with New Poems*. New York: Century Co., 1908.

———. *Remembered Yesterdays*. Boston: Little, Brown and Company, 1923.

———. *Songs of Liberty and Other Poems: Including Paraphrases from the Servian after Translations by Nikola Tesla, with a Prefatory Note by Him on Servian Poetry*. New York: Century Co., 1897.

John-Steiner, Vera. *Notebooks of the Mind: Explorations of Thinking*. Albuquerque: University of New Mexico Press, 1985.

Judah, Tim. *The Serbs: History, Myth and the Destruction of Yugoslavia*. New Haven, CT: Yale University Press, 1998.

Kent, David J. *Tesla: The Wizard of Electricity*. New York: Fall River Press, 2013.

King, Moses. *King's Photographic Views of New York: A Souvenir Companion to King's Handbook of New York City*. Boston: Moses King, 1895.

Lichty, Lawrence W., and Malachi C. Topping. *American Broadcasting: A Source Book on the History of Radio and Television*. New York: Hastings House, 1975.

Lomas, Robert. *The Man Who Invented the Twentieth Century: Nikola Tesla, Forgotten Genius of Electricity.* London: Headline Book Publishing, 1999.

Luria, A. R. *The Mind of a Mnemonist: A Little Book about a Vast Memory.* New York: Basic Books, Inc., 1968.

Madden, Thomas F. *Istanbul: City of Majesty at the Crossroads of the World.* New York: Viking, 2016.

Manchester, William. *A World Lit Only by Fire: The Medieval Mind and the Renaissance— Portrait of an Age.* Boston: Little, Brown and Company, 1992.

Marconi, Degna. *My Father, Marconi.* New York: Guernica, 1996.

Martin, Thomas Commerford. *The Inventions, Researches, and Writings of Nikola Tesla with Special Reference to His Work in Polyphase Currents and High Potential Lighting.* New York: D. Van Nostrand Company, 1894.

———. *The Inventions, Researches, and Writings of Nikola Tesla.* CreateSpace/Skytower Press, 2013.

———. *The Inventions, Researches, and Writings of Nikola Tesla.* New York: Fall River Press, 2014.

McAdams, Dan P. *George W. Bush and the Redemptive Dream.* New York: Oxford University Press, 2011.

McNichol, Tom. *AC/DCI: The Savage Tale of the First Standards War.* San Francisco: Jossey-Bass, 2006.

Morus, Iwan Rhys. *Nikola Tesla and the Electrical Future.* London: Icon Books, 2019.

Mrkich, Daniel. *Nikola Tesla: The European Years* (Part Three). Ottawa, ON: Commoners' Publishing, 2010.

Munson, Richard. *Tesla: Inventor of the Modern.* New York: W.W. Norton and Company, 2018.

Norwich, John Julius. *A Short History of Byzantium.* New York: Alfred A. Knopf, Inc., 1997.

O'Neill, John J. *Prodigal Genius: The Life of Nikola Tesla.* Albuquerque, NM: Brotherhood of Life, 1994.

Passer, Harrold C. *The Electrical Manufacturers 1875–1900: A Study in Competition, Entrepreneurship, Technical Change, and Economic Growth.* New York: Arno Press, 1972.

Perko, Marko. *Did You Know That . . . ?: "Revised and Expanded" Edition: Surprising-But-True Facts about History, Science, Inventions, Geography, Origins, Art, Music and More.* Fourth edition. New York: Open Road Distribution, 2017.

Ratzlaff, John T., and Leland I. Anderson. *Dr. Nikola Tesla Bibliography.* Palo Alto, CA: Ragusan Press, 1979.

Sacks, Oliver. *The Man Who Mistook His Wife for a Hat and Other Clinical Tales.* New York: Simon & Schuster, 1998.

Schultz, William Todd (ed.). *Handbook of Psychobiography.* New York: Oxford University Press, 2005.

———. *Tiny Terror: Why Truman Capote (Almost) Wrote Answered Prayers.* New York: Oxford University Press, 2011.

Seifer, Marc J. *Wizard: The Life and Times of Nikola Tesla: Biography of a Genius.* New York: Citadel Press, 1998.

Shapiro, James. *Contested Will: Who Wrote Shakespeare?* New York: Simon & Schuster, 2010.

Simmons, Michael W. *Nikola Tesla: Prophet of the Modern Technological Age.* Make Profits Easy, 2016.

Strevens, Michael. *The Knowledge Machine: How Irrationality Created Modern Science.* New York: Liveright, 2020.

Stross, Randall. *The Wizard of Menlo Park: How Thomas Alva Edison Invented the Modern World.* New York: Crown Publishers, 2007.

Tesla, Nikola. *Lecture before the New York Academy of Sciences*, 1897. Breckenridge, CO: Twenty-First Century Books, 1994.

———. *My Inventions: Nikola Tesla's Autobiography.* Edited, with an introduction, by Ben Johnston. Austin, TX: Hart Brothers, 1982.

———. *My Inventions: Nikola Tesla's Autobiography.* Las Vegas: Lits, 2011.

———. *My Inventions: The Autobiography of Nikola Tesla.* Eastford, CT: Martino Fine Books, 2018.

———. *Nikola Tesla: Colorado Springs Notes 1899–1900.* La Vergne, TN: BN Publishing, 2007.

———. *Tesla Said.* Compiled by John T. Ratzlaff. Millbrae, CA: Tesla Book Company, 1984.

———. *The Nikola Tesla Treasury.* Radford, VA: Wilder Publications, 2007.

———. *The Problem of Increasing Human Energy: With Special Reference to the Harnessing of the Sun's Energy.* Unabridged edition with photographs. Merchant Books, 2019.

Uglow, Jenny. *The Lunar Men: Five Friends Whose Curiosity Changed the World.* New York: Farrar, Straus and Giroux, 2002.

Wasik, John F. *Lightning Strikes: Timeless Lessons in Creativity from the Life and Work of Nikola Tesla.* New York: Sterling Publishing Co., Inc., 2016.

Wheatcroft, Andrew. *The Habsburgs: Embodying Empire.* New York: Viking, 1995.

Wilson, Ben. *Metropolis: A History of the City, Humankind's Greatest Invention.* New York: Doubleday, 2020.

Winchell, Mike. *The Electric War: Edison, Tesla, Westinghouse, and the Race to Light the World.* New York: Henry Holt and Company, 2019.

Wolfe, Tom. *Hooking Up.* New York: Farrar, Straus and Giroux, 2000.

Wootten, David. *The Invention of Science: A New History of the Scientific Revolution.* New York: HarperCollins Publishers, 2015.

Wright, Craig. *The Hidden Habits of Genius: Beyond Talent, IQ, and Grit—Unlocking the Secrets of Greatness.* New York: Dey St./William Morrow, 2020.

ARCHIVAL COLLECTIONS

American Philosophical Society, Philadelphia, PA

American Radio History, www.AmericanRadioHistory.com Archives

Cambridge University Library, Cambridge, UK, Houghton Library Repository

Columbia University Library, Rare Books and Manuscript Library, New York "Nikola Tesla Papers"

Edison Archives, http://edison.rutgers.edu/tesla.htm

Engineering and Technology History Wiki, https://ethw.org/Archives:Papers_of_Nikola_Tesla

Federal Bureau of Investigation, Washington, DC, Freedom of Information Act file on Nikola Tesla

Institute of Electrical and Electronic Engineers, IET Archives, Piscataway, NJ, Biographical Files

Institute of Electrical and Electronics Engineers (IEEE)

Library of Congress, Manuscript Division, Washington, DC

National Museum of American History, Archives Center, Smithsonian Institution, Washington, DC: Kenneth Swezey Papers

National Museum of American History, Dibner Library, Smithsonian Institution, Washington, DC: Manuscript Collections

New-York Historical Society Museum & Library, New York, NY

New York Public Library, Manuscripts and Archives Division: *The Century Magazine Collection*; Personal Miscellaneous Collection

Nikola Tesla Archive/2 https://en.wikipedia.org/wiki/Talk%3ANikola_Tesla/Archive_2 Nikola Tesla Museum, Belgrade, Serbia: Library of Nikola Tesla Scrapbooks of Clippings

Oxford University, Bodleian Library: Marconi Archives

Science Museum, London, UK

The Franklin Institute, Philadelphia, PA

"The Tesla Collection": University of Birmingham, UK: Cadbury Research Library, Special Collections

The Tesla Memorial Society of New York, www.TeslaSociety.com

Western Pennsylvania Historical Society, Pittsburgh, PA: Leland I. Anderson Papers [Anderson Collection] Historical Society of Western Pennsylvania, Pittsburgh

WEBSITES

https://www.archive.org

https://www.documentcloud.org

http://edison.rutgers.edy/searchsite.htm

http://edison.rutgers.edu

https://en.wikisource.org

https://www.hathitrust.org/

www.jstor.org

www.newspapers.com

http://patents.google.com

https://www.researchgate.net

https://scholar.archive.org

https://www.sciencehistory.org
www.theteslasociety.com
http://www.teslacollection.com
http://teslaradio.com
https://teslasciencecenter.org
https://teslauniverse.com
www.theteslasociety.com
https://timesmachine.nytimes.com/timesmachine
http://www.tfcbooks.com
https://teslaresearch.jimdofree.com
https://tesla-coil-builder.com
https://todayinsci.com
www.worldcat.org
https://z-lib.org/Z-Library

PERIODICALS/NEWSPAPERS

Colliers (An Illustrated Journal)
Colorado Springs Gazette
Electrical Engineer (London)
Electrical Review
Electrical Review (London)
Electrical World
Esquire Magazine
Harper's Weekly
IEEE Technology and Society Magazine
Journal of the Institution of Electrical Engineers
Liminalities: A Journal of Performance Studies
Nature (London)
New York Herald
New York Sun
New York World
Philadelphia Press
Pittsburg(h) Dispatch
Politika
Popular Science Monthly
Savannah Morning News
Science
Scientific American Supplement
The A.W.A. Review—First Edition
The Century Magazine (The Century Illustrated Monthly Magazine)
The Electrical Age
The Electrical Engineer

The Electrical Experimenter
The Electrical Review
The Electrician
The Electrician: A Weekly Journal of Electrical Engineering, Industry and Science
The Engineering Magazine
The New York Herald (Tribune)
The New York Sun
The New York Times
The New York World-Telegram
The Outlook: A Family Paper
Tribune (Chicago)
Western Electrician
World To-day (Heart's Magazine)

Index

About the Authors

Marko Perko is a graduate of the University of Southern California. He has always had an insatiable thirst for knowledge of all types, and as such, he is highly regarded as a modern-day Renaissance man, author, historian, polymath, and polemicist. He is the author of the critically acclaimed and wildly popular book *Did You Know That . . . ?* He is also a novelist and the writer of an international bestselling knowledge-based board game, as well as the creator of the Cultural Enrichment Programs education series, and a software developer.

The subject of Nikola Tesla has been a lifelong passion for Marko Perko. As a fellow Serb, Perko has lived in the then-Yugoslavia and spent decades studying Tesla's life. Over the years, he has lectured about Tesla as well as conducted personal interviews with some of Tesla's relatives, friends, and others who knew the man.

Perko has written for and edited numerous publications, and he has worked as a columnist, speechwriter, composer, musician, lecturer, and playwright. He is a member of the Authors Guild, the Biographers International Organization, the American Society of Composers, Authors and Publishers, Broadcast Music, Inc., the Institution of Engineering and Technology, the British Library, and the Organization of American Historians.

Presently, he is authoring new books as well as working on television, film, and internet projects based upon several of his intellectual properties. He lives in California with his wife, Heather, and their daughter, Skye Mackay Perko. Their son, Marko Perko III, lives in London, England.

www.MarkoPerko.com
www.TeslaBiography.com

Stephen M. Stahl, MD, PhD, DSc, has held faculty positions at Stanford University, the University of California at Los Angeles, the Institute of Psychiatry London, and the Institute of Neurology London. Currently he is a professor at the University of California at San Diego and an Honorary Fellow in Psychiatry at the prestigious University of Cambridge. Dr. Stahl serves as editor in chief of *CNS Spectrums*. Author of over five hundred articles and chapters, and more than sixteen hundred scientific presentations and abstracts, Dr. Stahl is an internationally recognized clinician, researcher, and teacher in psychiatry with subspecialty expertise in psychopharmacology. Dr. Stahl has written thirty-nine textbooks and edited thirteen others, including the best-selling and award-winning textbook, *Stahl's Essential Psychopharmacology*, now in its fourth edition, and the bestselling and award-winning clinical manual, *Essential Psychopharmacology Prescriber's Guide*, now in its fifth edition. Dr. Stahl has also published a novel, *Shell Shock*, a thriller that recounts the history of PTSD (posttraumatic stress disorder). Dr. Stahl is senior academic advisor and director of psychopharmacology for the California Department of State Hospitals (DSH), where he has a leadership role in addressing violence in the five-hospital, 6,500-patient DSH. He has been awarded the International College of Neuropsychopharmacology (CINP) Lundbeck Foundation Award in Education for his contributions to postgraduate education in psychiatry and neurology. His books have won the British Medical Association's Book of the Year Award and, recently, first prize for best digital medical book. Dr. Stahl is also the winner of the A. E. Bennett Award of the Society of Biological Psychiatry, the APA/San Diego Psychiatric Society Education Award, and the UCSD Psychiatry Residency Teaching Award, and he has been cited as both one of "America's Top Psychiatrists" and one of the "Best Doctors in America." He was honored with the Distinguished Psychiatrist Award of the APA and gave the Distinguished Psychiatrist Lecture for 2013. He was named the 2016 David Mrazek Award Winner of the American Psychiatric Association and gave the Mrazek Pharmacogenomics Memorial Lecture at the 2016 APA meeting. His alma mater, Northwestern University, honored him by naming the award for the most promising medical student to go into psychiatry the "Stephen Stahl Award." In 2018, he was awarded an Honorary Doctor of Science by Uskudar University in Istanbul, Turkey, for his lifetime of achievements in psychiatry.

When not traveling the globe as a highly in-demand medical lecturer, Dr. Stahl lives in Southern California and spends his creative time penning the "Gus Conrad" thriller series.

www.neiglobal.com
www.TeslaBiography.com